HOW CITIES WILL SAVE THE WORLD

How Cities Will Save the World
Urban Innovation in the Face of Population Flows, Climate Change and Economic Inequality

Edited by

RAY BRESCIA
Albany Law School, Albany, NY, USA

JOHN TRAVIS MARSHALL
Georgia State University College of Law, Atlanta, GA, USA

Routledge
Taylor & Francis Group

LONDON AND NEW YORK

First published 2016
by Routledge
2 Park Square, Milton Park, Abingdon, Oxon OX14 4RN

and by Routledge
711 Third Avenue, New York, NY 10017

First issued in paperback 2017

Routledge is an imprint of the Taylor & Francis Group, an informa business

British Library Cataloguing in Publication Data
A catalogue record for this book is available from the British Library

Library of Congress Cataloging in Publication Data
Names: Brescia, Ray, editor. | Marshall, John Travis, editor.
Title: How cities will save the world : urban innovation in the face of population flows, climate change and economic inequality / [edited] by Ray Brescia and John Travis Marshall.
Description: Farnham, Surrey, UK ; Burlington, VT : Ashgate, [2016] | Series: Urban planning and environment | Includes bibliographical references and index.
Identifiers: LCCN 2015034811 | ISBN 9781472450265 (hardback) | ISBN 9781472450272 (ebook) | ISBN 9781472450289 (epub)
Subjects: LCSH: Urbanization. | Urban policy. | Regional planning. | Regional economics. | Sustainable urban development.
Classification: LCC HT361 .H69 2016 | DDC 307.76—dc23

ISBN 13: 978-1-138-49032-1 (pbk)
ISBN 13: 978-1-4724-5026-5 (hbk)

Typeset in Times New Roman
by Apex CoVantage, LLC

Contents

List of Figures

List of Abbreviations

ABA	American Bankers Association
ABA	American Bar Association
AJC	Atlanta Journal Constitution
APFO	Adequate Public Facility Ordinances
ARC	Atlanta Regional Commission
ASMARE	*Associação dos Catadores de Papel, Papelão e Material Reaproveitável* or Association of Paper, Carton and Recyclable Material Pickers
ATM	Automated Teller Machine
BART	Bay Area Rapid Transit
BESE	Board of Secondary and Elementary Education
BRT	Bus Rapid Transit
CC&Rs	covenants, conditions, and restrictions
CCLRC	Cuyahoga County Land Reutilization Corporation
CCP	Center for Community Progress
CDBG	Community Development Block Grant
CDCs	Community Development Corporations
CDFIs	Community Development Financial Institutions
CDP	Urban Justice Center Community Development Project
CEA	Cooperative Endeavor Agreement
CIRC	Community Impact Report Card
CLC	Yale University Law School Community Lawyering Clinic
CMO	Charter Management Organization
CO_2	Carbon dioxide
CREDO	Center for Research on Education Outcomes
CRO	Chief Resilience Officer
CRPE	Center on Reinventing Public Education
CVH	Community Voices Heard
DIF	Developer Impact Fees
DHS	Department of Homeland Security
DNA	Deoxyribonucleic Acid
DOT	Department of Transportation
ECCO	Elm City Churches Organized
EDC	Economic Development Corporation
ELI	Environmental Law Institute
EPA	U.S. Environmental Protection Agency
FBI	Federal Bureau of Investigation
FCFC	First City Fund Corporation
FDA	Food and Drug Administration
FEMA	Federal Emergency Management Agency
FTC	Federal Trade Commission
GAO	Government Accounting Office
GDOT	Georgia Department of Transportation

GDP	Gross Domestic Product
GIS	Geographic Information Systems
GPS	Global Positioning System
GRAS	Generally Recognized as Safe
GSE	Government Sponsored Enterprise
GTIB	Georgia Transportation Infrastructure Bank
HCF	Household Carbon Footprints
HMPG	Hazard Mitigation Grant Program
HOLC	Home Owners' Loan Corporation
HUD	U.S. Department of Housing and Urban Development
ICE	United States Immigration and Customs Enforcement
ICLEI	International Council for Local Environmental Initiatives
ICMA	International City–County Management Association
IDEA	Individuals with Disabilities Education Act
IMF	International Monetary Fund
IPPUC	Institute for Research and Urban Planning of Curitiba
IRS	Internal Revenue Services
ITINS	Individual Tax Identification Numbers
JUNTA	Junta for Progressive Action, Inc.
KIPP	Knowledge Is Power Program
LAANE	Los Angeles Alliance for a New Economy
LEED	Leadership in Energy and Environmental Design
LGBT	Lesbian, Gay, Bi-Sexual and Transgender
LISC	Local Initiatives Support Corporation
MAPTAM	Modernization of the Territorial Public Action and Affirmation of Metropolitan Areas
MARTA	Metropolitan Atlanta Rapid Transit Authority
MBUF	Mileage Based User Fee
MFP	Minimum Foundation Program
MPG	Miles Per Gallon
NAF	National Arbitration Forum
NCIC	FBI's National Crime Information Center
NEPA	National Environmental Policy Act
Nexus	Nexus Research Group
NGOs	Non-Governmental Organizations
NHPA	National Historic Preservation Act
NHPD	New Haven Police Department
NORA	New Orleans Redevelopment Authority
NRA	National Rifle Association
NVPC	National Vacant Property Campaign
NVPC	National Volunteer and Philanthropy Centre
OECD	Organisation for Economic Co–Operation and Development
OPSB	Orleans Parish School Board
PAC	Political Action Committee
PB	Participatory Budgeting
PBS	Public Broadcasting Service
PHA	Project Home Again
PLCAA	Protection of Lawful Commerce in Arms Act

PRIs	Philanthropic Program-Related Investments
PTA	Parent-Teacher Association
PTO	Parent-Teacher Organization
REO	Real Estate Owned
RPPNM	Reservas Particulares do Patrimônio Natural Municipal program or Municipal Private Heritage Private Reserves
RSD	Recovery School District
SCCIC	Southern Connecticut Citizens for Immigration Control
SFALP	San Francisco Affirmative Litigation Project
Sgt	Sergeant
SHPO	State Historic Preservation Officer
SRAD	Special Resident Assessment District
SSTI	State Smart Transportation Initiative
St. Rose	St. Rose of Lima Catholic Church
SZEA	Standard State Zoning Enabling Act
the Board	New Haven legislative body, 30 member Board of Alders elected by the districts
the Card	Elm City Resident Card
TBD	Tax Benefit District
TIF	Tax Incremental Financing
TDR	Transfer of Development Rights
TOPS	Taylor Opportunity Program for Students
T-SPLOST	Transportation Special Purpose Local Option Sales Tax
TUF	Transportation Utility Fee
UK	United Kingdom
ULA	Unidad Latina en Acion
URA	Uniform Relocation and Real Property Acquisition Policies Act of 1970
URBS	*Urbanizacao de Curitiba SA*
USA	United States of America
USGS	United States Geological Survey
VAPAC	Vacant Abandoned Property Action
VAVT	Vehicle Ad Valorem Tax
VC	Value Capture
VITA	Volunteer Income Tax Assistance
VMT	Vehicle Miles Traveled
VPRO	Vacant Property Registration Ordinance
WB	World Bank
WPA	Works Progress Administration
YEI	Yale Entrepreneurial Institute

Contributors

Ray Brescia is an Associate Professor of Law at Albany Law School and the Director of the Government Law Center at that school. He also formerly served as a Visiting Clinical Associate Professor of Law at Yale Law School. His recent publications address a wide range of topics related to the financial crisis, including sub-prime mortgages and housing policy. Before joining the faculty at Albany Law, Professor Brescia was the Associate Director of the Urban Justice Center in New York City, where he coordinated legal representation for community-based institutions in areas such as housing, economic justice, workers' rights, civil rights, and environmental justice. In addition, he held a position as an adjunct professor at New York Law School from 1997 through 2006. Prior to Professor Brescia's involvement with the Urban Justice Center, he worked as a staff attorney at the New Haven Legal Assistance Association and The Legal Aid Society of New York, where he was a recipient of a Skadden Fellowship following his graduation from law school. He was also Law Clerk to the Honorable Constance Baker Motley, Senior U.S. District Court Judge for the Southern District of New York. Professor Brescia graduated from Yale Law School, where he was co-recipient of the Charles Albom Prize for Appellate Advocacy and served as a student director of several clinics, including the Allard K. Lowenstein International Human Rights Law Clinic and the Homelessness Clinic. He earned his B.A. from Fordham University.

John DeStefano, Jr. is Executive Vice President of Start Bank of New Haven, a community development financial institution of which he was an incorporating director and on whose board he continues to serve. Start is a locally owned, governed and managed bank dedicated to serving both its community development and commercial bank mission. Mr. DeStefano also holds appointments as an Instructor at Yale University and Southern Connecticut State University.

Mr. DeStefano served as the 49th Mayor of the City of New Haven, CT, from 1994 through 2013. Elected Mayor 10 times, Mr. DeStefano is the City's longest serving Mayor. During his tenure Mr. DeStefano led New Haven's nationally acknowledged school reform initiative, New Haven School Change, serving as a member of the school board and the appointing authority for the board's other members. The program is a collaborative, comprehensive and persistent effort among parents, the American Federation of Teachers, community-based and business organizations and Yale University, the principal funder of New Haven Promise, the City's merit-based full college scholarship program. Mr. DeStefano chaired the Citywide School Construction Program, a $1.5 billion effort that rebuilt or built new, virtually every school building in the city.

During his time in office the City of New Haven undertook a number of immigration initiatives including the Elm City Resident Card which, commencing in 2007, was made available to all residents of the city for identification and to access city services, regardless of immigration status. The City also initiated a Living (minimum) Wage program, a Domestic Partner benefits initiative and the State of Connecticut's first public financing program for elected officials during his tenure.

New Haven experienced robust economic growth throughout Mr. DeStefano's tenure as Mayor as the city solidified its position as a major educational and medical center. Collateral growth in life science and entrepreneurial business continue to flourish in the City. Currently New Haven experiences among the lowest vacancy rates in the nation for commercial occupancies. During his time in office Mr. DeStefano served as both the President of the Connecticut Conference of Municipalities and the National League of Cities, the oldest and largest association of America's cities and towns. The National Civic League named New Haven an All America City three times during Mr. DeStefano's mayoralty.

Renia Ehrenfeucht is a Professor and Director of the Community and Regional Planning Program at the University of New Mexico. Dr. Ehrenfeucht conducts research in two areas. First, she focuses on the production and meaning of public space and the specificity of sidewalks. This includes the politics of public space use and design as well as how institutions governing shared space—including the built form and municipal ordinances—and social norms interact. Another component of this work is the collective dimensions of private property and how public and parochial spaces shape property relations. Her second area of research is shrinking cities or cities that have sustained long-term population loss. The strands of this area include how planning processes function in the face of population loss, how cities address vacant land, and how residents respond to their changing environments. Dr. Ehrenfeucht has co-authored two books, *Sidewalks: Conflict and Negotiation in Public Space* and *Urban Revitalization: Remaking Cities in a Changing World*, Her work has also been published in the *Journal of Planning History, the Journal of Historical Geography, Urban Geography, Urban Studies, Environment and Planning, Planning Practice and Research*, *Cityscape* and the *Journal of Urban Design*. Dr. Ehrenfeucht worked in land use planning in Washington State for four years.

Hanna-Ruth Gustafsson is an associate attorney with Fried, Frank, Harris, Shriver & Jacobson LLP in New York City, where she concentrates on real estate law with a focus on land use. She graduated from Yale Law School in 2012, after serving as a student director of Yale's Community and Economic Development Clinic and an editor of the *Yale Journal of International Law*. She earned a B.A. in History, *magna cum laude*, from Yale University and worked for the Urban Design Division of the New York City Department of Transportation.

Julian Conrad Juergensmeyer is Professor and Ben F. Johnson Chair in Law at Georgia State University and Director of the College of Law's Center for the Comparative Study of Metropolitan Growth. Professor Juergensmeyer's teaching and research specialties are land use planning law, property law and comparative land use and environmental law and his books and articles on those topics are nearing 100 in number. They include a co-authored treatise/hornbook on Land Use Planning and Development Regulation Law that is widely used by law and planning practitioners and frequently cited by courts including the Supreme Court of the United States. He is considered a pioneer in the development of infrastructure finance-oriented regulations such as impact fees and has participated in developing growth management and capital improvement programs and litigation in regard to them in over 25 states. In recent years, he has also focused on smart growth regulations designed to encourage use of renewable energy and resource conservation. He currently serves as Adjunct Professor of City and Regional Planning at GSU's sister institution, Georgia Institute of Technology. He is also Professor of Law Emeritus of the University of Florida

College of Law. As a strong believer in the globalization of law and the legal profession, he has taught and lectured throughout Europe, North and South America, and Africa, and has held the title of visiting professor at such foreign universities as the Universities of Frankfurt, Aarhus, Warsaw, Strasbourg, Limoges and British Columbia.

Elizabeth A. Kelly is currently a Senior Policy Advisor for the National Economic Council. Prior to joining the NEC, Elizabeth clerked for the Honorable Stephen Higginson on the U.S. Court of Appeals for the Fifth Circuit and served as Special Assistant to the Assistant Secretary for Policy Development and Research at the U.S. Department of Housing and Urban Development (HUD). Ms. Kelly is a graduate of Duke University, the University of Oxford, and Yale Law School, where she was a student director of the Community and Economic Development Clinic. Ms. Kelly participated in the research and writing of this paper in her personal capacity as a student at Yale Law School from January to July 2012. The National Economic Council disclaims responsibility for any of the views expressed herein and these views do not necessarily represent the views of the National Economic Council or the United States.

Kermit J. Lind is Clinical Professor of Law Emeritus at Cleveland-Marshall College of Law, Cleveland State University, where for 16 years he supervised the Urban Development Law Clinic in providing legal services to neighborhood housing and community development corporations. Before joining the clinical faculty at the law school, Professor Lind taught history at Cleveland State, led non-profit advocacy organizations for 13 years, and practiced law in Cleveland for eight years. In 2005, he received the Michael R. White award for public service from the Cleveland Community Development Coalition. Professor Lind has tirelessly advocated for lender accountability during the mortgage crisis, successfully citing public nuisance abatement statutes in Cleveland courts to require big banks that own foreclosed and abandoned buildings to comply with local housing maintenance codes and to take responsibility for the harm caused to neighborhoods and communities by the deterioration of their properties. Professor Lind authored several law review articles prior to retirement, and now writes, consults, and lectures on community development law and public policy. He received his J.D. from Cleveland State University, M.A. from the University of Chicago, and B.A. from Goshen College.

John A. Lovett is the De Van D. Daggett, Jr. Distinguished Professor of Law at Loyola University New Orleans College of Law. He joined the Loyola law faculty in 2002, after five years in practice with the New Orleans law firm of Liskow & Lewis and judicial clerkships with the United States District Court, Western District of Louisiana, and the United States Fifth Circuit Court of Appeals. Professor Lovett has chaired the Property and Real Estate Transactions Sections of the American Association of Law Schools and has served as contributing editor to the ABA journal, *Probate and Property*. In the fall of 2009, Professor Lovett was a McCormick Fellow at the University of Edinburgh Law School. From 2012 to 2015, he served as Associate Dean for Faculty Development and Academic Affairs at Loyola. He is the co-author of *Louisiana Property Law: The Civil Code, Cases, and Commentary* (Carolina Academic Press 2014) and has authored numerous book chapters and law review articles on property law in common, civil and mixed jurisdictions. Professor Lovett received his J.D. from Tulane Law School, M.F.A. from Indiana University-Bloomington, and B.A. from Haverford College.

John Travis Marshall is an Assistant Professor of Law at Georgia State University and Associate Director of the College of Law's Center for the Comparative Study of Metropolitan Growth. He teaches land use and environmental law. His research focuses on private, non-profit, and government interventions to promote urban revitalization, including questions related to environmental justice, sustainable development, and long-term recovery from natural disasters. Professor Marshall joined Georgia State from Yale Law School where he was a Clinical Lecturer in Law and the Ludwig Community Development Fellow. He previously served as counsel to the New Orleans Redevelopment Authority (NORA), where he helped implement the Authority's post-Hurricane Katrina long-term neighborhood recovery strategy. Prior to joining NORA in 2007, Marshall was a partner with Holland & Knight LLP, specializing in land use, zoning, and real estate litigation. Professor Marshall graduated from the University of Florida College of Law. He has an M.A. in Government from the University of Texas at Austin, and a B.A. from the University of Notre Dame.

Kathleen Morris is an Associate Professor of Law at Golden Gate University School of Law. She is a former deputy city attorney at the San Francisco City Attorney's Office, where she headed up the Affirmative Litigation Task Force, a group of deputy city attorneys charged with developing and litigating public interest cases on behalf of San Francisco. She was a lead attorney in San Francisco's constitutional challenges to California's marriage laws and the federal Partial-Birth Abortion Ban Act. Her publications include *Expanding Local Enforcement of State and Federal Consumer Protection Laws*, 40 Fordham Urban L.J. 1903 (2013); *The Case for Local Constitutional Enforcement*, 47 Harv. C.R.-C.L. L. Rev. 1 (2012); and *San Francisco and the Rising Culture of Engagement in Local Public Law Offices*, in Why the Local Matters: Federalism, Localism, and Public Interest Advocacy (2010). Professor Morris has been a Visiting Lecturer at Yale Law School, a Lecturer at U.C. Berkeley Law School, and a Visiting Assistant Professor at Rutgers-Camden Law School. She has a law degree from the University of California, Berkeley; a Masters in Politics from the University of Edinburgh, Scotland; and a B.A. from California State University, Northridge. After law school, she served as a law clerk for Chief Judge Sidney R. Thomas, U.S. Court of Appeals for the Ninth Circuit.

Becht Neel was born and raised in Atlanta, Georgia. He left for four years to earn a B.B.A. in Real Estate from the University of Georgia in Athens, Georgia, and returned to attend Georgia State University College of Law from which he received his J.D., *cum laude*, in 2015. Becht's father is a practicing real estate attorney in Atlanta and his inspiration for pursuing a career in law.

Arthur C. Nelson is Professor of Planning and Real Estate Development in the College of Architecture, Planning and Landscape Architecture at the University of Arizona. He is also Presidential Professor Emeritus of City & Metropolitan Planning at the University of Utah where he was founding Director of the Metropolitan Research Center, Adjunct Professor of Finance in the David Eccles School of Business, and founding Co-Director of the Master of Real Estate Development program. Nelson has made significant contributions to the fields of: real estate analysis including the role of changing demographics in shifting long-term real estate development trends; urban growth management and open space preservation; central city revitalization; infrastructure financing; planning effectiveness; transportation and land use outcomes; metropolitan development patterns; the economic effects of facility location; the role of suburban redevelopment in reshaping metropolitan America; and the new "megapolitan" geography of the United States. His current work is on how changing demographics and preferences will reshape metropolitan America to mid-century.

Marla Nelson is an Associate Professor in the Department of Planning and Urban Studies at the University of New Orleans, where she serves as coordinator of the Master of Urban and Regional Planning program, the only accredited planning program in the State of Louisiana. Her areas of expertise include local and regional economic development, community development, and urban revitalization. Dr. Nelson's most recent research has focused on counteracting population decline in cities struggling with equity, efficiency, and environmental management in the implementation of redevelopment strategies to strengthen their infrastructures. She has also examined the locational preferences of socially motivated individuals, members of the "creative class," whose influx into New Orleans for the purposes of assisting in the rebuilding and recovery efforts following Hurricane Katrina helped the city to recover from a "brain drain" that had plagued it for decades. Dr. Nelson's research has been funded by the US Department of Housing and Urban Development (HUD), the National Science Foundation, the Lincoln Institute for Land Policy, and the Louisiana Board of Regents. In March of 2011, she received an Activist Scholar Award at the Urban Affairs Association's 41st Annual Meeting. She is the Vice President and Board Member of the Urban Conservancy, a non-profit organization dedicated to research, education, and advocacy that promote the wise stewardship of the urban built environment and local economies.

Ryan Max Rowberry is an Assistant Professor of Law and Associate Director for the Center for the Comparative Study of Metropolitan Growth at Georgia State University College of Law. He teaches Property Law, Natural Resources Law, Environmental Law, and Anglo-American Legal History. Professor Rowberry's research concentrates on cultural heritage, historic preservation, and natural resources law. He co-authored *Historic Preservation Law in a Nutshell*, a ground-breaking book that provides the first in-depth summary of historic preservation law within its local, state, tribal, federal, and international contexts. Rowberry graduated from Harvard Law School, where he was an Islamic Legal Studies Fellow, a Cravath International Fellow, and recipient of the Irving Oberman Award in Legal History. He subsequently practiced environmental and natural resources law at Hogan Lovells in Washington, DC. Professor Rowberry also briefed foreign judicial delegations on the American legal system and created a reference guide to foreign legal systems for American judges during his yearlong appointment with the Supreme Court of the United States as a Supreme Court Fellow. Professor Rowberry was selected as a Rhodes Scholar upon earning a B.A. in English from Brigham Young University, and he later received an M.Sc in Comparative Education Policy and M.St in Medieval British History from Oxford University.

Sarah B. Schindler is Professor of Law and Glassman Faculty Research Scholar at the University of Maine School of Law, where she teaches in the areas of property, land use, local government law, real estate transactions, and animal law. Prior to joining the Maine Law faculty in 2009, Professor Schindler clerked for Judge Will Garwood of the Fifth Circuit Court of Appeals in Austin, Texas, and practiced in the area of land use and environmental law at Morrison and Foerster in San Francisco. She was a Visiting Assistant Professor at the University of Georgia School of Law, and taught as a guest lecturer both at U.C. Berkeley School of Law (Boalt Hall) and U.C. Hastings College of Law. Professor Schindler researches emerging trends in municipal land use law, particularly those related to sustainable development. Her articles have been published in the *Yale Law Journal*, the *George Washington Law Review*, the *University of Chicago Law Review's Dialogue*, and the *Wisconsin Law Review*, among others. One of her recent articles was selected to be reprinted in the *2013–2014 Land Use and Environmental Law Review*, an annual, peer-selected compendium of the 10 best land use and environmental law articles of the year. She was also named as Pace Environmental Law Center's Distinguished Young Scholar of 2013. Professor Schindler received her J.D., *summa cum laude*, from the University of Georgia School of Law.

Celina Su is Marilyn J. Gittell Chair in Urban Studies and an Associate Professor of Political Science at the City University of New York. Her research concerns civil society and social policy, especially health and education. Her work focuses on how everyday constituents and citizens engage in and shape policy making and community development—via deliberative democracy, community organizations, and protest and social movements. Her publications include *Streetwise for Book Smarts: Grassroots Organizing and Education Reform in the Bronx* (Cornell University Press, 2009); *Our Schools Suck: Young People Talk Back to a Segregated Nation on the Failures of Urban Education* (co-authored, New York University Press, 2009); and *Introducing Global Health: Practice, Policy, and Solutions* (co-authored, John Wiley/ Jossey-Bass, 2013). She co-founded Kwah Dao/ Burmese Refugee Project (www.burmeserefugeeproject.org) in 2001, which employs participatory models to foster community development among Shan Burmese refugees in northwest Thailand, and has served on New York City's participatory budgeting Steering Committee since 2011. Her honors include the Berlin Prize and the Whiting Award for Excellence in Teaching. She earned her Ph.D. in Urban Studies from the Massachusetts Institute of Technology and her B.A. Honors in English and economics from Wesleyan University.

Foreword
Debunking Urban Myths about Cities

Overview

We humans are the world's first domesticated animals. Before domesticating other animals and crops we became social creatures assisting one-another in multiple tasks. As we discovered things about nature that improved well-being, those discoveries were communicated to future generations. Discoveries led to inventions that further improved well-being. Over eons, inventions made us increasingly productive to the point where not everyone was needed to track down or manage animals, or grow crops; hamlets, villages and cities emerged. The rise of cities was aided by inventing and reinventing social, political and administrative institutions. City-states became nation-states. But for millions of years most people did not live in cities. Only since 2006 have most of the world's people lived in them; by the end of this century more than two-thirds will. The world is coalescing around mega cities and larger urban mega regions. This is a good thing for humanity.

To be sure, cities historically were very, even extraordinarily, unpleasant places in which to live. An aristocrat in Pompeii lived next to a tannery in which urine was boiled to tan leather. A thousand years later, men would wear brimmed hats walking street-side protecting the women they escorted from the trash being thrown upon them from upper floors. In the 19th century, a large share of cities had open sewers; animal excrement coated (usually dirt) streets and fouled the air; coal-fired heat darkened the air and in many larger cities killed thousands during certain episodes; and the list goes on. From the beginning of cities, those who could chose lived away from them or vacationed outside them (Bruegmann). The rise of suburbia occurred as much as a reaction to the living conditions of cities as the ability to live away from them through advances in transportation technology.

This is changing. Edward Glaeser advances several arguments supporting this proposition in his 2011 book, the title of which says it all: *Triumph of the City: How Our Greatest Invention Makes Us Richer, Smarter, Greener, Healthier, and Happier*. This book, *How Cities will Save the World: Urban Innovation in the Face of Population Flows, Climate Change and Economic Inequality*, picks up where Glaeser's ends.

For one thing, we keep being reminded that as population density increases important economic benefits follow. A doubling of metropolitan area population increases jobs proportionately by about 15 percent (Bettencourt and West 52–53); a doubling of density increases jobs by 6 percent proportionately (Ciccone and Hall 54–70); and as density increases so does innovation as measured through patents (Bettencourt et al. 7301–7306). Incomes also rise. In a way, cities become their own self-fulfilling prophesies.

As cities grow birth rates also fall. Certainly one reason is that some of the largest and most densely settled cities, such as Tokyo, are viewed as too costly to raise families. But a larger reason is that unlike an agrarian population that depends on manual labor to survive—much of it derived from farmers' own children—an urban population has no such dependency. As cities grow the rate of the world's population growth falls. For instance, India's birth rate, which was about 5.2 live births per women between 15 and 45 in the 1950s, will fall to replacement rate or less (2.1 live births) before 2020 (Population Reference Bureau). Indeed, largely because of urbanization, the world's population growth rate will reach replacement, or less, well before this century's end, thus leading to stabilized if not reduced population thereafter (Yale Global Online).

Let me put this into context. Of the world's 10 billion people about mid-century, seven billion will be living in cities and their larger metropolitan areas—more than double the number in 2010. Does this mean the world's consumption of resources will increase by more than double with its proportionate effects on the environment? Does it also mean the ills of city-living will be accentuated? Urban myths to the contrary, the simple answer is "no."

An *urban myth* is modern folklore rooted in so-called common sense that is believable to its intended audience. Many urban myths characterize cities as places unworthy of human habitation compared to suburban and rural alternatives. I address such urban myths as crime and safety, personal health, air pollution, the so-called benefits of country living, and overall quality of life. There are many other urban myths about cities that need to be dispelled. This foreword addresses some of them while many others will be put to rest in *How Cities will Save the World*.

Crime and Safety

It is often alleged that US cities, especially large ones, are unsafe places to live and work. Myers et al. challenge this myth directly (408–418). In their analysis of 1,295,919 injury deaths in 3,141 US counties between 1999 and 2006, they found that the probability of death from violent crime and accidents is 22 percent higher in rural areas than large U.S. cities, with suburbs falling in between. While there may be more deaths in large cities than the rural countryside, only because populations are larger, the odds of being killed are higher in rural areas. I am reminded of what Sherlock Holmes says about this:

> It is my belief, Watson, founded upon my experience, that the lowest and vilest alleys in London do not present a more dreadful record of sin than does the smiling and beautiful countryside . .... The pressure of public opinion can do in the town what the law cannot accomplish. There is no lane so vile that the scream of a tortured child, or the thud of a drunkard's blow, does not beget sympathy and indignation among the neighbours, and then the whole machinery of justice is ever so close that a word of complaint can set it going, and there is but a step between the crime and the dock. But look at these lonely houses, each in its own fields, filled for the most part with poor ignorant folk who know little of the law. Think of the deeds of hellish cruelty, the hidden wickedness which may go on, year in, year out, in such places, and none the wiser (Doyle 1892).

To be sure, where poor populations are concentrated in urban areas—perhaps through a range of public policies that steer them into those areas—crime rates can be very high. Nonetheless, Kneebone and Raphael find that "[t]he gap between city and suburban violent crime rates declined in nearly two-thirds of metro areas. In 90 of the 100 largest metro areas, the gap between city and suburban property crime rates narrowed from 1990 to 2008. In most metro areas, city and suburban crime rates rose or fell together" (1).

There is an important theory underlying the reason why more densely settled areas should have less crime than less densely settled ones. Jane Jacobs hypothesized that through increased human interaction and with more "eyes upon the street" combined with buildings oriented towards the streets themselves, cities would be safer than other settled landscapes (35). Christens and Speer tested this hypothesis (113). They aggregated data to the census block group level to compare crime rates between the urban center of a city (Nashville, TN) with the central county (Davidson County) along with non-urban parts of the county. Using Jacobs' theory, they hypothesized that "population density at the census block level would negatively predict violent crime in the urban areas" (113). After controlling for a variety of socioeconomic and location characteristics, their analysis supported the hypothesis.

Personal Health

Writing for the *Wall Street Journal*, Beck observes that "[f]or many urban dwellers, the country conjures up images of clean air, fresh food and physical activities. But these days, Americans residing in major cities live longer, healthier lives overall than their country cousins—a reversal from decades past." The story reviewed a report published by Bennett, Bankole and Probst who studied all U.S. counties finding that rural residents are more prone to obesity, diabetes, stroke, heart attacks and high blood pressure than urban residents. In reviewing the details of their analysis, I find a continuum effect in which the higher the level of urbanity the greater the level of personal health. Earlier work by Ewing et al. shows that as counties become more sprawled the incidence of obesity with associated personal health problems increases (47–57).

Air Pollution

There is a myth that cities have more air pollution than their suburbs. This has been debunked by Jones and Kammen who use national household surveys to develop econometric models to estimate average household carbon footprints (HCF) for U.S. zip codes in metropolitan areas as well as cities, counties, and metropolitan areas themselves (895–902). They find lower HCF in urban core cities and higher carbon footprints in suburbs. For instance, for an "urban" zip code near downtown Atlanta, Georgia (30312), households generate the equivalent of 32.8 tons of carbon dioxide each but in "suburban" Sandy Springs (zip code 30338), about 16 miles north, households generate 62.9 tons each or nearly twice as much (Thorpe).

A common argument I have heard is that even if suburbs generate more pollution per household, suburban areas on the whole generate less pollution than downtowns and areas close to downtown because of the greater concentrations of activity towards the center than the periphery. This urban myth is not supported by the facts. These two zip codes noted above are substantially built out with not much new development possible. The urban zip code has a density of about 2,100 households per square mile generating nearly 69,000 tons of carbon per square mile. The suburban zip code had density of about 1,400 households per square mile generating more than 87,000 tons of carbon per square mile.

A common problem with air pollution is not what is generated within the downtown or closer-in, higher density areas, but rather what is generated in the lower-density, sprawling suburban areas and often times blown into higher density areas.

There is no question that high levels of human activity combined with high levels of impervious surface can create an "urban heat island" effect in which the ambient air temperature is higher than in suburban areas. At a certain threshold, higher temperature can generate more ozone, a form of air pollution. Yet, there are already simple solutions to reduce urban heat islands down to levels of moderate suburban density, and below. These include green roofs, street trees (especially well-placed ones), and high albedo (sun-reflecting) additives to asphalt used for pavement and roofs.[1] As roofs are replaced and street/parking surfaces replaced over their standard useful lives of about 20 years, and as street trees mature over the same period, the adverse effect of the urban heat island on ozone production can be all but eliminated.

The So-Called Benefits of Country Living

In *Triumph of the City*, Edward Glaeser confronts the urban myth that humankind is better off living Henry David Thoreau's *On Walden Pond* country life:

> Residing in a forest might seem to be a good way of showing one's love of nature, but living in a concrete jungle is actually far more ecologically friendly. We humans are a destructive species, even when, like Thoreau, we're not trying to be. We burn forests and oil and inevitably hurt the landscape that surrounds us. If you love nature, stay away from it (201).

Consider the U.S. context. Based on data published by the Economic Research Services of the United States Department of Agriculture, I estimate that 30 million acres of land were converted from natural resource uses to urban, suburban and exurban uses between 1974 and 2007— a 70 percent increase (Nickerson, et al. 5). In contrast, the U.S. population increased by about 40 percent. I use recent research to estimate the weighted average annual value of ecosystem services per acre at roughly $5,000 in the U.S. (DeGroot et al. 50–61; Costanza et al. 152–58). Using a 2.0 percent discount rate roughly reflecting high-grade borrowing rates in the middle 2010s, the fully capitalized intergenerational value of land converted to urban, suburban and exurban uses is about $50,000 per acre. American society thus lost about $1.5 trillion in ecosystem services over this time period. From an economic perspective, if the market for land internalized fully its ecosystem services value to society, less land would be converted, while that which is would be developed more efficiently.

Quality of Life

Mathis reports that 84 percent of suburban residents rated their communities as excellent or good places in which to live, in contrast to urban (75 percent) and rural (78 percent). While popular opinion suggests suburbs confer a higher quality of life overall than cities, it is cities that are becoming increasingly preferred after considering numerous tradeoffs. Writing for the Harvard Business Review, Ania Wieckowski observed that because of their improved accessibility to services and jobs, improved opportunities to walk, broader housing choices, higher degree of amenities, and other factors, businesses and individuals

---

1   For a review, see http://science.howstuffworks.com/environmental/green-science/urban-heat-island2.htm.

alike are shifting their preferences away from suburbs to cities. The article quotes Robert Fishman of the University of Michigan explaining the shifting preference as follows: "In the 1950s, suburbs were the future. The city was then seen as a dingy environment. But today it's these urban neighborhoods that are exciting and diverse and exploding with growth."

Concluding Observations

I conclude with insights from an emerging field of study called urban metabolism (Holmes and Prencitl). It measures the resources moving into cities (such as energy, water, and food), how they are "metabolized," and their byproducts (such as air pollution, solid waste). A growing body of research is showing that more densely populated areas (my interpretation of the data being above 15,000 persons per square mile) consume half the energy per capita as the average American suburban density (about 2,000 persons per square mile) and about a third of rural areas (less than 200 persons per square mile). If these relationships are confirmed through future research, a world in which 70 percent of a mid-century population of 10 billion live in cities of more than 15,000 persons per square mile (less than the 2015 densities of metropolitan London, Los Angeles, and New York) would consume fewer resources and generate less waste than the whole world did in 2015. Literature also suggests these people would be healthier, live longer, be more productive, and happier. *How Cities will Save the World* shows in much greater detail how this will be accomplished.

Arthur C. Nelson, Ph.D., FAICP
Professor of Planning & Real Estate Development
Associate Dean for Research & Discovery
College of Architecture, Planning and Landscape Architecture
University of Arizona

Works Cited

Beck, Melinda (2011). "City vs. Country: Who Is Healthier? Urban Areas Clean Up, Residents Live Longer, Stay Fitter; But Stress Is Less in Rural Regions." *Wall Street Journal* 12 July 2011. n. pag. Web. 23 Dec. 2014 <www.wsj.com/articles/SB10001424 052702304793504576434442652581806>.

Bennett, Kevin J., Bankole Olatosi and Janice C. Probst. *Health Disparities: An Urban—Rural Chartbook.* Columbia, SC: South Carolina Rural Health Research Center, 2008. Print.

Bettencourt, Luís M. A. and Geoffrey B. West. "Bigger Cities Do More with Less." *Scientific American* 305 (Sept. 2011): 52–53. Print.

Bettencourt, Luís M. A., et al. "Growth, Innovation, Scaling, and the Pace of Life in Cities." *Proceedings of the National Academy of Sciences* 104.17 (2007): 7301–7306. Print.

Bruegmann, Robert. *Sprawl: A Compact History.* Chicago: U of Chicago P, 2005. Print.

Christens, Brian and Paul W. Speer. (2005). "Predicting Violent Crime Using Urban and Suburban Densities." *Behavior & Social Issues* 14.2 (Fall/Winter2005): 113. Print.

Ciccone, Antonio and Robert E. Hall. "Productivity and the Density of Economic Activity." *American Economic Review* 86.1 (1996): 54–70. Print.

Costanza, Robert, et al. "Changes in the Global Value of Ecosystem Services." *Global Environmental Change* 26: 152–158. Print.

deGroot, Rudolf, et. al. "Global Estimates of the Value of Ecosystems and their Services in Monetary Units." *Ecosystem Services* 1 (2012): 50–61. Print.

Doyle, Arthur Conan. "The Adventure of the Copper Beeches." *Strand Magazine* (1892).

Ewing, R., et al. "Relationship Between Urban Sprawl and Physical Activity, Obesity, and Morbidity." *American Journal of Health Promotion* 18.1 (2003): 47–57. Print.

Glaeser, Edward. *Triumph of the City: How Our Greatest Invention Makes Us Richer, Smarter, Greener, Healthier, and Happier*. New York: Penguin, 2011. Print.

Holmes, Tisha and Stephanie Pincetl. *Urban Metabolism Literature Review*. Los Angeles: Center for Sustainable Urban Systems, U of California at Los Angeles, 2012. Print.

Jacobs, Jane. *The Death and Life of Great American Cities*. New York: Random House, 1961. Print.

Jones, Christopher and Daniel M. Kammen. "Spatial Distribution of U.S. Household Carbon Footprints Reveals Suburbanization Undermines Greenhouse Gas Benefits of Urban Population Density." *Environmental Science & Technology* 48 (2014): 895−902. Print.

Kneebone, Elizabeth and Steven Raphael. *City and Suburban Crime Trends in Metropolitan America*. Washington: Brookings Institution, 2011. Print.

Mathis, Sommer. "Overall, Americans in the Suburbs Are Still the Happiest: The First Results of Our State of the City Poll." *CityLab* 25 Aug. 2014: n. pag. Web. 25 Dec. 2014. <www.citylab.com/politics/2014/08/overall-americans-in-the-suburbs-are-still-the-happiest/378964/>.

Meyer, William B. *The Environmental Advantages of Cities: Countering Commonsense Antiurbanism*. Cambridge: MIT, 2013. Print.

Myers, Sage R., et al. "Safety in Numbers: Are Major Cities the Safest Places in the United States?" *Annals of Emergency Medicine* 62.4 (2013): 408–418. Print.

Nickerson, Cynthia, et al. *Major Uses of Land in the United States, 2007*. Washington: U.S. Department of Agriculture, Economic Research Services, Dec. 2011. Print.

Population Reference Bureau. "India: On the Path to Replacement-Level Fertility?" *World Population Data Sheet*. Population Reference Bureau, 2011. Web. 25 Dec. 2014 <www.prb.org/Publications/Datasheets/2011/world-population-data-sheet/india.aspx>.

Thorpe, David. "Suburbs Emit More Carbon Dioxide than Cities—Study." *Sustainable Cities Collective*, Jan. 8, 2014. Web. 25 Dec. 2014 <sustainablecitiescollective.com/david-thorpe/212726/suburbs-emit-more-carbon-dioxide-cities-study>.

Wieckowski, Ania. "Back to the City." *Harvard Business Review* May 2010. n. pag. Web. 25 Dec. 2014 <https://hbr.org/2010/05/back-to-the-city/sb1>.

Yale Global Online. "Global Population of 10 Billion by 2010? -- Not So Fast." 26 Oct. 2011. Web. 25 Dec. 2014 <yaleglobal.yale.edu/content/global-population-10-billion-not-so-fast>.

Preface

Recent global events have featured cities in stories of hope and despair, violence and peaceful political protest, crisis and celebration. Riots in Baltimore, xenophobic violence in Johannesburg, peaceful political encampments in Hong Kong, signs of rebirth in post-bankruptcy Detroit: these are all examples of cities serving as the locus of change, engagement, crisis and opportunity. Cities stand at the center of the challenges posed to our local, regional, national and global communities. They host not just the trials, tribulations, and triumphs of the human condition, but they also generate the innovations that are solving problems, renewing traditions, and marking new chapters in the human story.

While mindful of the formidable challenges that cities present for families, for civil society, and for government, we see these challenges as occasions for ingenuity, innovation, equity, foresight and, above all, pragmatism. Free from national and global politics, though always acting in their shadow, they are places where creative problem solving flourishes out of necessity, often apart from partisan rancor. Cities know how to get things done, and they are doing just that all across the world.

It is to cities that we must turn for solutions to the problems the world faces, because it is in cities where the challenges of the 21st century will play out: climate change, migration, and economic inequality. Cities are the future, and this volume explores just some aspects of a future dominated by cities, which will be the source, testing ground and locus of solutions to the problems of this century and the centuries to come.

Ray Brescia and John Travis Marshall

Acknowledgments

Ray Brescia

I am grateful for the support of Albany Law School in helping to give me the assistance and confidence to see this project through to fruition, most particularly President and Dean Alicia Ouellette and my former colleague Timothy D. Lytton, who has moved on to join the faculty of my co-editor, John Travis Marshall at George State University College of Law. Dean Ouellette and Professor Lytton have been tireless advocates for excellence and have encouraged me to pursue this project through their unwavering support.

I also wish to thank my colleagues at the Government Law Center at Albany Law School who make my job easy, and are creative problem solvers on issues great and small: Rose Mary Bailly, Robert Batson, Mary Berry, Emily Ekland, Amy Gunnells, Bennett Liebman, Barbara Mabel, Michele Monforte, Maureen Obie, Melissa Perry, and Lisa Rivage. My research assistants and Albany Law students, Cassandra Rivais and Carrie Dessereau, both supplied a level of grace, patience, attention to detail and skill that made invaluable contributions to this volume. I wish to extend my thanks to my law school classmate and friend, Paul Sonn, who reviewed prior drafts of some of the work contained in this volume. All mistakes and omissions are solely mine, however.

None of my own contributions to this work would have been possible without the encouragement of my spouse, Amy Barasch, a child of cities, whether it is her native New York or her adopted Paris, and the spirit of my son, Leo Brescia, whose ideas on our "Brainstorming Nights" help get me through the week.

A child of the suburbs myself, I fell in love with cities through the lens of the Bronx in the mid-1980s and I have never turned back. As a practicing attorney, I had the great honor of engaging in community organizing in Harlem, representing worker organizations in Chinatown, and forming and advising not-for-profits throughout New York City that struggled to take on the challenges all urban communities face. It is this work, though now mostly in my past, that animates the ideas that take shape in this book, and inspires my contributions to this volume.

John Travis Marshall

In September 2007, I resigned from my Tampa, Florida, law practice and moved to New Orleans to join the New Orleans Redevelopment Authority's growing staff. Arriving more than two years following the August 2005 levee breaches, I anticipated finding neighborhoods in the throes of revitalization. Instead, my colleagues and I found a city where vast stretches of once-densely populated neighborhoods were almost completely deserted. High grass masked front doors and windows. Only a handful of stores were open. New Orleans, one of the nation's oldest cities, was still teetering from Hurricane Katrina's blow, and it was not outwardly clear if the city could ever rebound from disaster.

Then I came to know the leaders of grassroots neighborhood redevelopment organizations, including the Broadmoor Improvement Association, the St. Roch

Improvement Association, the Lakeview Civic Improvement Association, and the Gentilly Civic Improvement Association. Although the federal funding to spur New Orleans's neighborhood revitalization was too little and trickled into the city at too slow a pace, the city's residents worked nights and weekends and spent their life's savings to press forward with the revitalization, house-by-house and block-by-block.

The mortgage foreclosure crisis may have reminded us that our neighborhoods can be torn apart by global financial decisions, but New Orleans's journey to recovery taught me that neighborhoods are a city's engine. When buffeted by fundamental challenges a city's neighborhoods are a locus of citizen innovation, cooperation, political power, and hope. In reading this book's chapters, I appreciate it is no coincidence that many authors highlight the importance of neighborhoods and their citizens as a city's firmest foundation. I thank those citizen leaders, the nonprofit developers, the foundation program officers, and the dedicated government staff—particularly at the New Orleans Redevelopment Authority—who taught me about mending an ailing city while also envisioning its promising new future. A Rockefeller Foundation Fellowship provided the opportunity to work in New Orleans. I thank the Rockefeller Foundation and the cohort of 27 mid-career professionals that relocated to New Orleans to play even a small role in helping the city's citizens travel the road to recovery.

I am grateful for the generous support I have received from my colleagues at Georgia State University's College of Law. Dean Steven Kaminshine ensured that I had two summer research grants and four wonderful graduate research assistants to assist me with every aspect of this book. Associate Dean for Research and Faculty Development, Wendy Hensel, continues to create great opportunities for me to present my cities scholarship at conferences and workshops. My colleagues and mentors Julian Juergensmeyer and Ryan Rowberry have given me the opportunity to travel to Asia, Europe, and Africa to research metropolitan response to crisis and disaster. They bring a stream of renowned cities scholars from across the country and the world to teach our law students and discuss the pressing problems facing urban communities. I am thankful for Julian and Ryan's thoughtful suggestions, support and encouragement regarding this book. I am indebted to them for generously agreeing to author chapters on topics for which they are widely known for their expertise. Ann-Margaret Esnard of Georgia State's Andrew Young School of Public Affairs has not only modeled what it means to be a cities scholar, but she has shown extraordinary generosity in sharing opportunities and contacts with me. Karen Johnston, Esq., Assistant Director of the Center for the Comparative Study of Metropolitan Growth, brings the Center's active Atlanta and international programming to life. Her diligent work and planning has helped make it possible for me to see firsthand the critical issues examined in this volume. I am also deeply appreciative of Karen Butler for her kind patience and great humor and for meticulously managing all range of administrative details pertaining to the Center's travels, meetings and visiting scholars.

My contributions to this book would not have been possible without excellent research and editing assistance from my team of research assistants during the 2014–15 school year, including Josh Joel, Peter Watson, Danielle Weit, and Joseph Collins. Josh, Peter, Danielle, and Joseph each showed me an attention to detail and follow-through that will make them wonderful attorneys. The College of Law's library team has also helped me at every turn in the writing and editing process, particularly Pamela Brannon and Austin Williams.

Although I grew up in Boston's suburbs, my parents, Marie Quigley Marshall and Willoughby Marshall, viewed cities as special places of beauty, culture, celebration, learning, opportunity, and reflection. Their work over the years as architect, environmental planner, non-profit executive, and school teacher required them to engage communities' challenges and dreams. I am grateful to my parents for making their children part of those civic engagements by bringing them to neighborhood meetings or by bringing their stories home to the dinner table. I have inherited their love of cities and I hope to be as good and caring a steward and advocate for cities as they have been.

Both Ray and John also thank Ashgate Publishing for its willingness to take on this project, and for all of the authors who contributed to this work, without whom it never would have come to fruition.

Introduction

Raymond H. Brescia and John Travis Marshall

The greatest challenges the world faces can be found, and will be solved, in its cities. Greenhouse gasses are disproportionately generated in cities, and the harshest effects of climate change hit cities the hardest. Economic inequality is concentrated in cities, but at the same time, economic growth is generated in cities. Given these phenomena, the social, economic and environmental solutions to the challenges the world faces in the coming decades—if they will be solved—will arise in and emanate from its cities. This volume not only describes how innovative cities from across the globe are addressing some of the hardest-to-solve problems the world faces today, it also presents novel and pioneering solutions that different cities can adopt to address these problems. It is a guide for policymakers, urban planners, elected officials and citizens who wish to confront some of the world's greatest challenges, challenges which will be solved by cities, if they are to be solved at all.

Cities are not only population centers. They are knowledge centers. They serve as first homes for immigrant families and young professionals alike, and often the permanent settling places for both. They are fertile environments for commerce and incubators for innovation. They protect a dense layering of a country's and a region's cultural, historic, and artistic stories. For millennia, cities have been a powerful beacon for people seeking opportunity. Just a few years ago the world reached a tipping point: a majority of the world's population now resides in its urban centers, a figure which is likely to reach 70 percent by the mid-point of the century.

Despite their prominence, however, and the likelihood that cities will be the locus of problem solving in the 21st century, cities are vulnerable. Their past accomplishments are not necessarily a prologue to their future successes. Three major sources of adversity have recently emerged that have challenged, and promise to challenge, the welfare and in some cases the basic viability of cities. Climate change, population shifts, and crises precipitated by economic inequality have etched in stark contrast cities' relative successes and failures at plotting a course for a sustainable future. We have firsthand experiences from cities here in the United States and across the world of the destructive and disruptive power of each of these forces. For examples of this power, we need look no further than the urban ruins left in the wake of Hurricane Sandy, the hemorrhaging of Detroit's population, and the vacant properties that litter cityscapes in the wake of the mortgage foreclosure crisis. We cannot escape the tragic stories and statistics that document a widening economic and achievement gap between low-income and upper-income families. We also cannot sit still. The stakes are only growing.

Climate change, population shifts and economic inequality are challenges so significant and fundamental that they demand new approaches—large and small—in how cities engage in local governance. In this volume, the authors not only capture recent and current challenges posed by climate change, population shifts and economic inequality, but they also frame concrete policy changes and new core capacities that cities must consider adopting to have the best chance at survival. This volume also analyzes the critical problems currently facing cities and provides specific recommendations and guidance on the range of interventions that 21st century cities would be prudent to consider in mapping their immediate responses to these critical problems. Taken together, these chapters describe not only the story of recent urban crises, but also the steps that cities have taken and ought to take to address these problems. The chapters that follow show how cities can use the law to manage the daunting challenges they face. The authors identify specific innovations that will help cities tackle continuing adversities caused by the forces unleashed by climate change, population shifts, and economic inequality.

Before we provide an overview of the upcoming chapters, we want to say a little bit about the subject matter of this work: urban innovation. Admittedly, many of the chapters, though not all, focus on innovations that are currently underway in the US. In this way, the book has a very US-centric outlook. While one could compile a work such as this on urban innovation that is taking place in six continents around the world, we focus on the US for several reasons.

Primarily, it is what we know. The authors come from many different disciplines but research, practice, and govern in the US. Some of them speak from personal experience in urban innovation, others have gained empirical knowledge about what works, and still others use a range of methodologies to understand local issues and local experimentation. But there are other reasons as well for why we chose to focus primarily on urban innovation that is occurring in the US.

While the world's population recently tipped the balance, and a majority now live in cities for the first time in human history, the US crossed that threshold roughly one century ago. As a result, it has a head start on much of the rest of the world in managing the transition from a primarily rural to a primarily urban population. But it also has a track record of urban innovation dating back to long before it became a nation where most of its citizens lived in cities. Boston, New York, Philadelphia: these were all urban engines of economic development, technological experimentation, and political ferment, even when the nation was primarily one made up of farmers.

In addition, the governmental structure in the US lends itself to urban innovation. Eighty years ago, US Supreme Court Justice Louis Brandies described the flexibility to innovate available to states in the American federal system as follows: the "state may, if its citizens choose, serve as a laboratory; and try novel social and economic experiments without risk to the rest of the country." Cities and local governments are the rooms inside those laboratories, where local experimentation can flourish just as far as the imagination, the political will, and the urban pragmatic spirit can take it.

Like English cities in the industrial revolution, and Paris in the *Belle Epoque*, urban centers in the US—from San Francisco and the surrounding regions, to Austin, Texas, and Boston, Massachusetts—are hotbeds of activity and innovation, in technology, culture and the arts, and are serving as economic engines that are fueling not just the US, but also the global economy. Its largest city, New York City, is the world's center of finance, arts and entertainment, fashion and the media.

While US cities serve as hubs of innovation and economic development, they are doing so despite the same kind of national political environment plaguing many nations across the globe. Like many countries today, national leaders in the US are caught in a political deadlock, swaying between austerity budgets and tax cuts, and not being able to pass meaningful legislation. Yet, despite this, pragmatic urban officials are exploring effective urban innovations and are doing more with less, simply out of necessity.

For these reasons, the innovations taking place in cities across the US can perhaps provide insights into how cities in other nations might address some of the problems we discuss here: climate change, economic inequality, and shifting populations. By no means is the US the only source of innovation in these areas, nevertheless, we have chosen to start here. We encourage others to join the debate and offer other examples of urban innovation from across the globe that are designed to address these and other looming crises that are currently gripping, or may yet grip, all of the nations of the world.

In light of these looming crises, we begin by examining urban crisis and response. Cities frequently face fundamental challenges to their continuing vitality and, in certain rare instances, their existence. An important investment for any city is to study the challenges that it should anticipate and the tools and systems necessary to address those challenges. Ray Brescia, John Travis Marshall, Ryan Rowberry, and Marla Nelson and Renia Ehrenfeucht examine fundamental threats to long-term metropolitan growth posed by natural disasters, climate change, and economic crises. Their pieces explore recent and historical catastrophes that have befallen cities and suggest how cities can and could respond to crisis and disaster.

Co-Editor Ray Brescia analyzes the causes of the Financial Crisis of 2008, including its under-appreciated roots in the deep economic inequalities that divide urban communities. Many have offered theories that attempt to identify the causes of the recent financial crisis. Some blame deregulation and a culture of greed on Wall Street. Others argue that lawmakers and presidents promoted homeownership too aggressively, sending mortgage credit to low-income communities to serve borrowers with poor credit and little likelihood of paying back their mortgage. Too often, various charges enjoy little empirical support. Taking a hard look at some of the economic indicators present in the buildup to the crisis, one fact stands out; prior to the crisis, the US experienced a stunning increase in income inequality.

There are several possible explanations for the potential connection between rising income inequality and the great strains on the economy it causes. Brescia offers one such explanation: that both income inequality and racial inequality created greater social distance and this social distance, in turn, led to greater predatory conduct. That predatory conduct turned a mortgage market into an economic killing field.

The review of the empirical information presented in his chapter uncovers critical information that may reveal new insights into the potential causes of the financial crisis. This analysis suggests not that low-income African-Americans are to blame for the foreclosure crisis, as some argue, but rather, that middle-class African-Americans were targeted for, and steered towards, loans on unfair terms, precipitating the foreclosures that are now concentrated disproportionately in communities of color.

Apart from the causes of the crisis, its effects are hitting hardest in the nation's cities, where foreclosed properties and abandoned homes mean cities collect fewer property taxes, but they also have to spend diminishing and precious city resources on policing such homes. Since cities bear the brunt of much of this economic fallout, they need tools to attempt to mitigate it. To address this issue, Brescia examines strategies cities are deploying, or are considering deploying, to deal with the impacts they are facing as a result of the Financial Crisis of 2008 and its lasting effects. Specifically, his chapter describes a range of these strategies, namely: litigation to combat discriminatory predatory lending, responsible banking ordinances, so-called "Community Impact Report Cards" for financial institutions, and the use of eminent domain to address underwater mortgages.

There is no way to predict disasters, but cities, states, and the federal government can take coordinated, concrete steps long before disasters strike to assess community vulnerability and create more resilient cities. In this next chapter, Co-Editor John Travis

Marshall describes the catastrophic toll that recent and widespread natural disasters have inflicted on cities and the factors contributing to a painfully slow path to recovery. He recommends creating a Resiliency Index as a tool to promote better local and national response to urban adversities, including crisis and disaster, and to help ensure quicker long-term urban recovery.

If states and the national government in the US can make informed predictions about local governments' challenges and competencies, then pre- and post-disaster planning for long-term recovery efforts can be more strategic and less a matter of guesswork. As the Gulf Coast continues its long-term recovery from Hurricane Katrina and as the tri-state area of New Jersey, New York, and Connecticut continues the journey toward recovery from Hurricane Sandy, a special opportunity exists to examine the weaknesses that leave communities susceptible to catastrophic loss and the strengths that allow cities to rebound more quickly.

Marshall calls for governments to develop a succinct evaluation of the range of community resources—governmental and non-governmental—that will figure critically in implementation of any long-term disaster recovery efforts. On completion, this assessment will provide such governments and the general public with a "snap shot" of a local government's resiliency: its capacity to bounce back from disaster. In particular, this chapter recommends creation of a Resiliency Index that evaluates community resources and capabilities critical to recovery.

Marshall notes that following disaster, the civic landscape, like the city's geographic landscape, is littered with obstacles. Each obstacle compounds problems associated with achieving long-term recovery. If local governments can remove obstacles from that landscape before disaster happens, then they can better focus on overcoming the adversities associated with disasters to rebuild neighborhoods and cities.

The story of catastrophic disasters fundamentally altering or substantially destroying cities is not new. It is age-old. Ryan Rowberry looks closely at disasters' impact on cities' historic and cultural resources. Cities support not only bustling commerce and vibrant social life but they also protect a layered living, and broad cultural record: architecture, art, monuments, music, industry, archaeological sites, education, religious observance, sports, ethnic immigration, and a palimpsest of successive city plans. These urban cultural resources are arguably what give cities their unique character, and they are without question economic magnets that draw significant numbers of people and a constant flow of money to urban areas. Rising sea levels, more frequent and more intense storms, and natural disasters threaten many aspects of urban communities, but none more so than cities' often fragile cultural resources.

In this chapter, Rowberry recounts the demise of the ancient city of Alexandria, Egypt. Once the world's preeminent city, Alexandria succumbed to two catastrophic disasters more than a century apart. Not only are Alexandria's streets and buildings largely lost to history, but its wealth of historical and cultural artifacts were destroyed or scattered. Rowberry cautions that it is critical that we find in Alexandria's fate something much more relevant than an ancient epic narrative of destruction. Like their ancient cousin, Alexandria, scores of major cities worldwide are historic cities caught in the crosshairs of potential disaster, ranging from earthquakes to changing tidal patterns causing increased erosion, rising sea levels, and major storms. This chapter will detail the risk factors threatening coastal cities and will evaluate programs and policies that national and local governments have established to protect urban cultural resources, particularly in the face of disaster.

How might cities that are recovering from major disasters, shrinking in size, or threatened by climate change encourage settlement patterns that allow for the most efficient allocation of urban resources and services? Marla Nelson and Renia Ehrenfeucht tackle a challenge that has largely confounded policymakers; they evaluate how a residential relocation program—in the form of a land swap program—might be designed to serve as a valuable tool for cities coping with hazards and population shifts.

As cities face a range of fundamental challenges to their continued viability, including increasing environmental vulnerability due to climate change, and population shifts that leave neighborhoods empty and with declining public services, residential relocation has become a critical site for action. In the US, the relocation of communities away from hazard-prone areas has become an important mitigation strategy. Meanwhile, in cities facing widespread abandonment, scholars and practitioners have advocated for the relocation of residents out of sparsely inhabited neighborhoods to more densely populated areas as a key component of "rightsizing" initiatives. Given the long history of negative outcomes associated with government-sponsored relocations, however, and the importance placed on private property rights, proposals involving the relocation of residents, whether forced or voluntary, are fraught with political, logistical, and financial obstacles.

Nelson and Ehrenfeucht's chapter examines the evolution of, and challenges to, voluntary relocation efforts in the United States. The authors begin with an exploration of key conceptual issues surrounding voluntary relocation programs—including the difficulty distinguishing voluntary from forced relocations—and identify factors that inform and constrain relocation decision making. From there they use the successes of Project Home Again, an innovative home building and land swap program developed in New Orleans after hurricanes Katrina and Rita, as a case study to show the possibilities and limitations of voluntary relocation efforts. They conclude with a discussion of how Project Home Again can inform pre- and post-disaster recovery planning as well as land reconfiguration efforts in legacy cities.

Catastrophic threats, such as hurricanes, keep mayors from sleeping and they spur cities' emergency managers regularly to drill their communities' response to crises. Cities' vigilance in preparing for these perils protects entire regions. It helps ensure that no disaster or crisis event could deliver a knock-out blow to a city's economy, historic fabric, culture, or neighborhoods. Fortunately, existential challenges to cities' welfare are rare; however, cities face a range of day-to-day challenges. These challenges can undermine the long-term ability of families to live safely and with access to basic resources and opportunities; these challenges also can impede businesses and their ability to compete. One of the ways cities solve problems is by considering challenges to their ongoing sustainability. Sarah Schindler, Hanna-Ruth Gustafsson and Elizabeth A. Kelly take on current and future questions concerning metropolitan sustainability.

Kelly and Gustafsson examine the renowned urban planning decisions that have made Curitiba, Brazil an international exemplar for sustainable urban development. Curitiba has experienced explosive population growth in the past half century, but has also achieved great success in mitigating the negative effects of urban growth, in large part due to its dedication to integrated planning. Curitiba has developed groundbreaking and cost-effective transportation models, flood prevention systems, and recycling programs that are primed for export, and have succeeded in large part due to close collaboration between the government agencies entrusted with developing and administering the component programs. At its core, Curitiba exemplifies how the integration of land use planning, transportation infrastructure, and environmental sustainability efforts can enable a city to meet the needs of its expanding population without accepting sprawl and environmental degradation as foregone conclusions. As such, both its innovations and the principles underlying them are deeply relevant to expanding metropolises worldwide.

Curitiba's story is particularly important to tell because cities in the US do not tend to look far for innovative solutions to urban problems. While the problems of urban growth are experienced worldwide, US policymakers tend to seek solutions largely from their constituent states or their Organization for Economic Cooperation and Development (OECD) neighbors. This narrow focus ignores sources of potentially valuable lessons, many of which may actually be more relevant to the US than those currently studied, as well as more cost-effective. Kelly and Gustafsson seek to turn the reader's attention from the oft-cited practices of European cities to an example further south. It highlights forward-thinking planning in Curitiba, Brazil that has been successful in alleviating the interrelated problems of urban growth: sprawl, environmental degradation, and economic inequality.

Taking stock of the profound questions that face so many cities relating to adequate water resources, Sarah Schindler takes a provocative and forward-looking view at policies that will help cities conserve valuable water resources. Recognizing their role in sustainability efforts, many local governments are enacting climate change plans, mandatory green building ordinances, and sustainable procurement policies. But as California and other western US states wilt under another extended drought, Schindler pursues the insight that, thus far, local governments have largely ignored one of the most pervasive threats to sustainability: lawns. Schindler's chapter examines the trend toward sustainability mandates by considering the implications of a ban on lawns, which are the single largest irrigated crop in the United States. She also considers municipal authority to ban or substantially limit pre-existing lawns and mandate their replacement with native plantings or productive fruit- or vegetable-bearing plants. Although this proposal would no doubt prove politically contentious, local governments—especially those in drought-prone areas—might be forced to consider such a mandate in the future. Furthering this practical reality, she addresses the legitimate zoning, police power, and nuisance rationales for the passage of lawn bans, as well as the likely challenges they would face. She also considers more nuanced regulatory approaches that a municipality could use to limit lawns and their attendant environmental harms, including norm change, market-based mechanisms such as progressive block pricing for water, and incentivizing the removal of lawns.

Tight municipal budgets mean that cities must stretch resources to clean streets, plow snow, and maintain school buildings. They also mean constrained capacity, as cities struggle to maintain staffs and fight attrition. In spite of these day-to-day challenges, many cities—and states in partnership with cities—manage to devise innovative solutions to persistent problems. These innovative cites test solutions to emerging policy challenges to ensure safer, more secure, and more vital neighborhoods for their residents. Ray Brescia, Mayor John DeStefano, John Lovett, Kathleen Morris, Kermit Lind, and Celina Su provide detailed accounts of how cities in the US have responded to the calls for livable minimum wages for their residents, better schools, a more supportive community for recent immigrants, safer housing conditions, greater transparency in decision-making regarding city spending, and a local marketplace free of deceptive or unscrupulous business practices.

The role of municipalities as "laboratories of democracy" has only grown in recent years. While localities have long been public policy innovators, the gridlock and other political process obstacles that impede innovation with respect to addressing economic inequality at the federal and state levels have made municipalities more important than ever as proving grounds for policy reform. Due to this gridlock at the national level in the

US, and in many state legislatures, local governments and advocates have taken the lead in experimentation around one policy reform designed to combat economic inequality: raising the minimum wage. Ray Brescia reviews the history of minimum wage experimentation at the state and local levels and shows that such experimentation has often led to policy reform, not just at the state and local levels, but also at the national level in the US. His chapter looks at strategies that have been deployed and are being deployed to raise minimum wage standards at the local level, and assesses the possibilities and pitfalls such experimentation can surface. It reviews the legal landscape for such experimentation and shows that when advocates and officials press for local minimum wage increases, broader political and legal change sometimes follows.

Similarly, political gridlock at the national level has also left critical questions unanswered about federal immigration policy. Many cities have simply clamored even more loudly for federal relief and guidance, but at least one city seized the opportunity to help provide for the safety and security of newcomers from abroad. Immigrants have been at the forefront of city building in the US for centuries. The current debate over immigration reform on the national level skips over the essential role that city policies play in dealing with recent immigrant communities. John DeStefano, former Mayor of the City of New Haven, Connecticut, explores the inclusive approach this city has taken towards its immigrant communities, including the issuance of a resident identification card that is offered to all city residents regardless of immigration status. This chapter recounts the ground-up approach of advocates in the city who pushed the mayor and his staff to adopt inclusive and welcoming legal and policy approaches toward immigrants, in the area of both public safety and government administration. What this chapter also recounts is how these inclusive approaches resulted in an aggressive backlash—whether intentional or unintentional—from national authorities. After New Haven instituted some of its policies towards immigrants, both documented and undocumented, national authorities stormed local homes of recent immigrants seeking, ostensibly, to execute on outstanding deportation orders. Mayor DeStefano recounts these experiences and the local response, providing insights into the role of local government officials and advocates in promoting an inclusive immigration policy.

While cities rarely craft policies or programs for immigrants because immigration matters are generally considered part of the federal domain, almost all cities find themselves at the center of a contentious and urgent discussion about school reform. Few cities have taken the type of leadership role assumed by New Orleans and the State of Louisiana. Years before the 2005 levee breaches decimated New Orleans, another tragedy began to unfold that jeopardized New Orleans' future. Public schools across the city were suffering from years of disinvestment, declining student achievement, and a continuing drop in white student enrollment. This educational catastrophe, whose wreckage consisted of persistently high drop-out rates and low college attendance among graduates, was not nearly as photogenic as the devastation wrought by Hurricane Katrina. But the long-failing status of New Orleans' schools had produced a landscape of despair almost as heartbreaking as the historic storm: a generation of children largely unprepared to thrive in an increasingly competitive global world.

The story of deeply underperforming and under-resourced city schools is not unique to New Orleans. US cities from Atlanta to Chicago to Houston to Los Angeles and dozens of urban schools districts in between have struggled for decades to provide children with an educational experience that promises a future of opportunity. US cities continue to pursue strategies to create better schools for their children. However, as John Lovett details, no city has implemented a plan for educational reform with the breadth and scope of the plan adopted by New Orleans following Hurricane Katrina. Over the course of just a few years, the State of Louisiana dismantled New Orleans' traditional public school district that largely assigned students to a school based on geographic zones and replaced it with a system comprised almost entirely of charter schools managed by their own boards and offering families a portfolio of school choice options stretching across the entire city. No other school district in the country has come close to matching the radical changes implemented in New Orleans. As Lovett notes, with 91 percent of New Orleans public school children attending charter schools, Detroit was a distant second with 55 percent of its public school children attending charter schools.

The State of Louisiana's effort to overhaul New Orleans's troubled public school system is widely recognized as one of the country's most far reaching educational reform initiatives. It may also be the most widely watched as its successes and failures will be used as a sword and a shield in ongoing battles about how best to improve our children's schools. Numerous studies continue to gauge almost every aspect of New Orleans's charter school transformation. Lovett examines recent scholarship studying charter schools' success at creating social capital to help improve student educational outcomes. Starting with the intellectual roots of the 20th century concept of social capital, Lovett describes why scholars and educators consider social capital surrounding schools important to students, families, and the larger city. Lovett's chapter helps all stakeholders to the education reform debate appreciate that public schools' impact on social capital creation is an important variable to consider in evaluating the contribution that charter schools make to comprehensive educational reform and to urban renewal.

While revitalizing public schools presents an enormous challenge for cities, this charge lies at the heart of a local government's responsibilities to its citizens. But what about individuals and companies who prey on city residents? What role can or should cities play in vindicating the rights of their residents?

Cities have historically sat on the sidelines, as state attorneys general or the US Department of Justice has sought redress for the types of harms that might generally impact urban residents. However, over half of the nation's population lives in urban areas and, according to a recent 2012 Pew Research Center poll, Americans trust local governments at a significantly higher level than both state and federal government. What are cities and their citizens missing by not pursuing claims to protect the rights of the people who live and work there, and what role could affirmative litigation play in making cities better places to live? In an era where large swaths of cities have been laid waste by deceptive lending practices and financial institutions that refuse to maintain foreclosed properties, important questions arise as to the redress available to cities. Kathleen Morris explains that a nascent template for local government public interest litigation is beginning to emerge as an area of legal innovation for cities.

Morris describes how the City of San Francisco, through its local civil law office, has aggressively pursued public interest cases on behalf of economically and socially vulnerable constituents. In essence, San Francisco has created an in-house public interest law firm. That firm has brought cases vindicating a wide range of public policy interests. They include the pursuit of a clean environment, protection of public health, access to health care, consumer protection (especially vis-à-vis banks and credit card companies), and defense of immigrant groups from economic fraud. But San Francisco is not the only city breaking new ground. Several other US cities have amassed public interest dockets but scholars have yet to document those efforts. Morris's chapter fills that gap by canvassing major

US cities—including New York City, Los Angeles, Washington, D.C. and others—and describes their efforts to employ litigation as a tool to vindicate the public interest.

As Morris describes, city law departments have rarely played the role of public advocate challenging unfair or deceptive business practices. But this does not mean that cities have abdicated all civil enforcement roles. Cities have traditionally attempted to protect neighborhoods and their residents through enforcement of public health and safety codes. In the wake of the Great Recession and the accompanying mortgage foreclosure crisis, code enforcement became every city's growth industry. Unfortunately, few if any cities were prepared to meet the challenge of skyrocketing rates of residential property abandonment and the rampant violations of city housing, health, and safety code that accompanied abandonment. In his chapter expounding strategic local government code compliance measures, Kermit Lind explains why cities were unprepared. Coping with declining local tax revenues and diminishing federal resources, many US cities were already understaffed or suffering administrative atrophy when the first waves of foreclosed properties swept across the nation's real estate markets. As a result, cities were fundamentally unprepared to triage and manage the thousands and thousands of real estate "casualties" of the subprime crisis.

According to Lind, cities' struggles continue. He urges that for some cities their failure to address major deficiencies in housing and building code compliance poses very real threats to their continued viability. Lind presses that communities must rebuild code enforcement capacity, highlighting three key capacities: (1) a core of committed community development and code enforcement leaders who collaborate across bureaucratic and institutional boundaries; (2) a comprehensive property information system that compiles data from public offices for use in data-driven study and decision-making; and (3) a strategic, proactive approach to the development and deployment of limited resources for specific results. In this chapter, Lind describes in detail why each of these three core capacities is essential to a city's successful long-term recovery from the vacant property crisis, and he tracks successes and challenges encountered by jurisdictions currently implementing public policy solutions to the vacant property crisis.

There is a critical need for cities to invest in improving their code enforcement programs and the bureaucratic and technical systems that support them, but it is relatively easy to tell when those systems are not operating well. City and neighborhood dysfunction that allows dilapidation and abandonment to multiply is difficult to conceal. Overgrown yards, crime, and vermin associated with blight is easy to spot from a car window. That is not necessarily the case with a city's decisions about how and where to spend its money. For years, city's spending decisions were largely made behind closed doors. That is changing. Over the last four years, participatory budgeting (PB) has spread from one to more than 45 cities in the US. It was also listed as a cornerstone of the Obama Administration's "Open Government" initiative last year, and it is touted as an increasingly popular model of "participatory democracy" in the US.

Drawing upon surveys, interviews, and observations, Celina Su writes about the New York City experience of PB, examining its contested role as an empowering tool for equity. Su details how, in New York's case, PB has successfully and dramatically broadened notions of stakeholdership and citizenship for many constituents (especially youth and undocumented citizens). Su also explains that the process of researching, collaborating with others, and deliberating local priorities and project proposals appears to have been transformative for many participants. Still, PB has not necessarily prompted a re-prioritization of budget allocations just yet. This is partly because of the limits of the funds allocated to PB thus far, and partly because the governmental bureaucratic structures that help to implement PB—such as city agencies—remain wary of participatory decision-making in local governance. Thus, in order to reach its transformative potential and address economic inequalities, PB processes must allow for coordination across neighborhoods so as to achieve economies of scale and true redistribution in racially and economically segregated landscapes. As Su describes, they must also provide training for both lay participants and city agency representatives, especially helping the latter to act as translators between local and technical knowledge, and to abide by the spirit (rather than the letter) of constituent proposals. Finally, many of the most exciting developments surrounding PB lie in spillover effects. There is evidence that the experience of participating in budgeting has educated, empowered, and even outraged constituents to demand more, and to hold government more accountable: not only regarding local discretionary expenditures, but regarding the municipal budget overall. These promising developments are likely to have ripple effects beyond PB, in other forms of public administration and governance.

Streets, sidewalks, highways, transit corridors, and bike lanes tie our cities together. They are the sinews that unite urban areas and they are the arteries that bring cities to life in the most basic and essential ways: carrying us to work or school and back, defining the supply chains that bring us food, and connecting us to our most critical needs and simplest pleasures. Transportation resources are just one class of infrastructure, but one of the most critical and cities will not thrive unless they are consistently maintained. The US entered the 21st century with essential transportation infrastructure barely adequate to meet current demands, not to mention unable to address future burdens associated with increased population, heavy continuing use, diminished government investment in upkeep, or climate change. The challenge of replacing or rebuilding this infrastructure looms larger every year.

In this volume's concluding chapter, Julian Juergensmeyer and Becht Neel grapple with what many consider the most vexing challenge facing 21st century cities: how to replace aging transportation infrastructure while funding new infrastructure to accommodate rapid urban growth? There are no easy solutions. But there are promising options available and they are options that cities must consider to realize the new road, highway, bridge, and mass transit systems for which cities are in dire need.

Current funding streams largely depend on local government property taxes and state support, and federal grants are too small to satisfy the need. It is imperative that local and state governments have appropriate financing tools at their disposal. Juergensmeyer and Neel outline the range of promising financing mechanisms that local and state governments must consider adopting to support infrastructure redevelopment and construction, including front-ending, mobility fees, and value capture requirements.

No single discipline or field of study yields all or even most of the cutting-edge policy ideas that make cities effective problem solvers. Transformative and innovative problem solving requires an interdisciplinary team armed with a range of experiences. The following chapters—written by urban planners, land use and public interest lawyers, current and former policy professionals from the national and local level, and a former mayor with 20 years' experience—come from seasoned practitioners and thinkers who have witnessed the strategic role cities play in crafting cutting-edge solutions to the most perplexing policy problems. We hope this volume stimulates continued discussions about the important role cities can play as 21st century problem solvers.

Chapter 1
Cities and the Financial Crisis

Ray Brescia

The Financial Crisis of 2008, and the foreclosure crisis that both preceded and followed it, resulted in financial hardship for many. Large financial houses closed or were absorbed into other financial institutions. Homeowners in the United States lost trillions of dollars in the collective value of their homes. Countless businesses shut their doors forever. Families were foreclosed upon and evicted. Many are still unemployed or have simply stopped looking.

Cities have borne the brunt of much of these aftereffects of the Financial Crisis. Indeed, it is in cities where the Financial Crisis's toll is, perhaps, the greatest. Cities often must deal with foreclosed and abandoned properties that scar the landscape and become magnets for crimes like arson, prostitution, and drug dealing. Shuttered businesses mean empty storefronts and reduced sales revenue. The precipitous drop in home values in many communities has meant a concomitant reduction in tax revenue for many cities. Couple that with the fact that abandoned properties rarely produce local tax revenues and one sees many cities hovering near the brink of bankruptcy, if they have not already hurtled into that financial abyss.

While much of the oversight of financial institutions in the United States occurs at the national level, and, to a lesser extent, in the states, cities are not powerless to combat the effects of the Financial Crisis on their bottom lines or community well-being. Some cities are fighting back, and many have the tools at their disposal to protect their citizens, strengthen their sources of revenue, and combat predatory behavior from financial institutions. This chapter will review both the causes of the Financial Crisis of 2008 as well as its consequences on US cities, with a particular focus on the connection between these causes and consequences and economic and racial inequality. It will then describe an array of tactics that cities are deploying to respond to the lasting effects of the Crisis. Specifically, this chapter will describe a range of these tools, namely: litigation to combat discriminatory predatory lending, responsible banking ordinances, so-called "Community Impact Report Cards" for financial institutions, and the use of eminent domain to address underwater mortgages.

These local regulatory tools offer cities and other local governments the chance to have a say in the way that financial institutions function and operate within their borders. They give these local governments a role in addressing some of the fallout from the Financial Crisis of 2008, and can help shape financial institution behavior moving forward. As this chapter explains, some of the harshest effects of financial crises can occur at the local level, where foreclosures—just one common by-product of the last such crisis—wreak havoc on local government functions and finances. Because of that, local governments need tools to address the fallout from such crises and the ability to try to prevent the next one. If the Financial Crisis of 2008 was the result of legal and regulatory failures at the national level, and this chapter will explore that question, cities, which are affected most dramatically by financial crises, should ensure they are doing all that they can to help shape financial institution behavior.

The Impact of the Financial Crisis on Cities

The main impact on US cities of the Financial Crisis of 2008 is the loss of property tax revenue. This loss of revenue comes about for three main reasons. First, foreclosures drive down property values in neighboring properties, as described below. This reduction in property values often results in a downward revaluation of properties for tax purposes with a corresponding reduction in the amount of taxes paid by homeowners. When a property is foreclosed upon, it is often sold at auction at a bargain basement price. In addition, in a so-called "short sale," a pre-foreclosure sale, a property is also sold at a reduced rate. Appraisers then use the ultimate sale price of these properties as "comparable" property values when appraising neighboring properties. These lower appraisals then can factor into the taxes homeowners pay their local governments. Second, when a property is abandoned, the owner of that property rarely pays taxes on it. Finally, when a community is particularly hard hit by abandonment, this drives down investment in the community and results in residential flight, further reducing the tax base. But not only does a municipality lose tax revenue, it also must often expend resources to maintain and police abandoned properties.

A number of studies over the last 15 years have attempted to assess the impact of foreclosures on the value of neighboring properties, and the picture that emerges is a bleak one. In the late 1990s, a group of researchers looked at the impact of foreclosures in the Chicago area and found that the value of each single-family home within one-eighth of a mile of a foreclosed property dropped by between 0.9 percent and 1.136 percent for each foreclosure (Immergluck and Smith 9). Each foreclosed property reduced the value of neighboring properties in the aggregate by as much as $371,000 (Ibid., 11). A recent study by the Federal Reserve Bank of Atlanta found that property values in the late 2000s, on average, were reduced by nearly one percent due to nearby foreclosures. The researchers attributed at least some of that reduction to disinvestment in properties in foreclosure and delays in the foreclosure process (Gerardi Kristopher, et al. 13–14, 33). While early studies of the current foreclosure crisis predicted a modest reduction in home values as a result of this crisis of between $356 billion to $1.2 trillion in home values (Pew Charitable Trusts 10; Global Insight 2), one current analysis is far more dramatic, as the US Government Accountability Office assesses the loss of homeowner equity as a result of the Financial Crisis at over $9 trillion (2013).

Foreclosures and short sales are not the only ways that neighboring homeowners lose value in their homes. And they are not the only ways that municipalities suffer when properties are foreclosed upon and/or are abandoned. When a property is neglected by the homeowner, or post-foreclosure, by the foreclosing entity that takes title to it, it can create an eyesore in the neighborhood or worse as those parties fail to maintain the lawns, facades, and other cosmetic elements of the property, or fail to ensure necessary repairs are made in a timely fashion. Once a property falls into disrepair, the cost of repairs can increase exponentially. A simple water leak or defective heater, if neglected, can lead to extensive and structural damage. Clearly, abandoned properties become a magnet for crime.

All of these forces align to not just lower property values, which has a ripple effect on city coffers, but also to force local governments to spend resources on property maintenance, emergency response to fires and injuries occurring on site, and policing these properties (Apgar, Duda and Nawrocki Gorey, 10–11). When homeowners in arrears on their mortgages also neglect to pay their property taxes, cities can choose not to expend resources to foreclose on their tax liens, instead choosing to sell such liens for a fraction of their face value (Ibid., 7). One study of the impact of the recent foreclosure crisis on states estimated that taxing entities in just 10 states stood to lose a total of $6.6 billion in tax revenue in 2008 alone (Global Insight, 5). Cities lose even when they decide simply to demolish abandoned properties rather than allow them to stand, thereby eliminating their attractiveness to criminals and vandals. The cost to municipalities to address vacant homes, whether through policing, response to a fire, or other interventions, can cost a local government as much as $34,000 per property (Apgar, et al., 23).

While many of these forces have aligned to diminish the tax base of municipalities hard hit by foreclosures, what they, together with other economic forces unleashed by the Financial Crisis of 2008, are also doing is threatening the population base of many cities, particularly those in the Northeastern United States. Of course, this threat is part of a trend that has taken place over the last 50 years, as many of the so-called "Rust Belt" cities have seen reductions in population of over 50 percent in the last 50 years, including Buffalo, NY; Cleveland, OH; Detroit, MI; St. Louis, MO; and Youngstown, OH.

It is evident from this review that US cities are bearing much of the financial brunt of the lasting effects of the Financial Crisis. But what were some of the causes of the crisis, and what are some potential strategies cities can deploy to remedy its effects? This next section explores the forces that coalesced to create the impacts cities continue to face as a result of the Financial Crisis of 2008. Subsequent sections will explore potential responses cities can undertake to address such impacts and prevent future crises.

The Forces at the Heart of the Financial Crisis

While it is said that success has many parents while failure is an orphan, the causes of the Financial Crisis can be traced back many decades. This brief overview highlights some of the causes of the Financial Crisis and then attempts to tie them back to the impacts of the Crisis on municipalities.

Deregulation of the US financial industry in the 1980s and 1990s cleared a path for the creation of the subprime mortgage products at the center of the Crisis (Mansfield, 493). At the same time, mortgage lenders began to automate the mortgage underwriting process. This allowed them to write more mortgages (Financial Crisis Inquiry Commission, 72). They also developed new mortgage products that became increasingly exotic, with new features that created new markets for borrowers previously excluded from home finance because they could not satisfy lenders' underwriting criteria (Moran, 21–22). While this criteria may have traditionally kept certain communities from gaining access to credit regardless of their members' creditworthiness, the new criteria often allowed those who could not repay a mortgage access to mortgage finance (Engel and McCoy, 33). Another development, namely, the ability to securitize mortgages, when coupled with new mortgage products and new markets, allowed financial institutions to increase the number of mortgages they could write. This new financial vehicle—the securitized mortgage—permitted mortgage lenders to convert what were historically the future income streams from mortgagor payments into liquid assets; these liquid assets could then fuel more mortgage lending (Ibid., 17–26). What is more, new compensation schemes rewarded mortgage brokers who would increase loan volume regardless of borrower creditworthiness. Similarly, mortgage lenders would receive fees from investment banks for packaging loans for securitization. This came to be known as an "originate-to-securitize" approach, where the goal of mortgage origination was simply to gather more loans for securitization and sale on the secondary mortgage market where investors were hungry for such products (Financial Crisis Inquiry Commission, 8).

Credit rating agencies also received fees for their work, blessing these mortgage-backed financial products with little regard for their future financial viability or the strength of the underlying mortgages (Financial Crisis Inquiry Commission, 43–44).

In the background of all of this financial engineering was a time of relaxed monetary policy that encouraged bank lending because those banks' costs to borrow were incredibly low (Bar-Gill and Warren, 3–5). Low relative returns on Treasury bills encouraged investors to seek the higher returns typically promised through investments in mortgage-backed securities (Financial Crisis Inquiry Commission, 85). These expected returns from mortgage-backed securities whet the appetite of investors to invest in them at precisely the time when mortgage lenders increased their home mortgage lending as described above, through new mortgage originations and mortgage refinancing. Outsized profits in the mortgage-backed securities market spurred investment banks to encourage mortgage lenders to generate more mortgages (Moran, 24–25). Once mortgage lenders had saturated the market through loans to all of those traditional or "prime" borrowers they could find, they used new products to reach less creditworthy borrowers with subprime loans (Engel and McCoy, 35–36).

This combination of forces—new sources of capital, new mortgage products, and lowered underwriting criteria—meant an influx of funds in the home mortgage market. This influx helped to create an asset bubble. Once that asset bubble popped, it generated a dramatic drop in home values. For borrowers in adjustable rate mortgages, reduced home values meant they could no longer refinance their mortgages to avoid the rising interest rates as mortgage lenders were no longer refinancing mortgages where the value of the home was less than the outstanding principal on the mortgage the borrower was trying to refinance (Ibid., 22–26). As borrowers lost the ability to refinance, they began to default on their loans, tipping the financial dominos that ultimately took down the economy.

Looking Deeper: Economic and Racial Inequality at the Heart of the Financial Crisis

This review of some of the forces behind the Financial Crisis of 2008 and its aftermath overlooks yet another potential cause of the Crisis, one that exists, not at the level of laws, regulations, and policies, but in the market, where bankers, brokers, and borrowers engaged in millions of transactions across the US. These transactions reveal perhaps another cause of the Crisis: social distance. Social distance is a term used to describe the differences we perceive between ourselves and others, between ingroup members and outgroup members, and status differences (Liberman, Trope, and Stephan, 357). A review of the lending practices that led up to the Financial Crisis and the subsequent path of the foreclosure crisis that followed reveals that social distance in two particular manifestations—economic inequality and racial differences—may have lurked beneath many of these transactions, resulting in trillions of dollars in home equity losses, foreclosures and evictions.

The height of the foreclosure crisis likely occurred in 2010, a time when income inequality had followed a decades-long stretch in which it increased considerably in the United States. One common measure of the economic inequality that exists in a society is the Gini coefficient or Gini index. As the US government's Census Bureau describes it, the Gini Index "varies from 0 to 1, with a 0 indicating perfect equality, where there is a proportional distribution of income. A 1 indicates perfect inequality, where one household has all the income and all others have no income." (Noss, 1.) Over the years, the Gini Index for the United States as a whole has fluctuated, but it has increased over the last few decades. When compared to other nations, the US Gini Index ranks the US among countries that are not typically associated with the US in terms of economic fairness, like Uruguay, Georgia, and El Salvador (World Bank, 2015).

An assessment of mortgage delinquency rates by states in 2010 reveals that states with higher levels of income inequality, as measured by the Gini Index, had, on average, higher delinquency rates. I use mortgage delinquency rates here as a proxy for predatory conduct because loans issued during the mortgage frenzy of the last decade that had predatory features were much more likely to enter into delinquency and default than loans without such features (Ding, et al., 2010). The figure below, which reveals this connections, plots delinquency rates by states in relation to each state's Gini coefficient. The higher the Gini score, on average, the greater the delinquency rate.

Racial distinctions can also create social distance. And if social distance based on race was one of the drivers of predatory mortgage lending, we would likely see the extension of credit on less-than-favorable terms along racial lines, which is precisely what happened in the middle part of the last decade. Research conducted by staff at the US Federal Reserve analyzed data available as a result of the requirements of the Home Mortgage Disclosure Act, through which financial institutions report their lending practices. This research reveals that, in 2006, nearly 54 percent of home mortgages extended to African-American borrowers had sub-prime features, while just 18 percent of loans to Whites had similar features, a nearly three-to-one ratio. Even controlling for a range of differences between Whites, African-Americans and Latinos, including income, these researchers found that over 30 percent of African-American borrowers, 24 percent of Latino borrowers, and just 18 percent of White borrowers received loans with predatory features. This research thus reveals that—even controlling for differences in income—African-American borrowers and Latino borrowers were 75 and 36 percent more likely, respectively, than White borrowers to enter into loans with subprime features (Avery, et al.).

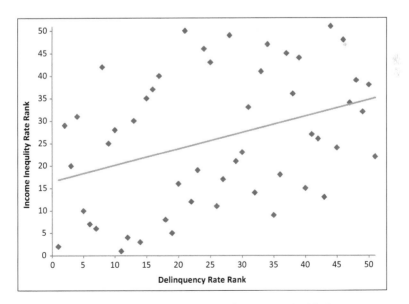

Figure 1.1   Delinquency Rates/Gini Index by State (Brescia, 2011)

The *New York Times* analyzed lending data in the New York City metropolitan region during the height of the subprime lending frenzy of the last decade and found similar patterns, particularly as such lending impacted the African-American middle class. This analysis yielded the following conclusion: "[T]he hardest blows rain down on the backbone of minority neighborhoods: the black middle class. In New York City, for example, black households making more than $68,000 a year are almost five times as likely to hold high-interest subprime mortgages as are whites of similar—or even lower—incomes." (Powell and Roberts, 2009).

But risky lending did not just impact borrowers adversely, as evidenced by mortgage delinquencies and foreclosures. Indeed, well over 100 mortgage lenders shuttered their businesses in the wake of the mortgage frenzy, and analysis of their lending patterns reveals that they, too, tended to lend disproportionately to borrowers of color, extending loans to them on unfair terms. Analysis of 2006 HMDA data from 167 financial institutions that closed their doors in 2007 shows that a disproportionate share of their subprime lending involved borrowers of color compared to industry averages. Seventy-four percent of the loans made to African-Americans by these failed lenders were subprime loans, and 63 percent of the loans made to Latinos had such features, compared to the industry average of 54 and 47 percent, respectively, as indicated in Figure 1.2.

While some have tried to blame government policies that promoted the extension of mortgage credit to communities of color as well as the borrowers of color who defaulted on their mortgages in disproportionate rates for the economic downturn that was spurred by the mortgage crisis, analysis of mortgage lending in the lead up to this crisis strongly suggests that it was not risky borrowers but risky lending on predatory terms that led to the downturn in the market and the broader financial crisis that followed. The Center for

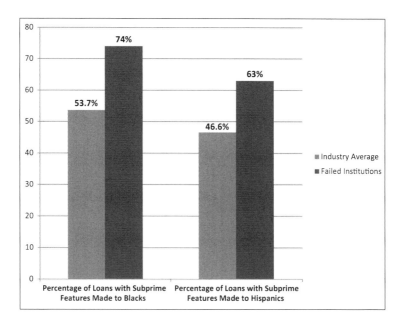

Figure 1.2   Data on 2007 Failed Institutions (Avery, et al, 2008; Avery, et al., 2007)

Community Capital at the University of North Carolina conducted a study comparing the performance of subprime loans to those using more traditional underwriting criteria. That study compared this performance using borrowers of similar economic profiles in each lending pool. The results reveal that subprime loans made in 2004 were four times as likely to enter into default as the traditional loans and loans made in 2006 were over three times more likely to do so (Ding, et al., 2010).

Another view of the reasons for the demise of subprime lending suggests that borrower fraud may have played a part in the poor performance of such loans, but what seems more likely is that lenders engaged in a degree of fraud to promote lending to unwitting and uncreditworthy borrowers. A Federal Bureau of Investigation study released during the height of the mortgage frenzy estimated that 80 percent of the losses stemming from mortgage fraud were the result of lender or mortgage broker misconduct (Federal Bureau of Investigation, 2005). Furthermore, an internal study conducted by Washington Mutual, one of the failed financial institutions that had expanded its subprime lending dramatically during the last decade, found that 83 percent of loans generated by just one of its own California branches were tainted by some form of fraud by bank officials (Levin, 2010).

What does all this data suggest about some of the additional causes of the foreclosure crisis? Can we tie social distance caused by racial and income inequality back to those causes? Social distance can lead to predatory conduct, as economic, racial and social differences can lead individuals to seek to take advantage of and prey on those not like them. As we have seen, higher delinquency rates appeared in states with higher income inequality. Is this an indication that economic differences led some to take advantage of others? Similarly, subprime lending was targeted towards borrowers of color,

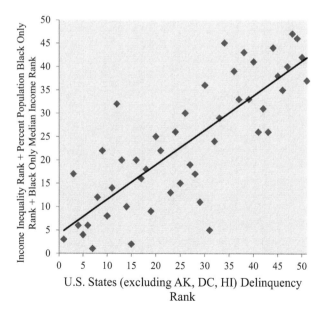

Figure 1.3   Targeting the Black Middle Class (Brescia, 2011)

particularly those in the middle class. Were the effects of economic and racial inequality compounded, such that one might see even poorer loan performance in areas with higher income inequality and greater concentrations of the African-American middle class? The preceding graph combined a number of data points by state: greater economic inequality, median income of the African-American community in a state, and percentage of the African-American community in a particular state. This data certainly suggests that states with higher income inequality and a larger and wealthier African-American middle class were also states where predatory lending (as indicated by higher delinquency rates) seems to have flourished (Brescia, 2011). If what this research suggests holds true, one might conclude that predatory conduct fueled by social distance may have played some role in the fallout from the aggressive predatory lending of the last decade.

Cities Respond

This information suggests that the Financial Crisis has its roots, at least in part, in economic and racial inequality, which were in some ways a product of the legal environment in which this inequality could flourish. A desire to participate in the homeownership market as a strategy for improving one's economic situation certainly led many to pursue home purchases. Communities of color that have endured a history of racial discrimination and fewer traditional mortgage options as a result of that history relied on unscrupulous mortgage brokers and banks that extended mortgage credit to such communities on discriminatory terms.

These are both national policies and trends, but they have had a significant impact on local economies. The fallout from the overextension of credit in the lead up to the Crisis has been dramatic, particularly in cities, where concentrated foreclosures have led to dramatic loss of property values and, in turn, loss of tax revenue for municipalities in the communities hit hardest by the effects of the Financial Crisis. While cities were not responsible for these national policies and trends, they are, in many respects, left to deal with their consequences. Furthermore, political gridlock at the national level means help is not on the way for municipalities dealing with distressed mortgages and economically distressed communities. Indeed, if anything, some elected officials in the national government are looking to weaken financial oversight, not strengthen it. All indications are that US cities cannot look to the national government for help to strengthen the regulatory infrastructure that will prevent the next financial crisis, or for strategies for dealing with the economic fallout from the last.

The remainder of this chapter is dedicated to ways cities can respond to the consequences of the Financial Crisis of 2008, consequences that still plague cities years after the peak period of fallout from the Crisis. It also offers ways that cities can serve as a check on predatory conduct, and look to lessen economic and racial inequality. What follows is a description of four somewhat distinct, somewhat interconnected strategies that cities have deployed or could deploy in the coming years to either remedy the harshest consequences of the Financial Crisis, or attempt to prevent similar crises from occurring in the future. They are the following: litigation, responsible banking ordinances, bank report cards, and eminent domain to address so-called underwater mortgages. Each will be discussed, in turn.

Litigation

In 2010, Congress passed the landmark legislation that has come to be known as Dodd-Frank, named for its chief sponsors, then Connecticut Senator Christopher Dodd and then Massachusetts Congressman Barney Frank, both Democrats. Through the law, Congress looked to close some of the weaker elements of the regulatory infrastructure upon which financial institution safety and soundness is founded. While one of the goals of the legislation was to restore trust in the financial system, and to prevent the next financial crisis, one of the things it failed to do was provide relief to homeowners who faced foreclosure, or lost their homes, and the municipalities that have been left to deal with a wake of abandoned homes, bank-controlled properties, and diminished property values. Just as law enforcement officials at the federal and state levels have attempted to use the courts to rectify some of the worst abuses that led to the Financial Crisis, cities, too, have gotten into the act and have used litigation to obtain some degree of financial relief from the worst fallout from the Crisis. The following discussion describes, first, the success of federal and state officials at using the courts to address this fallout, and then turns to the ways that cities are following suit.

The government response to just one bank, Bank of America, and its subsidiaries, the Countrywide family of companies, is perhaps the best example of federal and state government use of the courts to rectify the more abusive conduct that occurred during the lead up to the Financial Crisis of 2008. Countrywide Financial was one of the more aggressive home mortgage lenders in the early 2000s, and it had devised a dizzying array of mortgage products, each requiring less documentation and lower underwriting standards than the next. Like other subprime lenders in the last decade, Countrywide came up with a number of products that were designed for one purpose: to get more and more borrowers into mortgages with the bank. As the available supply of so-called "prime" borrowers dried up since many had already taken out mortgages, or refinanced existing mortgages, banks needed to continue to feed the desire of investors for mortgage-backed securities. In response, institutions like Countrywide found creative ways to generate new products that could cater to individuals who would not qualify for traditional mortgage products. Bank of America, seeking to bring this mortgage machine into its constellation of subsidiaries, purchased Countrywide and its affiliates in 2006. What it ended up buying was a lot of litigation.

In 2008, Bank of America entered into an $8 billion settlement with a series of state attorneys general over shoddy lending practices of Countrywide in extending subprime loans to borrowers. An example of the types of claims brought by these state law enforcement officials can be found in the complaint filed by California's then Attorney General, now Governor, Jerry Brown. The California complaint alleged that Countrywide used a range of tactics, like low "teaser" rates that would ultimately balloon into much higher charges that were designed to lure unwitting borrowers into loans regardless of their ability to repay them. The goal of the lending, Brown's complaint asserted, was to steer people into as many mortgages as possible so that Countrywide could resell those mortgages on the mortgage-backed securities market. Added to this $8 billion settlement with state attorneys general was a $16 billion settlement that Bank of America reached with the US Department of Justice to resolve claims related to the practices of the bank and its subsidiaries in the lead up to the Financial Crisis. All in all, according to one count, Bank of America has settled $50 billion in claims stemming from its conduct and that of its subsidiaries during the lead up to the Financial Crisis (*Economist*, 2013).

But Bank of America is not the only financial institution that has had to make billion dollar payments to settle claims and to stave off legal investigations for those institutions' conduct that led to the Financial Crisis. Here are some examples of the settlements that large financial institutions have reached to resolve claims that they engaged in illegal and fraudulent conduct in the mid-2000s. In November, 2013, the US Justice Department and other federal agencies entered into a $13 billion settlement with JP Morgan Chase to resolve claims that the bank's disclosures related to some of its mortgage-backed securities were flawed (Department of Justice, 2013). In a sweeping settlement with five of the largest financial institutions in the United States, the Justice Department and 49 state attorneys general settled claims for over $25 billion that these banks had engaged in faulty foreclosure practices, like fabricating court documents to move thousands of foreclosures through the legal system (Heavens, 2012). In July 2014, the Department of Justice reached a $7 billion settlement with Citigroup that resolves claims similar to those that have been raised against JPMorgan Chase and Bank of America (Narula and Stein, 2014).

While the US Department of Justice and state attorneys general have gotten most of the headlines for these multi-billion dollar bank settlements, city and county governments have also used the courts to try to rectify some of the worst financial institution abuses of the last decade. Following in the paths of litigation around the proliferation of illegal guns and climate change, local governments have attempted to use the courts to obtain some degree of relief from those practices that have left them responsible for dealing with abandoned and neglected properties, foreclosed homes, and diminished property values. In the area of predatory and discriminatory lending, local governments have filed lawsuits, and some have reached settlements, that might bring some relief to distressed municipalities and the homeowners who live in them.

Two cases that have had the most success to date have centered around claims of so-called reverse redlining on racial grounds. Reverse redlining is the term used to describe when a bank targets a certain community for loans on less favorable grounds on account of the demographics of the community. The term originally comes from traditional redlining: when a bank would choose not to lend to a particular community based on its racial or ethnic makeup. Two separate cases, one filed by the Mayor and City Council of Baltimore, Maryland, and a second, filed by the City of Memphis and Shelby County, Tennessee, alleged that Wells Fargo engaged in reverse redlining on racial grounds in those communities (Powell, 2009). The plaintiffs alleged that this practice in these communities resulted in those government entities suffering financial harm from diminished property values and increased blight (Ibid.). In addition to data that revealed what appear to have been discriminatory patterns of lending, the plaintiffs also produced sworn statements from former Wells Fargo employees that revealed that bank officials would refer to subprime loans as "ghetto loans" and borrowers of color as "mud people." (Ibid.) In May of 2012, Wells Fargo settled the claims against it in these communities by agreeing to make hundreds of millions of dollars of loans in Memphis and Shelby County (Campbell, 2012) and $7 million in loans in Baltimore (Relman, Dane, & Colfax, PLLC, 2012).

These two settlements will help to bring some degree of relief to borrowers in the communities these settlements reach. If the Federal Reserve analysis described earlier is to be believed, there are still millions of borrowers who were saddled with subprime loans on unfair terms because of their race. These settlements hardly begin to rectify what would appear to have been discriminatory conduct on the part of mortgage lenders that extended unfair credit on unfair terms to borrowers of color, regardless of their income. Does the success of the litigation of the type that Baltimore, Memphis and Shelby County prosecuted hold out hope that other communities could reap similar restitution for their residents? Unfortunately, cases such as these, addressing events that led up to the Financial Crisis of 2008, are likely a thing of the past. As described earlier, many of the subprime lenders that were so aggressive in extending subprime loans are now bankrupt, so it is unlikely that communities could bring affirmative cases such as these against such lenders and hope to achieve any real relief. Moreover, borrowers saddled with loans on unfair terms may have already been foreclosed upon, and bringing actions on their behalf at this point would face significant procedural barriers.

While the opportunity to bring these types of court actions to address racial discrimination in lending at the local level that occurred in the mid-2000s may have passed, local governments should keep this type of legal intervention in mind in the future, should financial institutions continue to engage in the types of behavior that can give rise to these legal claims. At the same time, these types of interventions are not the only ones available to cities, and the following discussion highlights some of the other, creative interventions cities are undertaking and considering undertaking.

Responsible Banking Ordinances

As the last discussion shows, select cities have taken at least one financial institution—Wells Fargo—to court for discriminatory acts that helped to fuel the Financial Crisis of 2008. In those cases, local governments have used laws on the books, most directly the federal Fair Housing Act, as the basis of those lawsuits. Claims of fraud and misrepresentation have also given rise to some of the other lawsuits described above. These claims mostly arise under state and federal law. What some intrepid local governments have done, some for a long time, is pass their own laws that have as their goal serving as an additional check on bank behavior. These types of laws, known generally as "Responsible Banking Ordinances," often attempt to use each local government's power of the purse—that is, the fact that such cities invest millions and sometimes billions of dollars with financial institutions—to try to influence bank behavior by directing where those local governments place their investment dollars: for example, pension funds, tax revenue, operating funds, and so on. In this way, these cities are taking advantage of the fact that they can be fairly lucrative customers for financial institutions; as such they can try to shape bank behavior by taking their business to those banks that seem to be engaging in what local governments see as responsible conduct.

These Responsible Banking Ordinances have manifested themselves in many different forms, but, for the most part, they are designed in such a way as to give a local government information about banking practices so that it can make a decision about where it can take its banking business. One approach that some localities take is to ask each financial institution that would like to do business with local government to report on its home mortgage and small business lending in the locality, with a particular emphasis on its activities in low- and moderate-income communities. Local government officials can compare bank performance in these areas and award city contracts to those institutions that seem to extend the highest volume of loans in underserved communities (Holeywell, 2012). Each city that has adopted such an ordinance has shaped it according to its own needs and interests.

Take as an example the Responsible Banking Ordinance of the first city to pass one, Cleveland, OH, which did so in 1991. Under Cleveland's Responsible Banking Ordinance, all banks that want to do business with the City must propose a four-year responsible banking plan. The City's Director of Community Development must approve the plan in order for the bank to be eligible to contract with that City. Local government officials then assess the bank's performance in light of its responsible banking plan: for example, how close has it come to meeting its goals, what is the number of branches it has opened in low- and moderate-income communities, and how successful it has been in hiring a racially and ethnically diverse staff. An important feature of the Cleveland ordinance is that the results of the City's assessment of bank performance are made available to the public.

When Cleveland first passed this ordinance, there was a fear that banks would not want to do business with the city, or would leave the area altogether rather than submit a responsible banking plan or subject themselves to an evaluation of their performance under such a plan. Despite these initial fears, few banks left the city, and the track record of bank service to low- and moderate-income communities in terms of extending banking services to them is better than the national average (Ibid., 2012).

It was not until 2002 that the next city, Philadelphia, followed suit and passed its own Responsible Banking Ordinance. While many aspects of the Philadelphia program are similar to Cleveland's, the results of the Philadelphia assessment of bank performance are kept confidential. At the same time, these ordinances have the same goal: that financial institutions that wish to seek city business will strive to meet the needs of all residents of those communities, regardless of their income.

While Cleveland and Philadelphia remained the only major municipalities that had adopted Responsible Banking Ordinances, and those ordinances certainly did not prevent the Financial Crisis from wreaking havoc on communities in those cities where subprime lenders engaged in risky lending, the Financial Crisis has spurred legislators in several major cities across the US to pursue the adoption of similar ordinances. It is clear that Responsible Banking Ordinances in their pre-Crisis incarnation—that is, where they would apply only to financial institutions that were hoping to do business with the local government—would do, and did do, little to prevent the Crisis from impacting those cities. A new round of Responsible Banking Ordinances is, perhaps, taking the lesson from Cleveland and Philadelphia's failure to prevent the Crisis from occurring in those cities despite the existence of responsible banking ordinances and putting more teeth into such ordinances moving forward.

Cities like Los Angeles, CA; New York City, NY; Boston, MA; Pittsburgh, PA; Seattle, WA; Portland, OR; and Kansas City, MO, have all either enacted Responsible Banking Ordinances or are considering them. In New York and LA, they have put their own unique twists on their ordinances. In LA, the legislation tracks bank service activity in the community, and then that information is made available to the public so its members can make decisions on where to bank, and the city government can decide with what financial institutions the city should do business. And New York City's ordinance creates an advisory council that reviews financial institution activity within city limits to assess how city government should invest its billions of dollars in pension funds and other financial resources. As this chapter goes to print, a federal judge has enjoined the operation of New York City's ordinance, stating that it is preempted by federal and state law; the City of New York is exploring its options, including whether to appeal the order.

While Responsible Banking Ordinances can leverage local government resources to try to shape bank behavior, there are some shortcomings to this approach that might weaken the impact of such legislation. First, only financial institutions that do business with cities where there are Responsible Banking Ordinances will care how they fare in the responsible banking system. Financial institutions that are not looking to pursue city business will not need to submit themselves to evaluation through the responsible banking system created by these ordinances. Few of the subprime lenders that were so active in the last decade in the lead up to the Financial Crisis did any business with local governments. As we saw in Cleveland and Philadelphia, the existence of responsible banking ordinances in those cities in the height of the mortgage frenzy did little to prevent its aftereffects from befalling those cities.

In order to ensure local communities have an ability to bring more oversight to local bank practices, approaches that review financial institution practices of all institutions that do business in a community (and not just with the local government there) might have a better chance of improving financial institution behavior in those communities. With this notion in mind, the next intervention—community impact report cards—holds out the promise of casting a wider net and shedding more light on a broader spectrum of banking behavior and a broader cohort of financial institutions. It is to this concept that I now turn.

Community Impact Report Cards

President Obama promoted and Congress passed post-Crisis financial legislation, mentioned above, in part, as an attempt to restore trust in the American financial system. The Great Depression, during which thousands of banks failed, and millions lost their savings, led to a severe lack of trust in the financial system at the time, until President Franklin Roosevelt helped to usher in a bank plan that would create savings deposit insurance and curbs on risky bank behavior. Dodd-Frank was supposed to accomplish that same sort of faith in the system. The legislation's limits on derivatives trading and minimum capital requirements for financial institutions, though important, are not the types of controls that the average consumer understands and might not accomplish the goal of restoring consumer faith in the financial system.

Building on the idea of market power that the responsible banking ordinances attempt to leverage, at least one municipality, the City of New Haven, CT, has taken an approach that aims to arm consumers with information that will help them decide with which bank or banks they should do business based on the quality and array of bank products available from those institutions. The Community Impact Report Card (CIRC) is an attempt by this city, and nonprofits across the country that are using similar methodologies, to give some power to consumers to shape bank behavior by giving them a method by which consumers can easily comparison shop between banks and decide which bank offers them the banking options that best suit their needs. By offering consumers an easy method for comparing bank products and services, the hope is that banks will compete to offer a wider array of consumer-friendly services. Using market power, then, the CIRC aims to shape bank conduct to make it attuned and responsive to community needs.

The CIRC is an index that ascribes different point values to a variety of bank products and services. City officials and volunteers gather information about each bank doing business in the New Haven area and then assess the activities of those financial institutions under each category of product or service measured through the system, awarding points based on bank performance under each such category. For example, through the 2012 version of the CIRC, banks were awarded three points if they had three or more branches in the area, two points for two branches and one point for one branch. Similarly, banks were granted three points for having highly flexible and accommodating branch hours, like evening *and* weekend hours; two points for somewhat flexible hours, like evening *or* weekend hours; and one point for non-flexible hours. Banks were also assessed for their relative performance in extending home mortgage loans in the community. A bank that exceeded the average loan volume by more than 20 percent in a particular category of loan—like loans to African-Americans—received five points. Those that were within a range of 20 percent above or below the average in each category received three points, and banks that failed to come within 20 percent of the average in a particular category of loan received one point. Other categories included the availability of free online banking, minimum deposit requirements, ATM locations and fees, language capabilities of bank tellers, and check overdraft fees.

After volunteers gathered the information regarding each bank's products and services, they tallied the points each bank received, and issued a final score for each bank. Conveniently, the points available added up to 100; if a bank received the highest possible score under each category, it would have received that perfect score of 100. In the first iteration of the CIRC, completed in 2012, no bank even scored in the high 70s. First Niagara, a regional bank operating in the area, received the highest score of 74, and Bank of America came in a close second, at 73. Wells Fargo came in last, earning 53 points.

Several nonprofits have utilized an index-like system for assessing and grading bank practices. The Association for Neighborhood and Housing Development conducts an analysis of bank reinvestment in New York City. The Maryland Consumer Rights Coalition recently utilized a bank scoring system for assessing the extent to which banks serving Baltimore, MD, are provided "age-friendly" services to older residents of that city.

The index approach used in New Haven, and in its progeny in other areas, could manifest itself in many different ways, with many different variations. For example, a tech-savvy municipality or nonprofit could create a user-directed interface that could grade banks along a particular consumer's specific interests. Once data is input into the system about bank products and services, a consumer could apply different weights to the different categories, eliminate certain categories, and highlight just those products and services that interest that particular consumer. That way, a consumer could create his or her own customized index. Several years ago, *New York Magazine* offered a web-based feature that allowed its readers to pick the neighborhood in New York City where they most wanted to live by permitting them to place varying weights on different features depending on their relative importance to those readers, like proximity to parks or public transportation. A similar approach, one that tailored a bank's score to each consumer's personal preferences, would empower consumers to comparison shop among financial institutions.

What impact these index systems might ultimately have remains to be seen. New Haven's iteration of its program has already gone through one update. It added features to the system that may have made it more complicated, which might diminish its use as a transparent method for assessing bank activities and their responsiveness to community needs. Unlike the responsible banking ordinances, the CIRC approach assesses the bank activity of every bank that serves a particular community, and gives consumers in those communities a means through which to utilize the services of those banks that seem to best meet their needs and the needs of those communities. The hope is that banks will care how they fare on these bank scoring systems, and perhaps improve their performance by improving the products and services they provide. Of course, the long-term impact of these systems remains to be seen, and bank profit from consumer services is but a small fraction of their overall incomes, so, it is unclear that these so-called impact report cards will have much of an impact on bank bottom lines. It is possible, however, that when a bank scores poorly under a CIRC-style system, it will diminish the public perception of the bank which will, in turn, encourage bank officials to improve the institution's practices if for no other reason than to improve the way it is viewed in the court of public opinion.

Eminent Domain

In September of 2013, the city council of the City of Richmond, California, approved a plan to use the city's power of eminent domain to seize what are known as underwater mortgages (Stanek, 2013). These mortgages are one of the lasting vestiges of the Financial Crisis. Borrowers who took on mortgage debt to purchase homes at the height of the mortgage frenzy of the last decade often found themselves with a mortgage where the principal still owed on that mortgage exceeded the value of the home on the open market. At one point during the height of the foreclosure crisis, the percentage of so-called underwater mortgages was nearly 25 percent of all outstanding home mortgage loans in the country.

This phenomenon was a result of several factors. First, borrowers often purchased homes at the high water mark of the market, as many buyers, often first-time home buyers, were eager to become homeowners for fear the housing boom would pass them by. Second, such buyers may have purchased their mortgages with little to no money down, taking out loans where the loan-to-value ratio was 95 percent or higher, meaning they had very little equity in the home. Indeed, some borrowers even had loans with negative equity, where they were given mortgages that exceeded the value of the home so that they could take out a loan against the house and even get "cash out" at the loan closing, meaning they did not just own a home and take out a mortgage, they also received money to spend on consumer goods, cars and vacations. The underlying premise of such lending was that home values would continue to increase indefinitely. Why worry about having little to no (or even negative) equity if the value of the home would continue to increase, while the principal of the mortgage would stay the same? Homeowners stand to gain considerably when home values increase; but the lender's interest remains the same, through good times and bad.

When those bad times hit, and property values fell, home owners with little or negative equity had little cushion to weather the impact of those falling home values. Their mortgage principal and monthly payments stayed the same, while the value of the home fell to below the outstanding principal on the mortgage. All in all, by one estimate, US homeowners lost $9 Trillion in home equity as a result of the Financial Crisis. Once the value of the home falls below the value of the outstanding principal, the borrower is underwater: he or she owes more than the home is worth.

In previous sections, I have reviewed some of the strategies that municipalities can deploy to attempt to remedy the worst effects of the Financial Crisis. What cities like California's Richmond have considered doing is using their eminent domain power to seize underwater mortgages. The eminent domain power pre-dates the signing of the US Constitution, and gives government the power to "condemn" property for public use, while paying the owner of that property fair market value for the property.

The US Supreme Court had the opportunity to review this eminent domain power in the middle part of the last decade, when it considered the constitutionality of a plan by the City of New London, CT, to seize property for what it described as economic development: taking the homes of middle class residents and other properties and making them available to the pharmaceutical giant Pfizer, among other uses, so that the city could lure that company and other businesses into the area. A deeply divided Court found that the city could seize property for economic development and transfer that property to a private concern while still staying within the requirements of the Constitution.

Some have argued that the Court's endorsement of broad government power under the Constitution to seize property for economic development means that a local government could seize underwater mortgages at a value that is in line with the value of the underlying property, and not the face value of the outstanding principal. What this would allow a city to do is essentially re-issue the mortgage to the homeowner at an amount that is more consistent with the value of the underlying property, regardless of the face value of the original mortgage. In one powerful stroke, municipalities could "right size" mortgages so their value would more closely correspond to the value of the underlying property thereby ensuring that homeowners were not saddled with debt that exceeds that value, and, perhaps, help them rebuild the equity that they lost in the fallout from the Financial Crisis.

In order for such a strategy to work, the government entity would give the bank no more than what the asset—the outstanding mortgage—would obtain at a sale on the open market. Such an approach is consistent with centuries of eminent domain jurisprudence. As holders of mortgage-backed securities know even now, a distressed mortgage sells for far less than 100 percent of the face value of the outstanding principal of the mortgage. Indeed, the federal government, in recent sales of distressed mortgages, sold such mortgages for roughly 60 percent of the outstanding mortgage principal.

Many homeowners, even those in communities where property values have fallen considerably, have not seen the value of their homes fall 40 percent, however. If cities could seize underwater mortgages and sell them on the open market, the city would likely obtain roughly 60 cents on the dollar of the face value of the outstanding principal of those mortgages. A city seizing such properties could thus take these mortgages and then repackage them in the form of a loan to extend to the homeowner, secured, once again, by the underlying property. The difference between the value of the seized mortgage and the value of the underlying property is what helps restore homeowner equity.

During the Great Depression, the Roosevelt Administration tried a strategy similar to an approach that would leverage the eminent domain power in the manner described above. In the early part of the 1930s, banks had billions of dollars in non-performing loans on their books. Instead of simply buying such unwanted debt, the federal Home Owners' Loan Corporation (HOLC) exchanged government-backed bonds for bank assets—nonperforming mortgages—and those banks could sell those bonds on the secondary market, where, if they had to sell their nonperforming mortgages, they would have had tremendous losses. The HOLC then turned around and reissued new mortgages, on reasonable terms, to the homeowners who were on the brink of default on their mortgages, making sure the borrowers had equity in the home (Home Loan Bank Board, 1952).

Cities looking to undertake an eminent domain program could finance it using an approach similar to that utilized by the HOLC; they could compensate financial institutions by offering them bonds equivalent to the value of the underlying mortgage were it to be sold on the open market, and then offer a mortgage back to the homeowner that would restore at least some of the equity in the home. While cities have been afraid to take the aggressive action of using eminent domain to seize underwater mortgages, perhaps some could attempt to institute pilot programs that could relieve banks of the burden of holding distressed mortgages, while bringing the debt on the home more in line with current property values.

The Role of Local Governments in Reducing Inequality Resulting from Financial Institution Misconduct

In this chapter, I have attempted to describe several of the steps that local governments have taken, or proposed taking, to combat some of the worst impacts of the Financial Crisis of 2008 and the foreclosure crisis that followed. Local governments are at the center of these impacts, yet layers of state and federal government oversight above them offer cities few traditional means of regulating bank behavior. Federal and state oversight of financial institutions means there is sometimes little municipal governments can do to rein in bank behavior through direct regulation of their conduct. But, as the preceding discussion shows, local governments have used a number of what I will call indirect ways of attempting to improve bank performance and combat the effects of the conduct that led to the Financial Crisis of 2008. These strategies—litigation under fair housing laws, Responsible Banking Ordinances, community impact report cards, and eminent domain—all offer potential strategies for local governments to address some of the harshest impacts of the Crisis, and the deepening divide in local communities that, as the discussion above attempted to show, has some initial basis in existing economic and racial inequality.

If some of the roots of the Financial Crisis are economic and racial inequality, will the local strategies laid out in this chapter help to combat inequality, and reduce the social distance that may have helped to create the conditions in which predatory conduct thrived? Certainly the use of litigation to rectify instances of racial discrimination that violate the Fair Housing Act is one strategy for addressing racial inequality. By forcing offending banks to compensate victims of racially discriminatory lending—whether it is traditional redlining, or the so-called reverse redlining through which banks were accused of directing lending on discriminatory terms to communities of color—municipal and private litigants can seek compensation for the victims of such practices, whether that comes in the form of monetary relief or increased lending on non-discriminatory grounds.

What about Responsible Banking Ordinances? With these, local governments use their consumer power to shape bank behavior. The extent to which financial institutions wish to do business with a particular municipality will dictate, to a certain extent, the level of impact this approach can have. A local government with little financial muscle might have little impact on bank behavior. Those communities might be able to get by on moral suasion, public relations, and the hope that financial institutions want to be seen as good citizens. But unless they have real financial might—like in New York City, Los Angeles or other large cities—it is unclear that responsible banking ordinances will have much bite. Another question for responsible banking ordinances is what type of behavior are they looking for from financial institutions? Here, the details of the particular responsible banking ordinance, and the type of behavior it seeks to measure, and, in turn, encourage, is what matters. Cities could choose to tailor their own ordinances to help combat economic and racial inequality by creating benchmarks for bank behavior that help ameliorate these inequalities: like the extent to which banks offer free or low-cost banking products, engage in meaningful lending in low- and moderate income communities and communities of color, and have hiring practices that encourage diversity at all levels of the institution.

Furthermore, a relatively weak banking ordinance is one that fails to set clear benchmarks for financial institution behavior and fails to encourage banks to take real steps towards delivering meaningful services to residents of the municipality. Publicizing the results of the municipality's research on bank behavior, using the findings for local governments to comparison shop between banks, conducting follow up meetings with bank officials to try to improve bank performance under the ordinance: these are all strategies local governments can use to ensure that their Responsible Banking Ordinances have the type of impact that matters, and can shape bank behavior.

Similarly, where municipalities utilize a community impact report card, the effect that strategy will have on bank behavior will hinge on a number of factors. Will the community adopt the scoring system when its members decide where to bring their banking business? A

CIRC system might influence financial institution behavior where community residents are made aware of the CIRC scoring system, they decide to use it to make decisions regarding where to bank, and banks are made aware that community residents are using the CIRC system in making such decisions.

Like with responsible banking ordinances, the degree of influence a CIRC may have on financial institutions may depend on the extent to which those institutions are concerned with their standing in the community as a result of the CIRC's analysis and its adoption by the community. Since consumer banking is not typically a profit center for commercial banks, banks may care little whether customers use the CIRC to bank with them. At the same time, most banks wish to be perceived as good citizens, and it is unlikely that a financial institution would want to score poorly in a CIRC report. As a result, the impact these systems may have may be primarily in terms of bank public relations, but the CIRC system gives consumers an objective and relative lens through which to gauge the extent to which banks are meeting consumer needs. Instead of just relying on banks' own public relations efforts to assess whether banks are meeting customer needs, consumers can use a CIRC system to gauge the extent to which banks are really meeting those needs.

As with responsible banking ordinances, CIRC-style systems can also be adapted to address economic and racial inequality. New Haven's system currently gauges bank performance on the extent to which financial institutions offer banking services for free and reduced costs, as well as their success in lending in communities of color. Advocates elsewhere could fine tune the New Haven system to address racial and economic inequality in new ways. As the saying goes, if it can be measured, it can be improved. Communities looking to rein in the extent to which bank practices can exacerbate racial and economic inequality can utilize a CIRC-style index to measure bank performance along metrics that directly impact racial and economic inequality as a way to encourage banks to improve their performance along those metrics.

While the adoption of CIRC systems have been slow, as knowledge about them grows, more communities may decide to utilize a CIRC-style index, or a national organization may undertake the task of grading the nation's largest banks along metrics like those developed in New Haven. When that happens, this local experimentation, first started in one college town in Connecticut, could have national impact.

In terms of eminent domain, to date, no municipality has taken the plunge and used this centuries-old power to rectify the problem of underwater mortgages. In terms of the sheer volume of underwater mortgages in the United States and the extent to which mortgage debt, particularly underwater mortgage debt, likely impacts wealth inequality, it is clear that this may be the strategy that has the most potential for shifting wealth imbalances. Individuals with outsized mortgage debt acting as a weight on their income and wealth can have their equity in their homes restored and the mortgage debt reduced through simple legal measures at the municipal government level.

One of the main reasons municipalities have not tried this strategy to date is that they fear what is, in effect, retaliation by banks for doing so. Financial institutions threaten that they would be hesitant to lend money to communities that are not afraid to use eminent domain to seize underwater mortgages. The mortgage those banks issue today could become underwater tomorrow, and the subject of an eminent domain action. Would banks really shun a city that was looking to take an active role in strengthening its financial health in ways that preserved home values throughout the community, having beneficial spillover effects for other homeowners, and, in turn, lenders throughout that community? Banks have long used similar arguments to ward off legal actions, stating that aggressive oversight can stifle financial innovation. Well, if the last decade is any indicator, sometimes financial "innovation" in the form of exotic mortgage products and opaque derivative investments are precisely the types of innovations we might want to stifle, since these were so central to the events that led the Financial Crisis of 2008. Furthermore, in recent years, as outlined above, many of the largest and strongest banks in the US have paid billions to settle legal claims as a result of their conduct in the lead up to the Financial Crisis of 2008, and yet not one of them has collapsed in the wake of such aggressive legal action.

Municipalities have borne the brunt of the worst impacts of the Financial Crisis of 2008. They have seen their tax rolls diminished due to foreclosures that have had economic and social ripple effects throughout communities across the US and the world. They have had to spend precious resources policing and even maintaining vacant buildings. The effects of these forces have driven some cities to the brink of bankruptcy, and others, like Detroit, Michigan, over that brink. Cities have few tools at their disposal to combat the national and even international forces at play that have caused the worst impacts of the Financial Crisis. In addition, where they have had some protections in place—like in Cleveland and Philadelphia where they had pre-Crisis responsible banking ordinances—such measures were not strong enough to ward off the Crisis and its impacts.

Cities are starting to recognize that they are not powerless to try to remedy the lingering effects of the Crisis, nor are they unable to adopt new measures that can try to prevent the next financial crisis, even in the face of national political gridlock. The measures described here—aggressive litigation to remedy mortgage discrimination, responsible banking ordinances with bite, banking report cards, and perhaps even eminent domain efforts—give local governments options for providing some oversight of financial institution behavior, to attempt, as much as local governments can, to influence bank behavior in a way that is positive for the constituents of those local governments.

Leaving matters to federal and even state officials saddled local governments with the responsibility for dealing with the worst effects of the Crisis. Cities, towns, counties and other local governments have a role to play in providing a check on financial institution conduct that affects those governments and the communities they represent. The tools described here are just some that cities are using, or contemplating using, and local governments should explore the extent to which these tools, and others they can devise, can ensure that local governments have a say in financial institution behavior that affects their communities. Since cities are bearing the brunt of that behavior, they should have a role in checking it.

As subsequent chapters will show, in many arenas, political gridlock at higher levels of government leaves cities no choice but to take aggressive and pragmatic approaches that help solve problems. Cities have a role to play in the financial system, out of necessity. The tactics described here are just some cities can deploy to ameliorate the lingering effects of the Financial Crisis of 2008 and prevent the next one from having, once again, such a disastrous impact on local communities.

Works Cited

Apgar, William C., and Mark Duda. "Collateral Damage: The Municipal Impact of Today's Mortgage Foreclosure Boom." Web blog post. Collateral Damage. Bear Market Economics Blog, 17 Apr. 2009. Web.

Apgar, William C., Mark Duda, and Rochelle Nawrocki Gorey. The Municipal Cost of Foreclosure: A Chicago Case Study. Rep. N.p.: Neighborworks America, 2005. Web.

"Baltimore Settles Landmark Fair Lending Case Against Wells Fargo." Web blog post. Relman, Dane & Colfax PLLC. N.p., n.d. 11 Sept. 2014. Web.

Bar-Gill, Oren, and Elizabeth Warren. "Making Credit Safer." 157 U. PA. L. REV. 1 (2008): n. pag. Web.

Brescia, Raymond H. "The Cost of Inequality: Social Distance, Predatory Conduct, and the Financial Crisis." *66 NYU Annual Survey of American Law* 641 (2011).

Campbell, Dakin. "Wells Fargo Pledges Investment to End Memphis Lawsuit." *Bloomberg.* 29 May 2012. Web.

Defaulting on the Dream: States Respond to America's Foreclosure Crisis. Rep. N.p.: Pew Charitable Trusts, n.d. The Pew Charitable Trusts. Web.

Ding, Lei, et al. "Risky Borrowers or Risky Mortgages: Disaggregating Effects Using Propensity Score Models." Univeristy of North Carolina, Center for Community Capital 1 (2010). Print.

Engel, Kathleen and McCoy, Patricia, "The Subprime Virus: Reckless Credit Regulatory Failure, and Next Steps." Oxford University Press (2011). Web.

Final Report of the National Commission on the Causes of the Financial and Economic Crisis in the United States. Rep. Vol. 72.: Financial Crisis Inquiry Commission. *The Financial Crisis Inquiry Report*, Authorized Edition. Print.

Financial Crimes Report to the Public. Rep. US Department of Justice—Federal Bureau of Investigation, May 2005. Web.

Gerardi, Kristopher, Rosenblatt, Eric, Willen, Paul S., and Yao, Vincent W. Foreclosure Externalities: Some New Evidence 4. 13–14, 33. Working paper no. 2012–11, 2012: *Federal Reserve Bank of Atlanta Working Paper Series*, 2012. Print.

"GINI Index (World Bank Estimate)." GINI Index (World Bank Estimate). N.p.: n.d. 3 May 2015. Web.

Global Insight. *The Mortgage Crisis: Economic and Fiscal Implications for Metro Areas.* Lexington, MA: Global Insight, 2007. Print.

Heavens, Alan J. "$25 Billion 'Robo-Signing' Settlement Reached with Five Banks." *Philadelphia Inquirer*, 9 Feb. 2012. Web.

Holeywell, Ryan. "Cities Using Deposits to Gain Leverage over Banks." Web blog post. Governing: The States and Localities., 2012. Web.

Immergluck, Dan, and Geoff Smith. *There Goes the Neighborhood: The Effect of Single-Family Mortgage Foreclosures on Property Values*. N.p.: n.d., 2005. Print.

Lesser Mansfield, Cathy. "The Road to Subprime 'Hel' Was Paved with Good Congressional Intentions: Usury Deregulation and the Subprime Home Equity Market." *S.C. L. REV 51.473* (2000): 492. Print.

Levin, Carl. Memorandum from Senator Carl Levin, Subcommittee Chairman, to Members of the Permanent Subcommittee on Investigations 1. N.p.: n.d., 13 Apr. 2010. PDF available at http://levin.senate.gov/newsroom/supporting/2010/PSI. LevinCoburnmemo.041310.pdf.

Liberman, N., Trope, Y., and E. Stephan. *Social Psychology: Handbook of Basic Principles*. Ed. AW Kruglanski and ET Higgins. Vol. 2. New York: Guilford, 2007. 353–83. Print.

Moran, Eamonn K. "Wall Street Meets Main Street: Understanding the Financial Crisis." *North Carolina Banking Institute* 13 (2009): 5, 21–22. Print.

"Mortgage-Related Bank Fines: Payback Time for Subprime." *Economist*, 16 Oct. 2013. Web.

Narula, Anthony, and Philip R. Stein. "Citigroup Settles with DOJ for $7 Billion." *National Law Review* (2014): n. pag. 23 July 2014. Web.

Powell, Michael. "Memphis Accuses Wells Fargo of Discriminating Against Blacks." *New York Times* 31 Dec. 2009, New York ed., US sec.: A15. Web.

Powell, Michael, and Janet Roberts. "Minorities Affected Most as New York Foreclosures Rise." *New York Times* 16 May 2009, New York ed., A1 sec. Print.

United States Department of Commerce. Census Bureau. Household Income for States: 2010 and 2011. By Amanda Noss. ACSBR ed. Vol. 11–02. N.p.: US Census Bureau, n.d. American Community Survey Briefs. Sept. 2012. Web.

United States Department of Justice. Justice Department. Federal and State Partners Secure Record $13 Billion Global Settlement with JP Morgan for Misleading Investors About Securities Containing Toxic Mortgages. Press Release. 19 Nov. 2013. Web.

United States. Federal Reserve. The 2006 HMDA Data, 93 FED. RES. BULL. A73, A95. By Robert B. Avery, et al., 2007. Web.

United States. Federal Reserve. The 2007 HMDA Data, 94 FED. RES. BULL. A107, A109. By Robert B. Avery, et al., 2008. Web.

United States. Government Accountability Office. Financial Regulatory Reform: Financial Crisis Losses and Potential Impacts of the Dodd-Frank Act. Vol. 21., 2013. GAO-13–180. Web.

United States. Home Loan Bank Board. Final Report to the Congress of the United States Relating to the Home Owners' Loan Corp., 1933–1951. Vol. 5.: Report to Congress, 1952. Web.

Chapter 2
Assessing Metropolitan Resiliency: Laying the Foundation for Urban Sustainability[1]

John Travis Marshall

Introduction

Horror stories of urban natural disasters were once the occasional subject of glossy National Geographic articles (Holm; PBS; Bourne 92; Fischetti 76; Larson). The articles described doomsday scenarios involving ravaged cityscapes, flooded subway tunnels, submerged iconic buildings, and scores of casualties. They also estimated multi-billion dollar price tags for disaster response and long-term recovery efforts (Bourne 92; Carroll 72; Gore 37; Parfit 2).

In the last decade, these articles, which seemed more science fiction than credible threats, proved prescient. In 2005, Hurricane Katrina and the catastrophic failure of New Orleans's floodwall systems reduced the city to a lifeless, nearly uninhabitable landscape for weeks. Katrina proved to be the third most expensive natural disaster in modern world history (Liu; *The Economist*). Ensuing storms proved Katrina was no fluke: Hurricane Rita (late 2005) and Hurricane Gustav (2008) nearly delivered a second knock-out punch to New Orleans and the Gulf Coast. The Atlantic Coast endured its own close calls with tropical storms over the next several years. In 2011, Tropical Storm Irene swiped New York City, causing billions in damage. Just a year later, Hurricane Sandy, a rare and powerful late season storm, decimated large swaths of the New Jersey and New York coastlines, including parts of the lower tip of Manhattan and New York City's densely populated coastal neighborhoods.

The 21st century has proved that weather-related urban catastrophes are no longer a matter of fantastic speculation. To make matters worse, the international financial market collapse, the American mortgage foreclosure crisis, and the Deepwater Horizon disaster came on the heels of Hurricanes Katrina, Rita, and Gustav. We have thus also learned that severe economic crises, human-caused catastrophes, and non-weather related natural disasters, such as earthquakes and tsunamis, can compound or mimic ferocious storms.

The federal government, along with many state and local governments, now has firsthand knowledge about how major disasters can cripple American metropolitan areas. Governments at all levels now face the daunting challenges of rebuilding cities quickly and ensuring they are stronger than before the disaster event (Becker).

1 The *Idaho Law Review* has kindly agreed to allow me to use portions of the article that Ryan Rowberry and I wrote for the *Law Review's* April 2014 Symposium, Resilient Cities: Environment, Economy, Equity. This article, "Urban Wreckage and Resiliency: Articulating a Practical Framework for Preserving, Reconstructing, and Building Cities" appeared in volume 50 of the *Law Review*. I am also grateful to Ryan Rowberry for working with me on that piece and for agreeing to allow portions of that article to appear in this chapter.

Not surprisingly, this has proven a tough task. The road to recovery for disaster-stricken cities such as Des Moines, Joplin, New Orleans, and New York has been slow and punctuated by adversity (Munson; McKelvey; Marshall 2014; Trevelyan). This challenge of rebuilding and, at the same time, cultivating more resilient cities has become the focus of engineers, architects, economists, planners, developers, lawyers, building construction specialists, philanthropic foundations, and policy makers (Rockefeller Foundation).[2] They are asking the same general question: how do we cultivate communities that can withstand disaster but also adapt to the many challenges posed by storms, earthquakes, economic crises, and/or dramatic population shifts (Zolli and Healy)?

This chapter suggests that governments use a City Resilience Index as a policy tool to measure cities' comparative resiliency. A City Resilience Index employs quantitative metrics that provide critical data to governments, allowing them to identify current shortcomings with a city's institutional or legal capacity, to track progress in improving that capacity, and to create more refined incentives for cities to incorporate specific tools, programs, and policies into their current and future planning (Sherwood). In the event of a major crisis or disaster, the Index also provides data for the formulation of more rapidly deployed, targeted responses (Sherwood). Devising and implementing a long-term recovery plan is a daunting process that leaves most states and cities flying blind (Stone). With the aid of an index, the stakeholders who are key players in developing resilient cities—governments, the private sector, nonprofit and philanthropic organizations, and most importantly, city residents—have a compass to guide disaster preparation and response, or simply to advocate for policy changes and investments to ensure the long-term vibrancy of cities (Resilience).

A broad range of potential city functions and services could be included in a City Resilience Index—from health care, to schools, to social services, to transportation infrastructure (Sandy Recovery Improvement Act; FEMA). This chapter looks only at housing and community development resources. Apartments, single-family homes, small rental properties, parks, libraries, and community centers are core housing and community development assets of any city's social, built, or economic landscape. These resources, which local governments often manage together under a single program, figure centrally in any long-term disaster recovery program. This chapter examines representative examples of local government institutions, constitutional provisions, statutes, ordinances, and policies critical to evaluating whether housing and community development resources have resilient characteristics. The chapter also suggests how these critical factors might be measured to create a City Resilience Index score.

Part I of this chapter describes what a City Resilience Index is and why it may be an effective public policy tool. Part II of this chapter explores the reasons why a City Resilience Index supplies governmental, non-governmental, and private citizens with a valuable guide to prepare for and overcome natural disasters and other challenges that threaten a city's vitality. Drawing on lessons from recent disasters in the United States, Part III describes some of the essential components of a local government housing and

---

2 In 2013, the Rockefeller Foundation has recently launched an initiative to select 100 resilient cities and provide them with financial support to create a chief resilience officer (CRO) along with technical support and resources to develop and implement plans for urban resilience over the next three years (Rockefeller). New York City and Boston have also recently conducted in-depth studies on the vulnerabilities of their respective cities and strategies to make them more resilient (Specter and Bamberger; Bloomberg).

community development program and, thus specifies several of the constituent parts of the City Resilience Index framework.

I. What is an Index and Why Should Policymakers and Scholars Consider Using One?

Indices, in one form or another, are part of the daily lives of millions of Americans. In this era of online ratings, imagine traveling to an unfamiliar city and selecting among restaurants for an important family or business meal without checking TripAdvisor or UrbanSpoon to evaluate the dining establishment's quality. Consider also the steps an individual or business might take when investing funds in a company. As part of its due diligence, the individual or business would likely consult the target company's bond credit rating or its Dun & Bradstreet rating. These rating systems serve as a kind of report card on a company's health. Employing a numerical scale or a classic schoolhouse "A" through "F" evaluation system (or another form of measurement), indices provide the general public with valuable information about particular characteristics that the index itself deems critical qualities as well as information regarding the graded or rated entity's success at meeting those characteristics (Calo 775–90; Gerken 5–9, 66–92; Ho 577–87).

Flip roles for a moment and put yourself in the position of the restaurant owner or the CEO of the company targeted for investment. Restaurants need happy customers to become repeat diners and to recommend the establishment to friends. If a restaurant sees that its overall composite restaurant rating is good, except for a poor score in the area of dining ambiance, chances are the owner will be inclined to invest time and money making the dining room more aesthetically appealing. Likewise, if you are the CEO of a company seeking capital to pursue a promising innovation, you will make sure you pay your corporate bondholders and other creditors on time, so that your bond and credit ratings are as strong as possible for potential investors to appreciate. In short, indices tend to prod the people who are being graded to take action (Calo 775–90; Gerken 5–9, 66–92; Ho 577–87).

Thus, broadly conceived, indices are commonly appreciated for being helpful in at least two ways. The first way is the role they serve in providing succinct information about the status or condition of something. They serve the function of providing us "notice" about the safety or quality of an institution or operation. The second way is that indices help motivate action. They can contribute to causing the restaurant owner and the company owner to take action on addressing or correcting factors on which the index places particular value. That is why indices are also recognized as providing a "nudge" toward action (Calo 775–90).

Currently, local, state, and federal government agencies do not have a concise framework for evaluating local government capacity for carrying out housing and community development work. This is an important oversight with potentially far-reaching consequences. Housing and community development work is a core element of any city's day-to-day responsibilities to its residents, and this work is also central to successful recent long-term disaster recovery programs. The objective of the City Resilience Index is to enlist an index's power to give notice to important constituencies and to nudge the graded or rated city to implement better urban development policy. The information the Index discloses about a city's current capacity to manage housing and community development work will let federal, state, philanthropic, and nonprofit stakeholders know exactly the local government capacities that must be nurtured to create a more resilient city. The Index will also exhort local governments to take initiative and improve its laws, institutions, and systems pertaining to housing and community development. The Index thus encourages cities not simply to ask whether the city is prepared to handle long-term recovery from disaster, but it also encourages consideration of the larger, forward-looking question of what the city and its stakeholders want their city to become.

II. The Need for a City Resilience Index

Imagine your hometown's cityscape following a disaster, such as an earthquake or hurricane. Or, consider what that city might look like with its streets pocked by vacant and abandoned homes and buildings. To understand an index's utility for framing and addressing urban development challenges, it helps to consider a common tool we currently employ to assess the condition of cities. A map might be a beneficial damage assessment instrument, enabling you to understand the scope of destruction or blight. The map might also identify potential targets for recovery and revitalization. We have seen such maps help tell the story of disaster and crisis on the pages of the *New York Times*, *The Wall Street Journal*, and other newspapers.

A typical map, however, provides relatively restricted data regarding the range of considerations critical to recovery. It cannot capture information about the potential capacities and limitations of a city's or state's staff, laws, policies, and housing and community development-related institutions. A map merely fixes physical targets and suggests potential strategies for redevelopment and revitalization. A map generally tells us very little about the critical consideration of whether the city has the capacity to carry out a recovery plan.

A City Resilience Index can fill the vital need for a tool that gauges a city's capacity to meet housing and community development challenges. It can serve as a means for assessing whether a city or state can deliver meaningful assistance to a target neighborhood, block, or community of people. Think, for example, about the adversity that low-income homeowners face in the wake of crisis or disaster. These homeowners' concerns should lie at the heart of any city's housing and community development programming. Families who own houses and earn less than the average area income are among the most vulnerable residents in times of crisis and disaster (Brescia, Kelly, and Marshall 328–29). Rebuilding the neighborhoods these families call home requires extensive and diverse resources. This major investment is essential to the city's post-disaster recovery. After all, these neighborhoods often are home to its teachers, first responders, service sector employees, and administrative support staff. But these are also the neighborhoods where families have limited savings and rely on each paycheck to meet their mortgage payment and other monthly bills.

Given the challenging economic circumstances faced by these families, a city cannot rely on the private sector to rebuild these neighborhoods. Further, given the exacting regulatory requirements that attach to federal housing and community development funds, it is also not sufficient for a city simply to make government grant funds available for nonprofit developers to rebuild homes. The local government, therefore, has a pivotal role to play in implementing the long-term recovery. The City Resilience Index can help highlight the range of considerations that must inform a city's successful efforts to revitalize low and moderate-income neighborhoods. These considerations begin with whether or not the city has a plan outlining goals for promoting affordable housing. The considerations also include whether the city's housing and community development staff has a track record for timely and effectively using government grant funds. Another consideration relevant to comprehensive urban revitalization efforts is what special powers it can deploy when low and moderate-income families abandon their homes, leaving their neighbors to cope with high grass, unsecured homes, and other public health and safety hazards. Under these circumstances it is crucial that state and/or local laws governing code enforcement and expropriation of private property allow the local government to take vigorous action to protect the neighbors who have chosen to stay and rebuild. Further, it is essential to know about the existence, or lack, of local government-sponsored institutions to help execute important aspects of any neighborhood revitalization plan, such as land banks, redevelopment authorities, and land trusts. Until now, local government housing and community development capabilities have largely been a matter of speculation. This lack of insight about local government capacity has had, and will continue to have, serious social and economic consequences in times of crisis and disaster.

The federal government's allocation of long-term recovery funds to the Gulf Coast, Louisiana, and New Orleans was unprecedented. Unfortunately, the federal government dispensed the long-term recovery funds without any objective tool to measure how effectively, efficiently, or equitably state and local governments could move the federal funds. In the heat of the continuing humanitarian, environmental, and political crisis, Congress and HUD did not have time to "kick the tires" or look "under the hood" of the Gulf Coast jurisdictions that would ultimately receive federal funds. They had little other than anecdotal information about local government legal and institutional capacity that would be vital in determining how a successful long-term recovery would proceed. They likely had no objective information on the legal and institutional questions that would ultimately have a major influence on the City of New Orleans's slow post-Katrina recovery, such as: did the local government have a fully functional redevelopment authority; did the city's community development agency have a procurement policy that meets local, state, and federal requirements; did city agencies have a history of complying with federal regulatory requirements for environmental review, wage and hour thresholds, relocation, or civil rights laws; did city staff have the knowledge and experience to acquire and dispose of large volumes of properties; and did the city support any local community development financial institutions devoted to helping provide financing for important community development projects to which private banks might not lend. The failure to know the answers to these questions contributed to slow and poorly executed neighborhood redevelopment programs that effectively denied recovery to families that did not have the resources to rebuild on their own (Marshall 2014, 101–02).

Designing and implementing long-term recovery efforts might be more effective and less a matter of guesswork if federal, state, and local governments were armed with a concise understanding of the housing and community-development related challenges and capacities of the local governments that they are assisting. It is not good enough to have an anecdotal understanding of the challenges faced by local governments. Prior to Katrina, there should have been a more detailed evaluation of the range of community resources—governmental and non-governmental—that would figure critically in implementation of any long-term disaster recovery efforts. Federal and state governments should have a pre-disaster "picture" of local government capacity.

A City Resilience Index can provide that valuable snapshot. The goal of this index project is to provide cities, states, and national governments with a well calibrated tool they can use to evaluate whether a city is in a position to pull the many levers of the machine of long-term recovery as opposed to having to endure the time-consuming, expensive, and frustrating process of inventing the long-term recovery machinery necessary to heal a city and help it thrive. A City Resilience Index could be composed of dozens of factors. In this chapter I sketch out how this Index might assess housing and community development factors critical to cities nationally and internationally.

Designing and producing a City Resilience Index represents a major undertaking. Think about cities and how they encompass a wide range of essential systems and services. Each system is critical to the daily lives of residents, businesses, visitors, and major stakeholder institutions. If you browse a local government's homepage listing for its major departments, the website covers many of these critical systems: cable television, fire and police, parks and recreation, planning and development, solid waste, wastewater, and water—just to name a handful. A thriving city sustains all of these systems—or it creates partnerships to sustain them. Failure to administer any one of these systems following a disaster critically impedes that city's long-term recovery from the crisis event.

Ideally, a City Resilience Index would take the pulse of each of these key city systems. Heather Gerken's efforts to lay groundwork for the Democracy Index illustrate the challenge of a resilient cities indexing enterprise. The Democracy Index represents a tool for improving just a single critical system: the government elections system. To meaningfully improve and inform local government election policies, Gerken suggests three separate index metrics. But, she emphasizes that the Democracy Index's power to spur election policy improvements depends largely on the finer points of defining and collecting appropriate data for these metrics (Gerken 110–33). In other words, the process and resources involved in building even a "single system" index (here, the electoral system) are formidable. Designing a City Resilience Index that covers multiple city systems looms as an enormous undertaking.

This project is worth pursuing because a City Resilience Index potentially provides a much bigger "carrot" and promises a much larger and more effective "stick" than even the Democracy Index. There are at least three critical properties that give the Democracy Index such powerful potential force to drive change: (1) the looming threat of exposing poorly performing local election operations (the "stick"); (2) the ability to highlight the work of effective local government election staffs (the "carrot"); and (3) the capacity to give the public easy-to-understand information that it can use to advocate for better election administration (Gerken 67–69). A City Resilience Index uses these same three vectors to propel change. But, it also taps at least two additional forces for change and better public policy.

First, the City Resilience Index helps measure a local government's likely aptitude for carrying out essential city building tasks. The City Resilience Index also gives the federal government important information about the type of technical assistance it may need to supply to assist local governments to develop essential city-building capacities. In the event of disaster, the measurement also gives federal and state governments critical intelligence about the relative strengths and weaknesses of local governments so that federal and state governments can calibrate their response to address needs they know local governments cannot handle.

Further, as was so well documented following Hurricanes Katrina and Rita, nonprofit and philanthropic funders are often the first entities ready to open their wallets to jump start long-term neighborhood and city rebuilding efforts. However, these nonprofit and philanthropic funders usually have minimal insight into the relative sophistication and functionality of the local government. They have no sense of whether they will be funding a short-term band-aid until a high-functioning local government assumes full control of long-term recovery efforts. Or, in the case of New Orleans, nonprofits and philanthropic groups have no appreciation for the fact that they may supply the principal boots-on-the-ground not just for weeks or months, but for a period of years following a disaster. Immediate disaster response and the challenging road to long-term recovery require deep, continuous, and far-reaching coordination among local, state, and federal government partners before crisis or disaster strikes.

It is extremely difficult to establish coherent boundaries between index categories that have a legal effect and those that may be less legal in nature. Considerations that might fit more naturally into the finance category of a resilient cities index have been excluded. This category would include indicators such as the local government's bond rating, the amount of its annual debt service as a proportion of its total annual budget, or the average length of time it takes a city to pay its outside contractors. A city that is unable to pay its contractors in a timely manner under normal, non-urgent, circumstances will face enormous challenges post-disaster complying with detailed federal requirements for documenting recovery work completed (Krupa; Public Strategies Group 2, 6, 8).

Indicators that would fall into the category of customer satisfaction or process improvement have also been excluded from the Index. It is not hard to see how a city's commitment (or lack thereof) to improving its public interface affects delivery of housing and community development programs. For example, does a local government offer one-stop shopping or streamlined permitting for building and other development permits? If they have not done so already, colleagues in the field of public administration might consider creating a City Resilience Index subcategory covering: public finance and public sector customer service. A city in which its residents are unhappy, its finances questionable, and its local business partners disgruntled will struggle. One scholar even suggests that people have a proclivity to choose where they live based on the package of services offered by local government (Gillette 945). Further, residents, businesses, and service providers operate as indispensable partners in helping cities run efficiently and rebuild quickly. If local government finance systems are efficient and its customer service systems are fair and timely, chances are that these stakeholders are more likely to improve their neighborhoods and seek out local government contracts (Public Strategies Group 8).

A smarter, more efficient, and more equitable recovery is possible if city stakeholders have access to information about the existence of laws, institutions, and staff capacities. A City Resilience Index is a tool for city redevelopment partners working on the frontlines of important initiatives to revitalize a city. Federal and state governments, as well as the philanthropic and nonprofit entities working to help communities recover from disaster or crisis could have more efficiently and expediently deployed assistance to Gulf Coast communities if they had basic information regarding the strengths and weaknesses of local government partners, including the City of New Orleans's capacity to promote housing and community development.

III. Building the Index: First Thoughts on City Resilience Indicators

Profoundly traumatic and destructive experiences associated with Katrina, Sandy, or any other major disaster are wasted if all that is expected of government-led relief is a band-aid to mend an injured community. Disasters generally expose what is not working correctly in a community. Although no two cities are identical, their general strengths and their dysfunctions likely share some common DNA. Thus, disaster recovery is not only about building cities back stronger than they were before a disaster event, but also about taking note of a city's relative strengths and weaknesses and spreading the word to other cities to help make them stronger—regardless of whether those cities should face any immediate peril from hurricane, tsunami, tornado, wildfire, or earthquake.

Disaster events such as Katrina and Sandy are not just cautionary tales. They distill the local government's experiences down to factors that helped the city adapt to and overcome adversity. At the same time, disaster events expose a city's unattractive underbelly. They frame the factors whose absence or near-absence may have hobbled the city before disaster and then left the city in a poor position to rebound post-disaster (Liu). If those critical factors can be isolated and their presence in cities can be meaningfully measured, then it is possible to create a tool that allows for a constructive dialogue about how to build and sustain resilient cities.

Drawing on lessons of recent catastrophic disasters, Part III of this chapter suggests critical legal indicators that can be used to build stronger cities. In particular, this Part outlines the legal index indicators that might be used to analyze the relative vitality of cities' housing and community development programs. The indicators are presented here in Figure 2.1.[3]

Policy Category	Subcategory	Indicator
Housing and Community Development	Local Government Housing Development Legal "Toolbox": *Planning*	Housing plan or strategy adopted by local government (housing plan developed in conjunction with local stakeholders and sets priorities for meeting a community's housing needs)
	Local Government Housing Development Legal "Toolbox": *Partnerships*	Duly procured agreements with Local or regional nonprofit (CDCs, philanthropic organizations, etc.) and/or for-profit housing developers to rehabilitate or construct affordable housing at scale (10 or more units)
	Local Government Housing Development Legal "Toolbox": *Property Acquisition, Disposition & Stewardship*	Duly authorized property acquisition through two or more legal tools for property acquisition, including private market purchases, eminent domain, code lien foreclosure, and land swaps (local government must follow local, state and federal requirements for acquiring property, including environmental review and appraisal requirements – at scale)

Figure 2.1   Examples of Index Categories and Indicators

A. Resilience Index Indicators for Housing

In this initial version of the City Resilience Index, the housing category includes just a single subcategory: the Local Government Housing Development "Toolbox." The housing category would likely be expanded to multiple subcategories as work continued on this Index. For instance, at least one of the additional housing subcategories would likely cover the legal landscape for redevelopment activities, which would assess the extent to

---

3   The Environmental Performance Index (EPI) (Esty and Levy 18–21) has strongly influenced formulation of this preliminary table of resilient city index categories and indicators.

which state constitutional and statutory law and local government ordinances and policies facilitate or impede neighborhood redevelopment activities (Marshall 2015).

The Housing Development "Toolbox" subcategory evaluates the important housing development "levers" or "pulleys" that a local government has, or should have, at its disposal. This subcategory focuses on legal strategies and partnerships that tend to sustain and promote safe and affordable housing. Keep in mind that the Index does not serve merely as an inventory or checklist. The Index aims to measure whether local governments use these important housing development tools as well as the level of sophistication and the capacity at which they are being used.

B. The Local Government Housing Development "Toolbox"

City dwellers want to go home to safe neighborhoods with well-maintained houses or apartment buildings. Although private real estate development interests drive a large share of a city's residential housing development, local governments in partnership with state and federal governments can play a vital role in promoting development and affordable housing options. Cities not only contribute land or vacant buildings for these housing initiatives, but they often make financial investments in housing development and redevelopment, providing the critical gap financing that allows the projects to proceed.

Local and state governments can play an even a larger role during long-term disaster recovery. A catastrophic disaster may destroy tens of thousands of homes. For example, Hurricane Sandy damaged or destroyed more than 650,000 homes (Hurricane Sandy Rebuilding Strategy 13). Many families will have no insurance or insufficient insurance coverage. In the 2005 hurricane season, more than 38 percent of damaged units either had no insurance or had property insurance but lacked flood insurance (Hurricane Sandy Rebuilding Strategy 13). Federal disaster block grant dollars supply states and cities with funds for rebuilding neighborhoods where there is an urgent need, such as a disaster, or the need to build housing for low and moderate income families. To receive such a grant, state and local governments must ensure that the projects they fund with CDBG dollars meet one of three required "national objectives": *activities benefiting low- and moderate-income persons, activities eliminating slum and blight, or activities addressing urgent community development needs* (24 CFR 570.208).

It is not safe to assume that different local governments possess comparable tools to promote neighborhood housing development. One local government's housing development experience may be largely limited to selling vacant properties to Habitat for Humanity for single-family housing construction. Another local government may have significant experience promoting construction of affordable homes by layering federal grants, with federal tax credits, with private foundation dollars. One city may not be able to find enough capable low- and moderate-income housing developers to spend the city's annual allocation of federal block grant funds, while a different city may enjoy intense competition for federal grant monies.

Under normal, non-urgent circumstances, it may make little difference to all but the city's poorest residents whether a local government effectively manages its federally funded housing programs. But the level of interest in the local government's ability to spend federal grant funds efficiently increases dramatically following disasters (Campbell). Suddenly, the fortunes of residents of every disaster-ravaged neighborhood, regardless of income, depend in some way on the aptitude and experience of state and local governments in facilitating housing development. A local government struggling for proficiency with only the most basic legal tools to promote housing development will be outmatched and fundamentally overwhelmed when disaster strikes and its neighborhoods must be rebuilt. To be sure, it is difficult to imagine that any local government can easily bear the stress of post-disaster rebuilding. All city residents affected by disaster will encounter tremendous adversity (Brescia, Kelly, and Marshall 328). But it is the very low-income, low-income, moderate-income, and even middle-income families that will have a great personal stake in the city's ability to design, implement, and manage critical facets of the neighborhood housing recovery (Marshall 2014, 93–106).

It is important for the federal government, state governments, nonprofit and philanthropic developers, and the general public to have the ability to gauge the capacity of the local government to promote sophisticated housing development. The following three housing index indicators are examples of the factors that would likely be important in determining whether a local government possesses the necessary legal "toolkit" to pursue the sophisticated housing development that promotes affordable and equitable housing opportunities under non-crisis circumstances. This same legal toolkit is essential for confronting the enormous problems that confront a community following a disaster. A more mature City Resilience Index would include a number of additional housing indicators.

Planning tools

When disaster strikes, almost everyone, including the chair of the planning board and board staff, becomes preoccupied with the steps that they must take personally to rebuild their own homes. This is not an ideal time for writing plans. Rather, this is the time to mine and implement existing plans, so that cities and their neighborhoods can address their essential housing needs and follow guidelines that have been carefully developed to make sure the next generation of housing development is stronger, greener, safer, and more affordable.

The housing category of the City Resilience Index should include at least one indicator for assessing whether local governments have plans in place to address the adversities presented by disasters. This index indicator will credit cities for adopting comprehensive plans and procedures for updating those plans on a regular basis. Enhanced index scores will be possible for jurisdictions that have adopted a comprehensive plan with a housing component that includes detailed goals and objectives for developing housing to serve the city's very low-income and low-income families. Such goals and objectives might include specific plans and strategies for moving families out of harm's way in the event of disaster and allowing for higher density development in areas that are considered safer (Nelson).

There are three reasons why it is critical for the City Resilience Index's housing category to include a planning indicator. First, it cannot be taken for granted that all local governments have housing plans or policies. When Hurricane Katrina struck in 2005, New Orleans did not have a housing policy; it did not even have a policy that could have been modified in the wake of Katrina (Transition New Orleans Task Force; Olshansky and Johnson 236–37). Second, plans stake out a path and priorities for orderly development prior to the chaos of crisis (Olshansky and Johnson 236–37). If a local government has adopted and updated a comprehensive plan, that document gives the local government basis for making tough or even unpopular redevelopment decisions following disasters. In the wake of a catastrophic disaster, the local governments implementing the recovery strategy generally should not be expected to make tough decisions regarding a rebuilding "triage" process. In other words, local governments often face political pressures not to adopt redevelopment priorities that make long-term recovery more sustainable and feasible by prioritizing or circumscribing neighborhood recovery (Olshansky and Johnson 37–71). The City of New Orleans issued building permits immediately after the storm for every neighborhood in the City, even the lowest-lying neighborhoods that sat in the shadow of failed floodwalls (Nossiter). Further, the City did not initially require homeowners to elevate their homes as a condition of receiving a building permit. Third, housing plans and policies should protect the interests of low- and moderate-income residents who are most vulnerable following a disaster and who face the greatest housing needs. In New Orleans, 55 percent of damaged homes were rental units and 20 percent of those units (16,000) were affordable to extremely low-income households (Rose 99, 113). Low-income renters are the people who have slim resources to ride-out a longer-term disaster recovery. Yet these low-income renters are also the people who most often fill service industry jobs critical to the community's recovery.

Partnership tools
Constructing or rehabilitating housing is an expensive endeavor requiring special expertise. To build housing at scale demands even more resources and greater skill. Generally speaking, if local governments aim to improve housing stock then they must prove capable of partnering with private and nonprofit housing developers. Knowledgeable affordable housing developers can, in turn, magnify the value and impact of local governments' housing investments by leveraging tax credits, philanthropic program-related investments (PRIs), bank loans, and private capital. It cannot be assumed, however, that all local governments have this important ability to broker sophisticated housing development deals.

The housing category of a City Resilience Index should include at least one indicator for assessing whether local governments have duly procured agreements with local or regional for-profit and nonprofit housing developers that have resulted in rehabilitation or construction of affordable housing at scale. This Index indicator would not give credit to cities for arriving at agreements that fail to yield occupied housing units. Instead, the Index would give enhanced credit to local governments for each different developer with which it partnered. A further boost in scoring could be made for the number of units developed as a proportion of the city's overall affordable housing units.

There are at least two reasons why it is critical for the housing index of a City Resilience Index to include an indicator that measures local government development partnerships. The first reason is that the private sector frequently spearheads long-term redevelopment efforts following a disaster. This is especially true of mission-driven nonprofit developers who will pursue projects even when national and regional economic conditions push private developers largely to the sidelines, as was the case for a long period during the Katrina recovery (Glauber and Zisser 375). A firm that has a good track record developing affordable housing under normal circumstances could emerge as a strong candidate for a development partnership when the local government is responsible for deploying tens of millions in housing recovery funds following a disaster. The second reason is that a city's failure to support and cultivate a community of public, private, and nonprofit housing development organizations—which would lead to a low score in that part of the index—represents a critical deficit of which federal and state governments should be aware. For instance, New Orleans was not well positioned to address demands of long-term recovery because the city could point to very few local or regional entities capable of doing housing redevelopment work (Rose 103). This is critical because, for redevelopment of low- and moderate-income housing following disasters, one needs community development entities which have some proven capacity to effectively use federal block grant funds as well as federal tax credit funding (Rose 103).

Property acquisition, disposition and stewardship tools

Cultivating safe and affordable housing depends not only on the local government's skill at cultivating partnerships with developers and leveraging funds, but also its expertise for assembling residential property, disposing of it, and monitoring its condition across a city.

The housing category of a City Resilience Index should include at least one indicator for evaluating whether local governments can effectively manage matters relating to residential properties, including public health and safety code compliance and lien foreclosure. This Index indicator would recognize local governments for such aptitudes as: (1) their demonstrated success in acquiring residential properties through timely and properly administered code lien foreclosure or eminent domain procedures; (2) the range of real estate disposition strategies that successfully return residential properties to commerce, including individual sales of multiple properties through requests for proposals; (3) and property swaps with private, nonprofit, or government entities. Additional index credits could be awarded to cities for volume of properties disposed. For example, the Index could award points on a sliding scale that measures a city's sales of property as a percentage of the total number of properties sold on the private market each year in that jurisdiction and/ or the value of properties sold or swapped as a percentage of the total value of real estate owned by the city.

There are three reasons local government property acquisition, disposition, and stewardship are critical to a resilient city. The first is that a basic function of local government is to intervene and protect citizens and the neighborhoods in which they live when private market forces cannot eliminate persistent blight or abandonment. For example, natural disasters and human-made catastrophes such as the mortgage foreclosure crisis unravel local real estate markets in ways that demand government intervention (Brescia, Kelly, and Marshall 305–307; Alexander 734). Other prime causes of these conditions are negligent absentee owners and so-called "heir properties," where multiple family members own fractional interests in a single property due to the family's failure to probate wills or otherwise administer estates (Way 117–19, 151–58). The second is that following disasters, cities suffer widespread problems with poor upkeep of properties (Alexander 730–31, 734). Widespread blight and dilapidation of property retards citizens' efforts to rebuild their neighborhoods and discourages new outside investment. Vacant and abandoned properties have a toxic effect on surrounding homes and businesses (Field Hearing). Not only do vacant properties push down surrounding home values, but they also can trigger higher insurance rates (Field Hearing). Further, when homes are abandoned and not occupied, businesses have little incentive to rent or buy properties to open stores and offices. High levels of neighborhood abandonment cause an economic drag on the community (Field Hearing).

The third reason that a local government's (or in some instances a state's) skill in managing real estate and overseeing code compliance is so important is that disasters force cities to employ many different real estate strategies across its recovering neighborhoods (Brescia 328–35). A homeowner buyout strategy for redeveloping a neighborhood directly impacted by disaster is different from that needed for a neighborhood where the disaster undermined an already weak real estate market, or where the disaster destroyed a stable real estate market (Brescia 330–331, 332–333). Resilient cities will have experience using a range of real estate acquisition and disposition techniques.

Conclusion

This chapter contributes to an ongoing and longer-term exploration of how an index can be used as a tool to build better cities and to prepare them to weather adversity. As a growing body of scholarly work examining urban resilience signals, it is more essential than ever that cities learn from disaster experiences to both nurture thriving cities and bolster their defenses to all manner of adversity. An index promises to serve as a transparent, data driven tool to assist in this effort.

Legal scholars and experienced legal practitioners can make particularly valuable contributions to the establishment of an index as policy tool. As most of the Index indicators discussed in this chapter show, proficiency at using legal tools and knowing how to navigate legal requirements are core competencies for city building and long-term disaster recovery. The City Resilience Index promises to support and focus the day-to-day work of federal and state lawmakers and policy administrators. The City Resilience Index also offers agencies at all levels of government the opportunity to advance significantly the way they think about crafting disaster response laws.

The federal legislative response to Hurricanes Katrina and Rita displayed a restrained and largely reactive view of the federal-state-local community development partnership. Under that view, the federal government's principal recovery role was to furnish money to the states, provide limited technical assistance to disaster stricken communities, and unleash programmatic audits to chase down expected non-compliance on the back end of disaster recovery projects. The City Resilience Index reinforces recent federal efforts to calibrate urban revitalization policy more effectively than was done for New Orleans and the Gulf Coast. This new federal approach emphasizes and demands coordination, cooperation, and communication between and among federal, state, and local governments. The federal approach to deploying disaster relief thus eschews the conception of the federal role in disaster recovery as one that delivers money to the states, allows local governments to flounder as they implement recovery plans, and then "catches" those local governments in frustration or failure through post-hoc audits. A City Resilience Index can be an integral part of a more collaborative way of implementing long-term disaster and urban revitalization policy. It can identify the coordinated, concrete and, thus, most cost-effective steps that cities can take—long before disasters strike or even if disaster or crisis never strikes—to neutralize critical community vulnerabilities and create more resilient cities.

Works Cited

24 CFR 570.208. 2012. Print.

Alexander, Frank S. "Louisiana Land Reform in the Storm's Aftermath." *Loyola Law Review* 53 (2007): 734–57. Print.

Becker, Christine. "Disaster Recovery: A Local Government Responsibility." *PM Magazine* Mar. 2009. Web. <http://webapps.icma.org/pm/9102/public/cover.cfm?title=Disaster%20Recovery%3A%20%20A%20Local%20Government%20Responsibility&subtitle=&author=Christine%20Becker>.

Bloomberg, Michael R. *PlaNYC: A Stronger, More Resilient New York*. 2013. Web. <http://www.nyc.gov/html/sirr/html/report/report.shtml>.

Bourne, Joel K. Jr. "Louisiana Wetlands: Gone with the Water." *National Geographic* Oct. 2004. Print.

Brescia, Raymond H., Elizabeth A. Kelly, and John Travis Marshall. "Crisis Management: Principles that Should Guide the Disposition of Federally Owned, Foreclosed Properties." *Indiana Law Review* 45 (2012): 305, 328. Print.

Calo, Ryan. "Code, Nudge, or Notice." *Iowa Law Review* 99 (2014): 773, 775–90. Print.

Carroll, Chris. "Hurricane Warning: In Hot Water." *National Geographic* Aug. 2005. Print.

"Criteria for National Objectives" 24 CFR 570.208. 2011.

Esty, Daniel C. et al. *Environmental Performance Index*. 2008: 19–21. Web. <http://www.yale.edu/epi/files/2008EPI_text.pdf>.

Fischetti, Mark. "Drowning New Orleans." *Scientific American* (Oct. 2001). Print.

Gerken, Heather K. *The Democracy Index: Why Our Election System is Failing and How to Fix it.* Princeton: Princeton University Press, 2009: 5–9, 66–92. Print.

Gillette, Clayton P. "Plebiscites, Participation, and Collective Action in Local Government Law." *Michigan Law Review* 86 (1988): 930, 945. Print.

Glauber, Diane and David Zisser. "Innovative Post-Disaster Community-Based Housing Strategies." *Building Community Resilience Post-Disaster*. Ed. Dorcas R. Gilmore. Chicago: American Bar Association, 2013: 375. Print.

Gore, Rick. "Wrath of the Gods: A History Forged by Disaster." *National Geographic* July 2000. Print.

Heath, Brad. "Rebuilt N.O. Homes At Risk Without Required Elevation." *USA Today* 19 Sept. 2008. Web. <http://usatoday30.usatoday.com/news/nation/2008–09–18-home-elevation_N.htm>.

Ho, Daniel E. "Fudging the Nudge: Information Disclosure and Restaurant Grading." *Yale Law Journal* 122 (2012): 574, 577–87. Print.

Holm, Erik. "A New York Hurricane: So Long, Subways." *Wall Street Journal* 1 Sept. 2010. Web. <http://blogs.wsj.com/metropolis/2010/09/01/hurricane-earl-new-york-hurricane-so-long-subways/>.

Krupa, Michelle. "Companies that Helped N.O. Getting Paid 2 Years Later." *Times Picayune* 29 Mar. 2008. Print.

Larson, Erik. "Hurricanes on the Hudson." *N.Y. Times* 25 Sept. 1999. Web. <http://www.nytimes.com/1999/09/25/opinion/hurricanes-on-the-hudson.html>.

Liu, Amy. "Rebirth on the Bayou: Lessons from New Orleans and the Gulf Coast." *New Republic*. 26 Aug. 2011. Print.

Marshall, John T. "Rating the Cities: Constructing an Assessment Tool for Evaluating the Effect of State and Local Laws on Long-Term Disaster Recovery." *Tulane L. Rev.* 90.1 (2015) (forthcoming fall 2015). Print.

———— "Weathering NEPA Review: Superstorms and Super Slow Urban Recovery," *Ecology Law Quarterly* 41 (2014): 81, 93–106. Print.

McKelvey, Tara. "Two Years After a Tornado, Joplin Struggles to Rebuild." *BBC News* 22 May 2013. Web. <http://www.bbc.co.uk/news/world-us-canada-22578180>.

Munson, Kyle. "5 Years Later: Remembering the 2008 Flood." *DesMoinesRegister.com* 8 June 2013. Web. <http://www.desmoinesregister.com/article/20130609/NEWS/306090021/5-years-later-Remembering-2008-flood>.

Nelson, Marla. "Using Land Swaps to Concentrate Redevelopment and Expand Resettlement Options in Post-Hurricane Katrina New Orleans." *Journal of the American Planning Association* 80.4 (2014): 426–37.

Nossiter, Adam. "Rebuilding New Orleans, One Appeal at a Time." *New York Times* 6 Feb. 2006. Web. <http://www.nytimes.com/2006/02/05/national/nationalspecial/05rebuild. html?pagewanted=all>.

Olshansky, Robert B. and Laurie A. Johnson. *Clear As Mud: Planning For The Rebuilding of New Orleans*. Chicago: American Planning Association, 2010: 236–37. Print.

Parfit, Michael. "Living with Natural Hazards." *National Geographic* (July 1998). Print.

Public Strategies Group. *City of New Orleans: A Transformation Plan for City Government*. 1 Mar. 2011: 2, 6, 8 <https://www.nola.gov/chief-administrative-office/documents/ nola_transformation_plan/>. Print.

"Resilience." *ResilientCity.org*. Web. 24 Apr. 2014 <http://www.resilientcity.org/index. cfm?id=11449>.

Robertson, Campbell. "A Race for a New Mayor; a Trial for an Old One." *New York Times*. 26 Jan. 2014. Web. <http://www.nytimes.com/2014/01/27/us/politics/a-race-for-a-new-mayor-a-trial-for-an-old-one.html?_r=0>.

Rockefeller Foundation. *"The Rockefeller Foundation 100 Resilient Cities Centennial Challenge."* Rockefeller Foundation. Web. 24 Apr. 2014 <http://100resilientcities. rockefellerfoundation.org/>.

Rose, Kalima. "Bringing New Orleans Home: Community, Faith, and Nonprofit Driven Housing Recover." *Resilience and Opportunity: Lessons from the US Gulf Coast after Katrina and Rita*. Eds Amy Liu et al. Washington: Brookings Institution, 2011: 99, 113. Print.

Sherwood, Christina Hernandez. "Ranking the 'Resilience' of Hundreds of US Cities." *SMARTPLANET* 20 July 2011. Web. <http://www.smartplanet.com/blog/pure-genius/ ranking-the-resilience-of-hundreds-of-us-cities/>.

Spector, Carl and Leah Bamberger. *Climate Ready Boston*. 2013. Web. <*www.cityofboston. gov/news/uploads/30044_50_29_58.pdf*>.

Stone, Adam. "Long-Term Recovery Planning: What You Need to Know." *Emergency Management* 15 May 2013. Web. <http://www.emergencymgmt.com/disaster/ Developing-Long-Term-Recovery-Plan.html>.

The Economist. "Counting the Cost." 21 Mar. 2011. Web. <http://www.economist.com/ blogs/dailychart/2011/03/natural_disasters>.

"The Man Who Predicted Katrina." PBS. WGBH, Boston, 22 Nov. 2005. Television.

"Transition New Orleans Task Force, Housing." Apr. 2010. Copy on file with author.

Trevelyan, Laura. "Superstorm Sandy: US Marks One Year Anniversary." *BBC News* 29 Oct. 2013. Web. <http://www.bbc.co.uk/news/world-us-canada-24721439>.

United States Cong. House. Committee on Financial Services. *Implementation of the Road Home Program Four Years After Hurricane Katrina: Field Hearing Before the H. Comm. on Fin. Servs., Subcomm. on Hous. and Cmty Opportunity*. 111th Cong. Washington: GPO, 2009. <http://www.gpo.gov/fdsys/pkg/CHRG-111hhrg53250/html/ CHRG-111hhrg53250.htm>. Print.

United States Department of Housing and Urban Development. *CDBG Disaster Recovery Assistance*. 28 Jan. 2014. Web. <http://portal.hud.gov/hudportal/HUD?src=/program_ offices/comm_planning/communitydevelopment/programs/drsi>.

——— *Current Housing Unit Damage Estimates: Hurricanes Katrina, Rita, and Wilma*. 2006: 23. Web. <http://www.huduser.org/publications/pdf/GulfCoast_Hsngdmgest. pdf>.

———— *Hurricane Sandy Rebuilding Strategy: Stronger Communities, A Resilient Region.*
19 Aug. 2013. Web. <http://portal.hud.gov/hudportal/HUD?src=/press/press_releases_
media_advisories/2013/HUDNo.13–125>.

Way, Heather K. "Informal Homeownership in the United States and the Law." *Saint Louis
University Public Law Review* 29 (2009): 113, 117–19, 151–58. Print.

Zolli, Andrew and Ann Marie Healy. *Resilience: Why Things Bounce Back*. New York: Free
Press, 2012: 6–8. Print.

Chapter 3
Avoiding Atlantis: Protecting Urban Cultural Heritage from Disaster

Ryan Rowberry[1]

I. Introduction

Rapidly rising sea levels claimed the ancient city of Atlantis in only a day. As Plato chronicled in *Timaeus* (360 B.C.), "in a single day and night of misfortune," violent earthquakes and floods caused Atlantis to sink "into the earth ... and disappear into the depths of the sea" (25). For millennia, Atlantis' watery demise has captured imaginations the way apocalypse movies like *Contagion, Armageddon, World War Z, Independence Day*, or *Dante's Peak* do today. Some historians, geologists, and anthropologists have hypothesized that an actual city named Atlantis was destroyed by earthquake, tsunami, and/or volcanic eruption, but no consensus has been reached on whether the city existed at all (Dušanić 25–52; Vitaliano 5–30; Balch 388–329). Though Atlantis may be only a myth, its real-life parallel exists in the ancient city of Alexandria, along with warnings of what costly consequences result when cities forget their own past and succumb to the urban hubris of indestructibility.

Described by foreign contemporaries as "the first city of the civilized world," Alexandria was unrivaled in its possession of "so many wonders ... [and] works that defy description" (Diordorus of Sicily; Herodotus xxxv). Scholars and laymen alike mourn the loss of Alexandria's famous library—"the world's first universal center of learning" (Abbadi 16, 81)—and ponder what secrets were stored in its estimated 700,000, carefully cataloged scrolls (Fraser 38–48; Macleod; Parsons). Nearby on Pharos Island in Alexandria's harbor stood the magnificent Pharos Lighthouse, one of the Seven Wonders of the Ancient World, which for centuries after its completion remained one of the tallest man-made buildings ever built.[2] But perhaps Alexandria's greatest monument was a living one: Cleopatra (69–30 B.C.). Though she was the last Ptolemaic Pharaoh, she is more often remembered as a shrewd ruler who captivated the two most powerful men on the planet (Julius Caesar and Marc Antony), and her legend has enthralled public imagination for generations (Schiff).

As a population center, cosmopolitan fashion mecca, international power, titan in global commerce, religious nexus, and repository for cultural heritage, Alexandria finds echoes in modern metropolises like Paris, London, Istanbul, New York, and Beijing (Rowberry and

---

1   I would like to thank the editors of this book for stimulating this project and Mitchell E. Jackson for his invaluable research assistance. Any infelicities in the chapter are mine alone.

2   The pyramids at Giza always eclipsed the Pharos Lighthouse in height, but are better termed structures rather than buildings, because they were not intended for human occupation or use. For additional information on the Pharos Lighthouse, see Judith McKenzie's *The Architecture of Alexandria and Egypt, c. 300 BC to AD 70* and Jean-Yves Empereur's "Underwater Archeological Investigations of the Ancient Pharos."

Khalil 88–90).[3] Alexandria's distinct skyline appeared on ancient souvenirs and artwork, much like those bought by tourists in New York, London, or Paris today (Philo 338; Vasunia). Unlike these modern cities, however, Alexandria was unrivaled in its own era: even the famous Roman orator, Cicero, conceded that Imperial Rome was little more than a provincial backwater compared to Alexandria (Cicero 44). But Alexandria and modern cities share another more somber similarity: like many residents of modern coastal cities, Alexandrians believed their city to be immune from the natural catastrophes that ultimately consumed it (Sunstein 61–108; Posner 23–24, 40).

Like the fate which befell its fabled cousin Atlantis, Alexandria was struck by a devastating earthquake and subsequent tsunami in 365 A.D and again in 1375 (McKenzie 8–16). In the centuries before and after, the city's wonders were further damaged by fire, war, and pillage (McKenzie 8–16). As earthquakes and shifting tides gnawed away the foundations of the city, its nucleus, the Royal Quarter, crumbled slowly into the sea (McKenzie 8–16). Over time, remnants of the ancient city became scattered through sale and plunder amongst museums and palaces continents away. Far more often, however, the magnificent vestiges of Alexandria were demolished for raw materials, repurposed haphazardly in new buildings and walls, or abandoned altogether, to lie forgotten beneath the sands of Egyptian desert or the waves of the Mediterranean (McKenzie 9; Empereur 36–43).

The first step in becoming a forgotten city is when a city forgets its past. With scarcely a trace of its Hellenistic heritage intact, Alexandria, whose government-sponsored studies produced the first scientific discoveries on water levels and currents, failed to adapt to changing tides and its urban glory was largely washed away (Plini CC.li.60; Lewis 105–15; Achilles IV.II.-15). What battered remnants of Alexandrian tangible heritage remain rest more than 20 feet below the surface of the modern harbor. Only in the past two decades have underwater archeologists begun to explore the weathered sphinxes and pocked palace walls poking through the sandy ocean floor, hoping the ocean will yet yield some secrets of a city that forgot itself (Empereur 36–43). Though few relics of Alexandria remain to testify to its former glory as an axis of global influence, tourists are nonetheless captivated by the story of a once-powerful city that slid into the sea, visiting in dive suits and glass-bottom boats to peer through murky depths at the submerged ruins.

In an era when coastal cities face natural disasters like those which devastated ancient Alexandria with increasing frequency, it is necessary think critically about how modern cities can avoid similar destruction (Farber 206; Braine 4; Mileti 66). Will future generations visiting Manhattan or New Orleans gaze at their monuments beneath the water clutching souvenir guidebooks?

---

3 Coptic Christianity began in Alexandria, and Biblical papyri in the Coptic language predate the oldest authoritative Greek versions of the Scripture. Moreover, the Catechetical School established by Christian scholars circa 200 AD in Alexandria generated the first systematic Christian theology that laid the foundation for the later ecumenical movement under Constantine (306—337 AD), the first Christian ruler.

II. Why Protect Cultural Heritage?[4]

As leading population centers that hold a tremendous share of global cultural heritage, "coastal cities"—cities located on rivers, seas, or oceans—should be accorded special significance by policy makers (UNESCO 7).[5] Humans have traditionally settled in close proximity to large bodies of water, and a significant proportion of the world's population now lives on the borders of major rivers, seas, or near oceans in some form of urban agglomeration (Farber 23–24). Cultural heritage comprises a significant, core aspect of all cities, and historic resource preservation's "matrix of laws, incentives, [and] policies . . . has become a fundamental tool for strengthening . . . communities" (Rypkema vi). Cities did not somehow emerge fully formed; they developed incrementally, usually in oscillating, uneven lurches of development over time. A simple stroll around any city reminds us that urban areas are vibrant, living landscape palimpsests of our past—pockets of which have been preserved, rehabilitated, or revitalized. These pockets of tangible memory—buildings, parks, and single-family homes to name a few—are critical to a city's overall resilience and should be zealously protected. Indeed, many city dwellers live in historic structures or housing made possible through historic preservation programs.

Scholars have long noted the numerous benefits that flow from preserving historic and cultural heritage (Bronin and Rowberry 12–16). One critical factor from an urban resiliency standpoint is that historic resource preservation is a powerful, broad-based impetus for economic development. Recent statistics from the National Park Service reveal that, in 2012, economic impacts related to the federal historic preservation tax credit—a 20 percent credit for qualifying rehabilitation expenditures—accounted for the creation of approximately 58,000 jobs, generated $3.4 billion in gross domestic product (GDP), and produced over $2.5 billion in income (National Park Service). Just as important as the dollar figures, many of the 744 certified rehabilitated buildings that leveraged this credit in 2012 were "abandoned or underutilized, and all were in need of substantial rehabilitation to return them to, or for their continued, economic viability" (National Park Service 1). Many of these rehabilitated buildings reside in older urban cores and have breathed new life into once derelict domains (National Park Service 15).

Detailed research regarding the economic benefits of cultural heritage preservation in states and localities also reveals that it is a potent economic driver. In Georgia, for example, heritage tourism sustains 117,000 jobs, generates roughly $204 million in wages, and levies $210 million in local taxes on an annual basis (Rypkema and Cheong 2). For the decade 2000–2010, the rehabilitation of historic resources in Georgia (usually buildings) resulted in 10,168 local jobs and over $420 million in income for Georgia workers and proprietors (Rypkema and Cheong 2). Analyses of the economic impacts of historic resource preservation for Utah, Connecticut, Delaware, and Florida tell similar stories: historic preservation creates local jobs; revitalizes older neighborhoods; enhances local sustainability measures; adds needed affordable housing; boosts local taxes; and provides a powerful source of local revenue (Rypkema; Rypkema and Cheong; McLendon). Historic

---

4 Portions of the following two sections have been excerpted and adapted from "Urban Wreckage and Resiliency: Articulating a Practical Framework for Preserving, Reconstructing, and Building Cities," co-authored with Professor John Travis Marshall (50–83).

5 There is no universal definition for cultural heritage. For the purposes of this chapter, cultural heritage is defined broadly as tangible movable or immovable property that exhibits great historical importance (e.g., historic buildings, sites, artifacts, landscapes).

resource preservation is thus a key component in spurring the economic revitalization and resilience of older communities in metropolitan areas, cities, and towns.

Historic resource preservation also has a powerful, positive effect on mental health and the ability of people to cope with change—traits that are desperately needed during a disaster event. Pioneering studies in England have found that adults and teenagers who live in areas with higher concentrations of historic buildings are more likely to have a stronger sense of place (Bradley 5). This reinforced sense of place has many positive benefits on self-esteem, identity, and health, which in turn lead to stronger, more civically engaged communities (Bradley 8; Sandifer 1–15). English researchers also discovered a significant, positive link between the historic environment and social capital—the bonds that connect groups and individuals (Bradley 8). Adults and teenagers who visited historic properties or could identify a local building or monument as being unique or special are likely to have a higher level of social capital, an important element in individual health as well as strong, resilient communities (Bradley 3, 8). These conclusions about the importance of historic resource preservation to the mental health and resiliency of individuals and communities echo the results of research in other fields that display the positive power of connecting present generations with the past: psychologists are discovering that children who know about their family's history (good and bad) are more resilient because they can better moderate the effects of stress (Rai); and military academies have learned that "teaching recruits about the history of their service increases their camaraderie and ability to bond more closely with their unit" (Feiler).

III. Why Preserve Urban Cultural Heritage in Coastal Cities Now?

So, why should policy makers worry about preserving historic resources in coastal cities now? Coastal cities today face an unprecedented risk from both immediate and attenuated risks. Rapid disaster events such as hurricanes, earthquakes, and volcanic eruptions are increasing in frequency and ferocity and disproportionately affect coastal metropolises like San Francisco, New Orleans, New York, and Istanbul, as well as smaller urban coastal communities in Central America, the Caribbean, and the Pacific (Cai 132–37; Rowberry 196–97; Blake; Knabb; Lopez-Marrero and Wisner 129–68; Inter-Agency Subcommittee). These quick and volatile disasters are not isolated events; each brings a litany of devastating secondary effects—fire, flooding, tsunamis, mudslides, public health crises, abandonment—that are often overlooked but continue to grind down weakened urban infrastructure, leaving swathes of cities desolate for years (McNeill and Nelson; Pan-American Health Organization; Stone and Kerr 1602).

Furthermore, though natural disasters focus popular attention on the devastating damage delivered by swift storms or sudden earthquakes, the gradual degradation and dissipation of the coastal environment poses an equal, if not more daunting threat to coastal cities. Once regarded merely as an ominous possibility, mounting tides pose a very real long-term danger to the integrity of coastal communities. And regardless of how intensely scientists and policymakers debate the purported causes of global warming and climate change—topics omnipresent in today's news—evidence consistently shows that ocean levels are rising and coastal floods in the United States are increasing at an extraordinary pace (Melillo 1–62; McNeill and Nelson; Intergovernmental Panel).

Less researched, but just as important to coastal city resilience is the steady degradation of marine ecosystems caused by coastal development. Coasts, rich in biodiverse estuaries yet often lacking in freshwater and rife with disease like malaria, initially offered few places naturally suitable for long-term human habitation. Early settlers claimed the best sites: high ground, usually well above storm surge levels. Twentieth century advances in disease control, transportation, building materials (e.g. the widespread use of concrete), and the ability to pump freshwater to distant areas enabled an explosion in coastal development within the past 50 years.

Given the scenic beauty of estuaries, floodplains, beaches, and sea-islands, most of this growth has occurred in these fragile ecosystems (Farber 205–06; Hoge 34–35). These new coastal developments not only endanger many faltering marine species, they also tend to obliterate critical natural protective features of coastal landscapes like sand dunes, maritime forests, and coastal vegetation that can serve as incredibly effective buffers to sea-level rise and storms (Danielsen 643). Reinvigorating historic coastal cities helps mitigate the proliferation of environmentally insensitive and socially exclusive "resort style" coastal communities—particularly those in the southeastern United States—by concentrating coastal population growth in safer, established settlements (Campanella 351–54; Griffin 290–94, 353–56; Coclainis; Goldfield). Such concentration leaves protective marine ecosystem features intact and increases the resiliency of coastal communities by reducing the social vulnerability of the poorest populations that have traditionally suffered the most from disasters (Katz; Tonay 209–32).

IV. Protecting Urban Cultural Heritage in Coastal Cities

Our incredible capacity to predict and track natural disasters far outstrips current policies in place to deal with them. There is an urgent need for practical legal tools and policies that are both effective and fiscally viable. Far too often, legal avenues available to communities recovering from a catastrophic event are clogged with the detritus of myopic planning and stale policies, providing inadequate mechanisms for efficiently expediting actions required to stabilize vulnerable cultural heritage resources. The following three areas comprise some common programmatic deficiencies encountered in disaster planning and recovery with respect to preserving historic resources and provide policy makers with some powerful opportunities for positive change: (A) Historic Resource Inventories, (B) Adaptation, (C) Streamlined Review Processes.

A. Historic Resource Inventories

A fundamental principle for any effective historic resources management program is simple in theory yet bedeviling in practice: *know what you have.* It is particularly difficult to know all of the historic resources located in a city because there are so many forms they can take—commercial buildings, archaeological sites, residential homes, public buildings, parks, monuments, battlefields, museums—and each year more and more resources may be classified under law as historic. A resource must typically be 50 years old and meet requirements for significance and integrity before it may be designated as historic (Bronin and Rowberry 37–68). Nevertheless, most states, and some local governments, have historic resources statutes that require them to catalogue and protect historic and cultural resources (Bronin and Rowberry 37–68). To fulfill this obligation, many governments around the world utilize an online, publically accessible, searchable inventory using geographic information systems (GIS), which visualizes, analyzes, maps, and interprets data related to physical geography, including historic resources in urban areas. But unfortunately, most historic resources databases are woefully incomplete. For instance, prior to Hurricane Katrina, the Louisiana State Historic Preservation Office had "nineteen thousand [19,000] resources in its statewide GIS" (McCarthy 17). Following the disaster response, another fifty-five thousand (55,000) historic resources were added in the New Orleans area alone (McCarthy 17).

There are several reasons why GIS-based historic resources inventories are critical to preserving coastal city resilience. Here, I mention only two. First, GIS databases are already a widely used planning tool that can be relatively inexpensive to establish and maintain (Arches). In fact, the Getty Conservation Institute and the World Monuments Fund have just rolled out an open-source, geospatial software system—ARCHES—that is purposefully built to help inventory and manage all kinds of immovable heritage to internationally adopted standards (Arches). The ARCHES system is free; any organization can download, install, and customize it (Arches). Currently, Los Angeles city officials are using ARCHES in a program called SurveyLA to create the first comprehensive inventory of Los Angeles' historic resources (SurveyLA).

Second, GIS databases have the capability of layering information spatially on a digital map (Arches). This allows governments to perform a range of activities that are important to a city's long-term resilience on an everyday basis and that are even more critical following a disaster (Arches). Some of these activities include pinpointing individual resources or grouping historic resources by zip code, county, or neighborhood; determining the needs and priorities for investigation, research, conservation, and management of historic sites in targeted areas or by type of resource; formulating management plans for investigating and/or conserving and leveraging historic resources; creating risk maps for particularly vulnerable historic resources; and raising awareness among the public and other authorities about the types and condition of historic resources in their areas (Arches).

But how are cash-strapped governments supposed to keep their inventories up to date? A low-cost, powerful technological tool state and local governments should wield to ensure the accuracy and coverage of historic resources inventories is online crowdsourcing. Simply put, online crowdsourcing allows someone to obtain needed services and/or content by soliciting voluntary contributions from the online public community rather than hiring employees or paying contractors. It has been an extremely effective, low-cost tool for preserving historic resources in many countries. The National Library of Finland, for instance, is using online crowdsourcing to index its scanned archives (De Benetti). Similarly, the University of Cape Town in South Africa is using online crowdsourcing to transcribe collections containing Bushman language, stories, and way of life (Munyaradzi). The National Geographic Society is using online crowdsourcing to analyze millions of satellite images of Mongolia showing potential archaeological sites in the hopes of discovering the tombs of Genghis Khan and his descendants. And an English non-profit organization has utilized online crowdsourcing and online crowdfunding—funds donated by the interested public online—to provide both finances and labor for an expert-led excavation of a Bronze Age causeway composed of millions of timbers in the Cambridgeshire fens (Palmer; Carvajal).

There are two primary reasons why online crowdsourcing would be a fantastic aid to preserving urban historic resources in coastal areas. The first is scarce government resources. City and/or state authorities responsible for historic resources never have enough time, money, and staff to document and catalogue all known historic resources. It would be relatively easy to create an online portal attached to a state or local historic resources inventory website. This portal could offer training modules to citizens on historic resources recording practices and standards and afterwards ask them to collect and upload descriptive information, statistics, pictures, videos, and maps on historic resources in their neighborhoods. While prominent historic resources are likely to have been catalogued, online crowdsourcing can be extremely useful for recording smaller-scale historic resources (*e.g.*, façades) that deserve cataloguing and protection but are low priority. To ensure quality control, any information uploaded to this portal could be screened and vetted by the appropriate authorities before adding it to the inventory. In this way, cities and/or states could gather and preserve vast amounts of data related to their historic resources in a short period of time and at minimal cost.

Second, online crowdsourcing fosters civic pride, a sense of community, and a deeper, more tangible connection to the city's past, particularly for those of younger generations who are adept at using technology (Newman). The social effect of such participation is an increased resilience to economic shocks or natural disasters because community participants in historic preservation crowdsourcing become intimately invested in the future of the city (Rai 2). Additionally, this strategy offers governments and communities peace of mind knowing that, should a disaster occur, as many historic resources as possible have been recorded and preserved for future generations.

B. Adaptation

The second major area underutilized by policy makers to address disasters is the physical adaptation of cities. Measures taken to fortify historic resources in cities vary greatly according to political will, probability of risk, and finances. Adaptation can take many forms. Among the most well-known physical adaptations of the environment to protect historic urban areas are the systems of dykes and levees perfected by the Dutch (Schuetze; Tung 212; Bray 1). And the Experimental Electromechanical Module project in Venice—nicknamed "MOSES" for the biblical figure who parted the Red Sea—is among the most striking (and costly) efforts undertaken to protect urban cultural heritage by adapting the physical environment through human engineering.[6] Others believe, however, that restoring the natural environment of a coastal area may ultimately prove more successful in protecting cities than modifying the urban footprint (Danielsen; Schwab and Brower 283–301).

While physical adaptation or environmental restoration alone can mitigate predictable, immediate natural risks, neither fully addresses the need for gradual change required for adequate long-term preparation against natural disasters. As the levees of New Orleans famously proved after Hurricane Katrina, physical adaptations may ultimately offer far less protection from nature's wrath than is widely believed (Farber and Chen 204–05). One

---

6 Increased flooding, erratic tides, and rising sea levels critically threaten the globally cherished buildings found along the winding canals of Venice. In an effort to ensure the long-term survival of the city, Venetian and Italian policy makers have implemented a radical plan to thwart any watery hazards by controlling the level of the Venetian lagoon. Engineers have devised a series of 78 hollow 300-ton panels hinged to concrete bases that lie on the ocean floor along the mouth of the lagoon. When pumped full of air, these panels rise above the ocean to create a wall that insulates the Venetian lagoon from the currents of the Adriatic Sea. Though impressive, solutions like MOSES are only viable solutions to sea level rise under very specific geographic conditions and may result in an adverse environmental impact. Furthermore, the $8.8 billion the Italian government estimates it will cost to complete MOSES will dissuade resource strapped cities from adopting similar measures (Ammerman 1301–1302).

part of addressing this long-term aspect of disaster preparation lies in creating policies that more efficiently reuse extant historical structures in coastal cities.

Governments at all levels possess scores of vacant or underutilized historic buildings that can be adapted for a range of purposes from public housing to businesses to storage facilities. In the United States, hundreds of historic federal post offices in urban areas are redundant or underused (Advisory Council on Historic Preservation). The Federal Government also owns several neglected historic buildings within coastal National Parks that hold exciting potential for rehabilitation. Similarly, in the United Kingdom (UK) the central government owns "hundreds of other properties across the country which are underused or lying vacant" (Gov.uk). And Barcelona boasts a bevy of vacant or mothballed historic warehouses, factories, and industrial sites (Vilaseca 55).

To more effectively use historic government buildings, coastal cities should follow the lead of the UK, which recently established a policy called the "Right to Contest." The "Right to Contest" provides a swift mechanism for anyone—businesses, NGOs, community groups, individuals—to seek the sale of surplus or redundant publicly owned land or buildings where they could be put to better economic use. Parties may also challenge government buildings that are currently in use, as long as the building's operations could feasibly be removed to a different location. To contest the use of government property, historic or otherwise, applicants need to fill out an application form showing why the property in question is potentially surplus or redundant and offer supporting evidence (e.g., photographs, location plan or a description of the buildings involved). Once an application is submitted, the department holding the building and/or land will be asked if they agree to sell the site or to justify why it needs to keep the property. If the department agrees to sell the property, the case is closed, and the property is released on the open market. Alternatively, if the department argues to keep a property, the matter is referred to a committee of governmental ministers to make a decision, typically within six weeks.

Since 2010, the UK central government has raised over £1 billion pounds by selling over 770 buildings (many of them historic), thus reducing its deficit while providing increased space and financing for numerous businesses and tens of thousands of housing units, including public housing. Coastal states and the cities within them could benefit greatly by fashioning their own "Right to Contest" policy. This would decrease sprawl, concentrate people in traditionally safer areas where historic buildings tend to be located, and provide a needed fiscal injection to reduce government debt or shore up critical infrastructure.

C. Streamline Historic Resources Review Processes

Finally, policy makers should make every possible effort to streamline the review process for historic resources in order to protect them. During times of disaster, lengthy environmental and historic resources review processes can jeopardize the integrity of historic resources, keep residents in historic structures from rehabilitating their homes, and stop local governments from restoring critical historic areas.

State and local environmental and historic resources review processes are usually modeled on two federal statutes: the National Environmental Policy Act (NEPA) (42 U.S.C. 4332(C)–(D)); and Section 106 of the National Historic Preservation Act (NHPA) (16 U.S.C. 470f). Most state environmental protection statutes closely track NEPA by requiring an environmental review when a proposed agency action significantly impacts, or is likely to significantly impact, the environment (Cal. Pub. Res. Code. 21100(a); Conn. Gen. Stat. Ann. 22-a-1b(c); Ga. Code Ann. 12–16–4). Similarly, the NHPA requires the governor of every state to appoint a State Historic Preservation Officer (SHPO) to administer a preservation program in the state (16 U.S.C. 479a(6)). The SHPO consults with federal agencies when a federal undertaking has an effect on a state cultural heritage resource that is listed on or eligible for listing on the National Register of Historic Places as well as the State Register (16 U.S.C. 470a(b)(3)(F)). Such federal undertakings include federal permits, licenses, or funding that state or local governments need to begin rehabilitating or protecting historic resources (36 C.F.R. 800.17(y)). Thus, prior to the granting of federal, state, or local permits and funds for the rehabilitation of historic resources, both an environmental *and* historic resources review must be completed for each individual historic resource.

As implementation of the environmental and historic review processes taught the United States post-Katrina, inartful implementation of these processes can impede important long-term recovery efforts (Holdeman). Federal long-term recovery monies cannot be dispensed to reimburse state and local governments for recovery work until the environmental and historic reviews are completed (24 C.F.R. 58.22(a); Mandelker 7:10). This means homeowners repairing their residences (whether historic or not) following a disaster event and seeking reimbursement through a state or local government's federally-funded home rehabilitation project cannot receive reimbursement for repair costs until an environmental review has been performed on the home. Ironically, this could potentially delay repairs to historic properties. Unnecessary delay in protecting and rehabilitating historic resources after a disaster may be avoided by creating a regulation streamlining the environmental and historic resources review processes during times of disaster (United Nations Office for Disaster Risk Reduction; Natural Hazards Center).

There are several reasons why state and local governments should streamline regulations for environmental and historic review processes. The first is time. Following disaster, there is no time for legislators to devise speedy alternatives to the normal review processes; they are busy tending their families, homes, and devastated communities. Such streamlined regulations may take many forms. One possibility is for state and local governments to integrate their environmental and historic resources review processes, much like the federal government has recently done. This helps to avoid duplicative review efforts, saving time and resources. Another possibility is for states and/or localities to sign a programmatic agreement with the Federal Emergency Management Agency (FEMA) "to exclude specific routine activities from Section 106 review and streamline project evaluation during all phases of emergency response" (Advisory Council on Historic Preservation). Prototype programmatic agreements are available online are designed to be customized to local situations (Advisory Council on Historic Preservation; FEMA).

The second reason is money. As noted above, environmental and historic resources review processes must be completed *before* the disbursement of moneys to facilitate the protection or rehabilitation of historic resources. Put simply, not a dime of federal, state, or local moneys can flow to restore or repair historic resources until these reviews are completed (Louisiana Land Trust). But, if states and/or cities have streamlined regulations and a programmatic agreement in place before disaster strikes, these reviews can be finished efficiently so that money can be released to help rehabilitate historic buildings and homes. Otherwise, many historic resources in coastal cities may be in danger of festering in mold or mildew, or falling due to prolonged structural instabilities.

V. Conclusion

From the destruction of ancient Alexandria to the incessant winds and waves that caused the *Grote Mandrenke* (Great Drowning of Men) in 1362 that forever changed the northwest coastline of mainland Europe to Super Storm Sandy (2012) that wreaked havoc on a regional scale in the northeast United States, natural disasters have long devastated coastal communities and cities (Scott 187–190). Severe disaster incidents along with more attenuated, yet dangerous, sea level rise are increasing in frequency (Marshall). Concurrently, the majority of the world's cultural heritage—including most of the properties inscribed on the World Heritage List—are located in "historic cities, urban sites, or edifices located in cities," leaving urban cultural heritage in coastal cities to face extreme risks in the near future (UNESCO 7).

Coastal cities in particular need to cast off typical policy-making peredinations by immediately implementing preparatory policies and programs to save the souls of their cities (Mayes). Creating and managing an online historic resource inventory, adapting vacant or underused historic government properties for reuse by the wider community, and streamlining historic resource review processes are simply a start. These suggestions must become part of a more comprehensive disaster mitigation plan to be truly effective. Fortunately, we now have tools to make this a reality.

The FEMA has developed a handbook that offers cities, municipalities, and smaller communities step-by-step guidance on how to: (1) develop and implement pre-disaster planning strategies for historic properties and cultural resources; and (2) integrate cultural heritage into hazard mitigation planning (FEMA Handbook). The problem remains, however, that almost all cities and communities fail to act.[7] While some coastal cities and their cultural heritage may yet share the same fate as ancient Alexandria, these tragic losses will more likely be due to forgetfulness and inaction than ignorance or incapacity.

Works Cited

16 U.S.C. Sec. 470a. 2000.

24 Code of Federal Regulations Sec. 58.22(a). Print.

36 Code of Federal Regulations Sec. 800.16(y). Print.

42 U.S.C. Sec. 4332(C)–(D). 2012.

Achilles Tatius. *Leucippe and Clitophon.* Trans. S. Gaselee. Loeb Classical Library 45. Cambridge, MA: Harvard U P, 1969: IV.II-15. Print.

Advisory Council on Historic Preservation. *Federal Emergency Management Agency Model Statewide Programmatic Agreement.* Web. 21 Sept. 2010 <http://www.achp.gov/fema-pa.html>.

Advisory Council on Historic Preservation. *FEMA Prototype Programmatic Agreement.* Web. 18 Dec. 2013 <http://www.achp.gov/fema_prototype_pa.html>.

Advisory Council on Historic Preservation. *Preserving Historic Post Offices: A Report to Congress.* 2014. Web. 17 Mar. 2015 <http://www.achp.gov/historicpostoffices.pdf>.

---

7 Although the FEMA handbook was published nearly a decade ago, to my knowledge, only Annapolis, Maryland has actually begun following the steps outlined in the handbook to develop a Cultural Resource Hazard Mitigation Plan.

Alexander, Frank S. "Louisiana Land Reform in the Storms' Aftermath." *Loyola Law Review* 53 (2007): 730–31, 734. Print.

"Atlantis Beneath the Vineyard," *Science News* 92.6 (1967): 125–26. Print.

Ammerman, Albert J., and Charles E. McClennan. "Saving Venice." *Science, New Series* 289.5483 (2000): 1301–302. Print.

Angel, Shlomo. *Making Room for a Planet of Cities*. Lincoln Institute of Land Policy: 2011. Print.

Aston, Michael. *Interpreting the Landscape: Landscape Archaeology and Local History*. New York: Routledge, 1997. Print.

Balch, Edwin. "Atlantis or Minoan Crete." *Geographical Review* 3.5 (1917): 388–92. Print.

Blake, Eric et al. "Tropical Cyclone Report Hurricane Sandy (AL182012)." Working Paper, National Hurricane Center, 2013. Web. 15 Mar. 2015 <http://www.nhc.noaa.gov/data/tcr/AL182012_Sandy.pdf.>.

Bradley, David et al. *5395 Assessing the Importance and Value of Historic Buildings to Young People: Final Report to English Heritage* (2011): 5. Web. 13 Mar. 2015 <http://www.english-heritage.org.uk/publications/historic-buildings-young-people/importance-value-historic-buildings-young-people.pdf>.

Bradley, David et al., *Sense of Place and Social Capital and the Historic Built Environment: Report of Research for English Heritage* (2009): 2, 8. Web. 12 Mar. 2015 <http://hc.english-heritage.org.uk/content/pub/sense_of_place_web.pdf>.

Braine, Theresa. "Was 2005 the Year of Natural Disasters?" *Bulletin of the World Health Organization* 84.1 (2006): 3–8. Print.

Bronin, Sara and Ryan Rowberry. *Historic Preservation Law in a Nutshell*. St. Paul, MN: West Academic, 2014: 12–16. Print.

Cai, Wenju et al., "Increased Frequency of Extreme La Niña Events Under Greenhouse Warming." *Nature Climate Change* 5 (2015): 132–37. Print.

California Public Resources Code Sec. 21100(a). Print.

Campanella, Richard. *Bienville's Dilemma: A Historical Geography of New Orleans*. Lafayette: Center for Louisiana Studies, U of Louisiana at Lafayette, 2008: 351–54. Print.

Carvajal, Doreen. "More Paris Institutions Turn to Crowd-Funding." *New York Times* 7 Oct. 2014: n. pag. Print.

Center for Progressive Reform. "An Unnatural Disaster: The Aftermath of Hurricane Katrina." Washington, DC, 2005: 34–35. Print.

Cicero. *Pro Quinctio. Pro Roscio Amerino. Pro Roscio Comoedo. On the Agrarian Law*. Trans. J. H. Freese. Cambridge, MA: Harvard U P, 1930: 44. Print.

Coclanis, Peter A. *Globalization and the American South*. Ed. James C. Cobb and William Stueck. Athens, Georgia: U of Georgia P, 2005. Print.

Connecticut General Statutes Annotated Sec. 22a-1b(c). Print.

Diordorus of Sicily. *Library of History*, XVII.52.4. Trans. C. Bradford Welles. Cambridge, MA, Harvard U P: (1990): 267, 269. Print.

Danielsen, Finn et al. "The Asian Tsunami: A Protective Role for Coastal Vegetation." *Science* 310 (2005). Print.

David, Scott et al. *Coastal Wetlands of the World: Geology, Ecology, Distributions and Applications*. Cambridge, United Kingdom, 2014: 187–90. Print.

De Benetti, Tommaso. *Digitalkoot: Crowdsourcing Finnish Cultural Heritage*. Crowdsourcing.org. 8 Feb. 2011. Web. 17 Mar. 2015. <http://www.crowdsourcing.org/document/digitalkoot-crowdsourcing-finnish-cultural-heritage/9397>.

Dušanić, Slobodan. "Plato's Atlantis." *L'Antiquité Classique* 51 (1982): 25–52. Print.

Empereur, Jean-Yves. "Underwater Archeological Investigations of the Ancient Pharos." *UNESCO—CSI (Coastal Regions and Small Islands) Environmental Development in Coastal Regions and Small Islands*. Web. 13 Mar. 2015 <http://www.unesco.org/csi/pub/source/alex6.htm>.

——— "Diving on a Sunken City." *Archaeology* 52.2 (1999): 36–43. Print.

Farber, Daniel. "Chapter 2: Why Things Go Wrong: Causes of Disaster, B. Population Growth and Demographic Trends." *Disaster Law and Policy*, 2nd ed. Frederick, Maryland: Aspen Publishers, 2010: 23–24. Print.

Farber, Daniel and Jim Chen. "Prevention and Mitigation." *Disasters and the Law: Katrina and Beyond*. New York: Aspen, 2006: 204–05. Print.

Farber, Daniel et. al. "Social Vulnerability." *Disaster Law and Policy*. Frederick, Maryland: Aspen Publishers, 2010: 205–06. Print.

Feiler, Bruce. "The Stories That Bind Us." *New York Times* 15 Mar. 2013: n. pag. Web. 5 Mar. 2015 <http://www.nytimes.com/2013/03/17/fashion/the-family-stories-that-bind-us-this-life.html?pagewanted=all&_r=0>.

FEMA. *FEMA Handbook, Integrating Historic Property And Cultural Resource Considerations Into Hazard Mitigation Planning: State and Local Mitigation How-To Guide*. May 2005. Web. <https://www.fema.gov/pdf/fima/386–6_Book.pdf>.

Fire Following Earthquake. Eds. Charles Scawthorn, John Eidinger, Anshel Schiff. Reston, VA: American Society of Civil Engineers, 2005. Print.

Fraser, I. "Culture and Power in Ptolemaic Egypt: The Museum and Library of Alexandria," *Greece & Rome* 42.I (1995): 38–48. Print.

Georgia Code Annotated Sec. 12–16–4(a). Print.

Goddio, Franck. *Alexandria: The Submerged Royal Quarters*. London: Periplus, 1998. Print.

Goldfield, David R. *Cotton Fields to Skyscrapers: Southern City and Region*. Baton Rouge: Louisiana State U, 1989. Print.

Griffin, Larry J., et. al. "Gated Communities." *The New Encyclopedia of Southern Culture*. Ed. Charles Reagan Wilson. Vol. 20 Social Class. Chapel Hill: U of North Carolina, 2012: 353–56. Print.

Griffin, Larry J., et. al. "Urbanization in the New South." *The New Encyclopedia of Southern Culture*. Ed. Charles Reagan Wilson. Vol. 20 Social Class. Chapel Hill, NC: U of North Carolina, 2012: 290–94. Print.

Herodotus, *The Histories*. Trans. George Rawlinson. New York, Knopf: 1997. Book II, xxxv. Print.

Holdeman, Eric. "Hurricane Katrina and the Lessons Learned from Mississippi's Recovery." *Emergency Management*. 29 August 2012. Web. 17 Mar. 2015. <http://www.emergencymgmt.com/disaster/Hurricane-Katrina-Lessons-Learned-Mississippis-Recovery.html>.

Hoge, Patrick. "Homes in Flood Catastrophe Zones: A Booming Demand For Housing Has Developers Building New Earthen Levees Inside the Old Ones in the Delta Floodplain," *San Francisco Chronicle*. 30 Jan. 2006. Print.

Intergovernmental Panel on Climate Change. "Climate Change 2001: Working Group II." *Impacts, Adaptation, and Vulnerability*. 2001. Web. 17 Mar. 2015 <http://grida.no/publications/other/ipcc_tar/>.

Katz, Bruce. "Concentrated Poverty in New Orleans and other American Cities." Brookings Institute. 4 Aug. 2006. Print.

Knabb, R. D. et al. *Tropical Cyclone Report Hurricane Katrina 23–30 August 2005.* National Oceanic and Atmospheric Administration, National Hurricane Center, 2011. Web. 15 Mar. 2015. <http://www.nhc.noaa.gov/data/tcr/AL122005_Katrina.pdf>.

LaRiche, William. *Alexandria: The Sunken City.* London: Weidenfeld, 1996. Print.

Lazarus, Amber. *Relationships Among Indicators of Child and Family Resilience and Adjustment Following the September 11, 2001 Tragedy.* Working Paper. 2004: 12. Web. 15 Mar. 2015. <http://www.marial.emory.edu/pdfs/Lazarus_36_04.pdf>.

Lewis, Naphtali. *Greeks in Ptolemaic Egypt.* Oxford: Clarendon P., 1986: 105–15. Print.

López-Marrero, Tania and Ben Wisner. "Not In the Same Boat: Disasters and Differential Vulnerability in the Insular Caribbean." *Caribbean Studies* 40.2 (2012): 129–68. Print.

Los Angeles Department of City Planning, Office of Historic Resources. *SurveyLA.* Web. 17 Mar. 2015 <http://www.preservation.lacity.org/survey>.

McKenzie, Judith. "How Ancient Alexandria Was Lost." *The Architecture of Alexandria and Egypt, c. 300 BC to AD 700.* New Haven, Yale University Press: 2007. Print.

MacLeod, Roy. *The Library of Alexandria.* London, Tauris: 2000. Print.

Marshall, John Travis and Ryan Rowberry. "Urban Wreckage and Resiliency: Articulating a Practical Framework for Preserving, Reconstructing, and Building Cities," *Idaho Law Review* 50 (2014): 50–83. Print.

Mayes, Tom. *Why Do Old Places Matter? How Historic Places Affect our Identity—and our Wellbeing.* National Trust for Historic Preservation, Washington D.C., 2015: 1–114. Print.

McCarthy, Deidre. "The Importance of Heritage Inventories in Preparation and Response." *Conservation Perspectives: The GCI Newsletter.* 2013: 17. Print.

McLendon, Timothy et al. *Economic Impacts of Historic Preservation in Florida: Update, 2010* (2010). Web. 13 Mar. 2015 <http://www.law.ufl.edu/_pdf/academics/centers-clinics/centers/executive_summary_2010.pdf>.

McNeill, Ryan and Deborah Nelson. "Coastal Flooding Has Surged in the U.S." *Scientific American.* 10 July 2014. Print.

Melillo, Jerry M., Terese (T.C.) *Climate Change Impacts in the United States: The Third National Climate Assessment.* Eds. Richmond, and Gary W. Yohe. U.S. Global Change Research Program, 2014: 1–841. Print.

Mileti, Dennis. *Disasters by Design: a Reassessment of Natural Hazards in the United States.* Washington, D.C.: Joseph Henry, 1999: 66. Print.

Munyaradzi, Ngoni. "Crowdsourcing to Preserve Bushman Heritage." *Crowdsourcing. org.* 14 Nov. 2012. Web. 17 Mar. 2015 <http://www.crowdsourcing.org/article/-crowdsourcing-to-preserve-bushman-heritage/21527>.

National Park Service, United States Department of the Interior. *Annual Report on the Economic Impact of the Federal Historic Tax Credit for FY 2012.* Washington, D.C., National Park Service, 2013: 3, 5. Web. 15 Mar. 2015 <http://www.nps.gov/tps/tax-incentives/taxdocs/economic-impact-2013.pdf>.

Newman, Mark et. al., "Understanding the drivers, impact and value of engagement in culture and sport: An overarching summary of the research." *Gov.uk.* July 2010. Web. 17 Mar. 2015: 28 <https://www.gov.uk/government/uploads/system/uploads/attachment_data/file/71231/CASE-supersummaryFINAL-19-July2010.pdf>.

"NOVA—Transcripts—Treasures of the Sunken City—PBS." Pbs.org. 18 Nov. 1997. Web.

"Over 500 Dead, $200,000,000 Lost in San Francisco Earthquake." *The New York Times.* 18 Apr. 1906. Print.

Palmer, Jason. "Flag Fen Hosts 'Crowdsourced' Bronze Age Archaeology Dig." *BBC News: Science and Environment.* 13 Aug. 2012. Web. 17 Mar. 2015 <www.bbc.co.uk/news/science-environment-19192220>.

Pan-American Health Organization, Regional Office of the World Health Organization. *Health Cluster Bulletin: Cholera and Post-Earthquake Response in Haiti.* Port Au Prince: World Health Organization, 11 Oct. 2011: No. 28. Print.

Parsons, Edward. *The Alexandrian Library: Glory of the Hellenistic World.* New York: Elsevier, 1967. Print.

Philo, "On the Embassy to Gauis," *Book XLIII.* C.D. Trans. Yonge, *The Works of Philo.* Peabody, MA: Hendrickson Publishers (1993): 338. Print.

Plato. *Timaeus. Critias. Cleitophon. Menexenus. Epistles.* Trans. R. G. Bury. Loeb Classical Library No. 234. Cambridge, MA: Harvard U P, 1929: 25 c–d. Print.

Pliny, *Natural History*, Trans. H. Rackham. Loeb Classical Library 330. Cambridge, MA: Harvard U P, 1938: X.li.60, V.x.58. Print.

Posner, Richard. *Catastrophe.* Oxford: Oxford U P, 2004: 23–24. Print.

Rai, Tage. *Mental Resilience and Narratives: Physiological Stress Responses to Media Coverage of 9/11.* Working Paper. Apr. 2006: 2. Web. 15 Mar. 2015 <http://www.marial.emory.edu/pdfs/tage%20rai%20wp51.doc>.

Rowberry, Ryan, "Anchoring Memory in the Face of Disaster: Technology and Istanbul's Cultural Heritage Preservation Regime." *Bahçeşehir University Law Review* 8 (2014): 196–97. Print.

Rowberry, Ryan and John Khalil. "A Brief History of Coptic Personal Status Law," *Berkeley Journal of Middle Eastern and Islamic Law* 3 (2010): 88–90. Print.

Rypkema, Donovan and Caroline Cheong. *Good News in Tough Times: Historic Preservation and the Georgia Economy* (2010): 2. Web. 12 Mar. 2015 <http://georgiashpo.org/sites/uploads/hpd/pdf/Economic_impact_study.pdf>.

———— *The Delaware Historic Preservation Tax Credit Program: Good for the Economy, Good for the Environment, Good for Delaware's Future* (2010). Web. 13 Mar. 2015. <http://history.delaware.gov/pdfs/rypkemaReport.pdf>.

———— *Investment in Connecticut: The Economic Benefits of Historic Preservation* (2011). Web. 12 Mar. 2015. <http://www.cultureandtourism.org/cct/lib/cct/Economic_Impact_Study_(Final_6–2011).pdf>.

Rypkema, Donovan et al., *Measuring Economic Impacts of Historic Preservation: A Report to the Advisory Council on Historic Preservation by PlaceEconomics* (2013): vi. Web. 13 Mar. 2015 < http://www.achp.gov/docs/Economic%20Impacts%20v5-FINAL.pdf>.

———— *Profits Through Preservation: The Economic Impact of Historic Preservation in Utah* (2013). Web. 13 Mar. 2015 http://www.placeeconomics.com/wp-content/uploads/2011/03/profits-through-preservation_utah-shortreport.pdf>.

Sandifer, Paul et al. "Exploring Connections Among Nature, Biodiversity, Ecosystem Services, and Human Health and Well-Being: Opportunities to Enhance Health and Biodiversity Conservation." *Ecosystem Services* 12 (2015): 1–15. Print.

Schiff, Stacy. *Cleopatra: A Life.* New York: Brown and Company, 2010. Print.

Schueteze, Christopher F. "Awakening the 'Dutch Gene' of Water Survival." *The New York Times* 19 June 2014. Web. 17 Dec. 2014 <http://www.nytimes.com/2014/06/30/world/europe/netherlands-water-management-system-global-climate-change-sea-level-rise-dutch-gene.html?_r=1>.

Schwab. Anna K. and David J. Brower. "Increasing Resilience to Natural Hazards: Obstacles and Opportunities for Local Governments Under the Disaster Mitigation Act of 2000." *Losing Ground: A Nation on Edge.* Eds. John Nolon and Daniel Rodriguez. Washington, D.C.: Environmental Law Institute, 2007: 281, 283–301. Print.

Stone, Richard and Richard A. Kerr. "Girding for the Next Big Wave." Science 310 (2005): 1602. Print.

Sunstein, Cass. "Probability Neglect: Emotions, Worst Cases, and Law." *Yale Law Journal* 112 (2002): 61–108. Print.

Tonay, Stewart E. "The African American 'Great Migration' and Beyond." *Annual Review of Sociology* (2003): 209–32. Print.

Tung, Anthony M. "Preservation and Social Conscience." *Preserving the World's Great Cities: The Destruction and Renewal of the Historic Metropolis.* New York: Clarkson Potter, 2001: 212. Print.

United Nations. Inter-Agency Standing Committee, Office for the Coordination of Humanitarian Affairs. *Inter-Agency Real- Time Evaluation of the Humanitarian Response to the Earthquake in Haiti, 20 Months After.* New York: United Nations, Jan. 2012. Print.

——— UNESCO. *Developing Historic Cities: Keys for Understanding and Taking Action,* 2014: 7. Print.

United States. Cong. House Committee on Financial Services, Subcommittee on Housing and Community Opportunity. *Implementation of the Road Home Program Four Years After Hurricane Katrina.* 111th Cong. Washington: GPO, 2009: 95. Print.

Vasunia, Phiroze. *The Gift of the Nile: Hellenizing Egypt from Aeschylus to Alexander.* Berkeley: U of California P, 2001. Print.

Vilaseca, Stephen. *Barcelonan Okupas: Squatter Power!* Madison, New Jersey: Fairleigh Dickenson U P, 2013: 55. Print.

Vitaliano, Dorothy B. "Geomythology: The Impact of Geologic Events on History and Legend with Special Reference to Atlantis." *Journal of the Folklore Institute* 5.1 (1968): 5–30. Print.

Chapter 4
Moving to Safety? Opportunities to Reduce Vulnerability through Relocation and Resettlement Policy

Marla Nelson and Renia Ehrenfeucht

Introduction: Increasing Options in Hazardous Places

In 2014, the Louisiana coast was disappearing at the rate of 25 to 35 miles per year resulting in thousands of residents living closer and closer to open water in the hurricane-prone area. In 2014, aerials of Detroit, Michigan, showed more downtown parking lots than buildings and a few houses scattered among green swaths where dense neighborhoods once stood. As sea levels rise, as cities shrink, and as regional economies restructure or retract, people's circumstances decline or grow more hazardous. In situations where residents become more vulnerable when they stay put, residential relocation has become a critical site for action. Yet, in the United States, urban and regional institutions and policies have not been designed to effectively enable timely relocation and resettlement when it becomes necessary.

While helping residents relocate can prevent the loss of life and private property (FEMA; Perry and Lindell; Schwab et al), reconfiguring urban and regional land use including resettling residents is rare in the U.S. (Olshansky, Johnson and Topping). Instead, mitigation measures commonly help people stay in place. Nevertheless, continuing to live in areas that become more hazardous comes with significant financial and emotional costs to households and communities. Direct expenses are also borne by all as public entities step in to better protect vulnerable areas or help when disasters strike.

Although it is critical to rethink policies designed to reduce vulnerability, it has proven difficult. After all, more than a century of government policies have promoted the broadest possible footprint for urban growth and development. There have been myriad public sector actions from local governance structures to national housing and infrastructure policy that created habitable, well-serviced urban regions. These were tremendous urban innovations that allowed people to live in various circumstances from the high densities of Manhattan to Atlanta's low-density regional form to the urban fringe and rural areas that are well connected to urban systems. These institutions, however, were developed in contexts where managing growth and making land habitable for development were primary concerns. Contemporary circumstances including sea-level rise and dramatic population loss were not the focus. National and local policies are effective at developing infrastructure and facilitating development, but not well suited to moving people out of harm's way and reconfiguring established settlements. As a result, it has become necessary to adapt urban institutions in order to coordinate citywide and regional action. Relocation policies have also been controversial, however, and it is imperative that resettlement initiatives enable better options for residents to move to less hazardous circumstances without forcibly displacing residents or resulting in increased vulnerability for residents who remain.

This chapter examines how relocation and resettlement policy can provide people more options to reduce risk and build more resilient communities. We begin by explaining the need for relocation policy as well as the histories that foster resistance to relocation efforts. Next we reconceptualize relocation to account for the iterative and interdependent decision-making processes that shape household and community scale action. We then discuss three types of relocation and resettlement options that have been tried in the United States. We conclude with some of the ongoing challenges and ways to move forward.

The Urgent Need for Effective Long-Term Action

Relocation is often seen as a last-resort solution that residents often resist. Nevertheless, the potential damage from failing to develop viable relocation and resettlement strategies is staggering. Eight of the twenty cities worldwide at greatest risk from coastal storms are in the United States (NRC 2014). Five million Americans live where they are at increased risk from coastal flooding as the sea level rises. Half those residents live in Florida and Louisiana while New York and New Jersey residents are also vulnerable (Strauss, Tebaldi and Ziemlinski).

The majority of funding to address coastal risk is spent on disaster response and recovery rather than reducing risks (NAS 2014). Data indicate that in the US, major weather and climate disasters are increasing in both frequency and aggregate losses (Smith and Katz). Thus, disaster costs are likely to continue to increase. While disaster policy typically addresses short-term displacement, it overlooks long-term relocation and resettlement issues (Comerio; Levine, Esnard and Sapat; Peacock, Dash and Zhang; Zhang and Peacock). And because post-disaster recovery funds are tied to extreme events, they are ill suited or unavailable to respond to gradual environmental change.

Relocation and resettlement policy is also needed for shrinking or legacy cities facing long-term population loss, economic decline, and property abandonment. Despite its depopulation, 700,000 residents still live in Detroit, and millions live in depopulated cities and neighborhoods throughout the country with the accompanying abandoned buildings, fire, and worsening services. In the US most federal urban assistance and public revenues are dependent on population. As populations decline, resources to adapt to the depopulation shrink, making it impossible for shrinking cities to effectively adapt to the changing circumstances.

In areas growing more hazardous, market driven outmigration reduces vulnerabilities for those who leave while increasing vulnerabilities for those who remain (Warner et al). Oftentimes, the most advantaged residents migrate seeking better opportunities elsewhere, leaving behind residents who are more likely to be poor and elderly. Population loss can drive down property values and the quality of services, worsening conditions for those who stay behind. Low value properties also provide remaining residents few resources for potential future moves if they sell. As the population declines, financing regional mitigation becomes more expensive per capita at the same time that fewer tax revenues are collected in an area.

Although the urgency is great, the time needed to resettle both urban and rural communities is long. Kiruna, Sweden has decided it must move because the mine that provides its livelihoods is also undermining the stability of the ground underneath the town. Even though the mine will fund the move, relocating a city of 23,000 will take decades (O'Sullivan). In many cases, the public sector has a necessary role in funding relocation and resettlement initiatives, and, in doing so, reducing the costs of disaster response. It also plays a critical role in creating tools and governing structures necessary to facilitate long-term change.

Relocation as a Policy Framework Between Migrating and Staying Put

Relocation policy can be seen as a way to strategically bridge residents staying put in hazardous conditions and their long-distance migration. Global migrations are a defining characteristic of the 21st century. Consider that people situated as differently as the transnational capitalist class and those pursuing work in an expanding service sector move internationally for work (Brash; Sassen). These migrants will be joined by as many as 200 million environmental refugees who lose their homes or livelihoods by the mid-21st century (Campbell). In response to growing risk that accompanies sea level rise and extreme weather and in anticipation of more global migrations, researchers have asked, "where will people go?" (Oliver-Smith 2009).

Little is known about climate-induced migration and how environmental factors interact with economic and political factors to shape migration decisions, but the social costs of migration are high (Warner et al; McLeman and Hunter). In the US, 37 percent of Americans continue to live in their hometowns and another 20 percent have never moved from their home states. They identify the feeling of belonging as a major reason. Of the residents who have continued to live in their hometowns, 74 percent identify family ties as the reason they have stayed (Cohn and Morin). In some cases, people are unable to migrate or are involuntarily immobile. This is often due to immigration restrictions, but other factors such as available resources or social obligations can also hold people in place (Carling).

People respond in varied ways to adverse changes. They can do nothing, accepting declining life quality or increased risk (Bailey, Grambling and Laska). More often, when residents do not consider moving an option—or prior to their decision to move—they work to manage the adverse impacts in ways that would allow them to stay. To mitigate the changes, residents take varied, intentional actions at the parcel and neighborhood scale, as well as advocating for community or citywide mitigation measures and political and policy change (Warner et al; Bailey, Grambling and Laska; Kefalas; Woldoff; Kinder).

Because people are actively working to stay, it is important to recognize this as an influential phenomenon (Hanson). Relocation and resettlement policy, therefore, can be seen as a set of tools and processes intended to facilitate both strategic migration away from areas where staying becomes impossible or conditions will worsen through time and resettlement in areas that best suit the migrants. In these circumstances, moving becomes a better option for people than staying in place, but long distance migration is often undesirable or unnecessary.

Destructive Legacies that Continue to Shape Perceptions About Relocation

Envisioning a proactive relocation and resettlement policy framework is politically fraught because relocation in the U.S. invokes almost a century of public displacement. "Slum clearance" efforts began in the 1930s as part of Works Progress Administration (WPA) programs intending to provide work for unemployed men (Leighninger). Title I of the Housing Act of 1949 authorized the most infamous urban redevelopment program in the

US. This program, called "urban renewal" in the Housing Act of 1954, authorized the federal government's efforts to stimulate redevelopment of urban areas by razing thousands of neighborhoods, forcing over 300,000 families to relocate (Schwartz). Rockefeller Center in New York is an example of a redeveloped urban renewal site. In other cases, housing—including public housing—was built although few former neighborhood residents moved into the new housing. In some situations the land sat empty. The development of the national interstate freeway system also bulldozed thousands of neighborhoods. Between 1957 and 1968, at least 330,000 housing units were destroyed, and in the 1960s, highway construction displaced on average 32,400 families per year (Mohl).

Although the destructive legacy of urban renewal and freeway construction has been widely recognized, federal programs continued to facilitate displacement in the 1990s. As national commitment to maintaining public housing diminished, the US Department of Housing and Urban Development's HOPE VI program provided over $6 billion in assistance to redevelop public housing into mixed income developments (US HUD, n.d. (a)). Despite residents' and their allies' protests against displacement, only a small percentage of former residents returned to the redeveloped neighborhoods (Vale). In all the former cases, African-American residents were disproportionately displaced.

Forcible displacement hurt the residents who had to move. It disrupted the affective bonds between people and places and destroyed social connections and community ties that provided stability, familiarity, and security, causing feelings of loss and depression (Altman and Low; Fried; Gans; Marris). Relocated residents faced additional costs when they moved to an area where travel to work and school became difficult, or when they lost interdependent connections with family and neighbors for childcare and transportation.

The legacy of displacement has hampered relocation and resettlement initiatives intended to improve conditions for residents who lived in depopulated neighborhoods. This has been evident in shrinking cities where consolidating settlement could result in stronger neighborhoods and better services for remaining residents. In 1976, Roger Starr, at the time serving as New York's Housing and Development administrator, proposed planned shrinkage as a way to abandon depopulated neighborhoods and move remaining residents into a smaller area. The immediate opposition made it impossible to consider and Mayor Abraham D. Beame disavowed the idea (Rybczynski and Linneman; Lambert). Residents in Detroit in the 1990s and New Orleans in the 2000s also explicitly rejected shrinking the city's footprint or decommissioning neighborhoods as a solution to population loss and property abandonment (Ehrenfeucht and Nelson; Hackworth; Oosting). In Philadelphia in the early 2000s, the city's Neighborhood Transformation Initiative also included planned relocations as part of its ambitious anti-blight strategy, but residents, responding to the lack of clear guidelines and parallels to earlier urban renewal policies, rejected the initiative (McGovern). Even in Youngstown, Ohio, the US city that planned extensively for a smaller population through a participatory process, residents opted against incentives to move from empty neighborhoods (Schatz; Joffe-Watt). This visible opposition to resettlement makes relocation policies politically unpopular.

Reconceptualizing Relocation as Dynamic Decisions in Dynamic Environments

It would be a mistake, however, to assume that people oppose relocation in principle. Instead, people take reflective action that responds to their specific situations and their knowledge about likely future conditions, including when to accept or oppose relocation.

Both people's situations and perspectives change over time forcing them to make decisions in dynamic circumstances. An accurate understanding of the dynamic decision-making in which individuals, households, and communities engage to increase resilience and reduce risk breaks down the distinction between voluntary and involuntary relocation.

Although population movements are conventionally conceptualized as "voluntary" or "forced," migration should be understood along a continuum instead of as a voluntary-involuntary binary (Hugo 1996). Different forms of pressure, like the lack of viable options, pressure from government agencies, or resource constraints, can be brought to bear on relocation decisions, blurring the distinction between volition and compulsion, choice and obligation (Schmidt-Solatu and Brockington). Because relocation is triggered by an extreme event, affected residents must make a resettlement decision that would not have occurred under normal conditions (Hugo 1996). When the town of Allenville, Arizona, was relocated to the new town of Hopeville after devastating floods in 1978, for instance, few residents developed serious opposition to the relocation plan, but the majority of property owners felt that they had no viable alternatives (Perry and Lindell 52). Similarly, research on the Federal Emergency Management Agency's (FEMA) floodplain buy-out program indicates that many flooded property owners who took a FEMA buy-out reported that they had no choice. Even though they were under no obligation to sell their home, many felt that the government pressured them to participate in the program through the buy-out process (FEMA; de Vries and Fraser).

It is also important to understand decisions to migrate or stay as iterative processes where households and, in some instances, communities have changing perspectives that respond both to shifting circumstances and to an ongoing reflection on likely futures. Decisions may respond to a succession of events rather than a single happening (Black et al) and sudden-onset versus slow-onset changes also could influence decision making differently (Mclean and Hunter). People also must come to terms with the change and loss, a process that occurs gradually through time (Marris; Gans). Residents who have stayed in Detroit's neighborhoods until problems become insurmountable feel a sense of ambiguous loss, an experience when the object of mourning such as the neighborhood and associated life in that neighborhood is not gone even if the situation has changed so dramatically that it is no longer what it was (Weber). This leads to conflicting responses, and residents often resist moving until the moment they decide to do so.

A framework that understands the decision to stay or migrate as iterative allows for multiple, sequential decisions, such as both moving from and returning to areas that are growing more hazardous. It accommodates conflicting views on the best course of action. It also avoids presupposing a given decision-making moment about staying or leaving.

The dynamic circumstances also influence household decision-making and a region's trajectory. Sea-level rise, declining water security, changes in agricultural production and health impacts due to the changing disease vectors or increased heat-related diseases are potential local environmental changes that will be produced by global warming (Campbell). These will impact people and regions at differing rates and within a given city or region, some people will be affected sooner or have more resources to protect themselves (De Sherbinin, Schiller and Pulsipher; Huq, et al.).

Migration has both temporal and spatial dimensions. It can be a permanent or temporary measure that varies in duration. It also can vary in spatial scale from an intraregional move to moving across the globe (McLeman and Hunter). Resettlement decisions have similar temporal and spatial dimensions that are a basis for establishing an effective relocation and resettlement policy framework.

Relocation and Resettlement Options for Diverse Circumstances

In limited circumstances, efforts that attempt to facilitate equitable relocation that address both moving from one area and resettling in another have been attempted in the United States. These initiatives highlight both the possibilities for and ongoing challenges associated with developing effective relocation and resettlement policies. Here we focus on three relocation and resettlement options. The first are individual buy-outs where a government entity purchases properties from private individuals who choose where to relocate. Second are community resettlement efforts, initiatives in which an entire community or a substantial part of it relocates together to a new site. A third option, concentrated relocation, can create opportunities for individual households to move to a nearby relocation site through land swaps or another mechanism. Each approach has distinct advantages and disadvantages and is applicable in specific contexts.

Individual Buy-Outs

The most common strategy has been the relocation of individual households through property acquisition or buy-out programs. In these cases, the emphasis is on helping residents move from an unsafe area. Since the 1990s, FEMA has incorporated relocation into its national mitigation strategy through floodplain "buy-outs," or home acquisition and homeowner relocation programs (FEMA; deVries and Fraser; Schwab et al). FEMA's Hazard Mitigation Grant Program (HMGP) is among the largest buy-out programs. Through the program, FEMA provides grants to states and communities to acquire, elevate, or relocate homes to reduce or eliminate the losses from future disasters. Once the community buys the private property at pre-disaster market value, it acquires title to it, clears it, and maintains it as open space in perpetuity. Since the program's inception in 1993, states, local governments and non-profit organizations have leveraged FEMA funds to buy out over 20,000 homes (FEMA).

State and local governments have incorporated buy-outs into disaster recovery programs. The State of Louisiana, for instance, used disaster Community Development Block Grant (CDBG) resources to fund buy-outs as part of its comprehensive housing recovery initiative, the Road Home Program. The State of New York provided a buy-out incentive to homeowners in the most vulnerable neighborhoods with low rates of population recovery in its Hurricane Sandy recovery strategy (Binder).

Buy-outs allow the household the maximum flexibility in deciding where to relocate and are the least complicated relocation schemes for government agencies to manage. Shifting development patterns can provide opportunities for residents who want to relocate locally but the numbers of impacted people are too great or the relationships too varied to reasonably organize community-based relocation. However, low- and moderate-income residents who participate in buy-outs are often priced out of their communities. They have fewer resources with which to relocate, and the monetary compensation they receive is often insufficient to allow them to purchase property nearby (FEMA). If buy-outs are targeted to a particular community or neighborhood without providing designated relocation sites, residents may be dispersed to neighborhoods throughout the city and beyond, disrupting social networks that a community values and on which it may depend.

An additional critique of buy-out programs and of "voluntary" programs more broadly is that without full participation, some households remain in vulnerable situations and local governments must continue to provide services to those that remain (FEMA). And, while the situation for residents who take buy-outs and move outside of the area may improve, by departing they can contribute to decline in their original communities. They may be leaving those neighbors who decide to stay put or rebuild in place worse off over time.

Community Resettlement

In some instances, entire communities have relocated to sites near the original settlement. Community resettlement can improve the safety and living conditions for residents while maintaining social relationships and social support networks. Although some residents will choose not to participate in the community relocation effort, instead settling outside of the planned community, for small tight knit communities, the community-moving approach can mitigate some of the adverse social and psychological impacts of relocation, and thus has distinct advantages over individual buy-outs. However, community relocations are complex and costly and often take multiple years to complete. Developing the political will and reaching near consensus within the community as well as acquiring adequate financing can be challenging. Because of the cost and complexity, FEMA discourages such efforts (Bronen and Chapin).

The aforementioned resettlement of residents from Allenville, Arizona, to the new town of Hopeville following devastating flooding in 1978 is one example of the community resettlement approach (Perry and Lindell). In that case, government entities, including the Army Corps of Engineers and the Arizona Division of Emergency Management, worked closely with organized residents to relocate 35 of approximately 50 households to the new town. The community relocation preserved social ties and networks of the tightly knit community and also enabled some Allenville renters to use the financial incentives tied to the relocation plan to purchase property (Perry and Lindell).

In 1993, after Pattonsburg, Missouri, flooded twice, 90 percent of the declining farm community's 400 residents voted to relocate to higher ground. The town flooded 33 times in the 20th century, and residents had previously discussed relocating. In 1993, the town received $12 million in federal disaster assistance to help finance the transition (Greenberg et al; NCAT).

After the 1993 Mississippi River flood, Valmeyer, Illinois's 900 residents also chose to relocate (Hunter; NCAT). The levee protecting the community breached and, months later, the river still flowed through town. Working with the county's regional planning committee, community members and community-based businesses chose to relocate to a 500 acre nearby bluff instead of rebuilding in place or receiving individual buy-outs. The first property owners moved to the new neighborhood in 1995. In addition to single-family houses, a senior complex, school, commercial space, and governmental facilities were rebuilt.

Although in most situations, small, rural communities have been relocated in their entirety, urban communities also have been shifted and reconfigured. In 1986, the 200-household Crest Street neighborhood in Durham, North Carolina, shifted after residents organized to oppose a freeway extension. Lawyers for the Crest Street residents successfully argued that the preservation of social ties and social support networks among Crest Street residents was an important characteristic of replacement housing. As a result, rather than dispersing residents throughout the city, the North Carolina Department of Transportation used last resort housing funds provided for under the Uniform Relocation

Act to relocate Crest Street residents as a community (Rohe and Mouw).[1] To achieve this, the proposed expressway had to be realigned, 65 houses were moved, 178 new housing units were constructed, 12 housing units were rehabilitated in place, and senior housing was constructed. In addition, new infrastructure, including two parks and a community center, was built.

For some communities already affected by land loss and climate-induced sea level rise, relocation has become necessary. Native Alaskan villages and Native Gulf Coast communities have begun developing relocation strategies because adapting in place is no longer a viable option due to rising sea levels, thawing permafrost, land subsidence, and extreme weather events (Bronen and Chapin; GAO; Maldonado et al).

In Alaska, at least 12 threatened villages have elected to explore relocation, and the Yup'ik Eskimo village of Newtok is among the furthest along in their relocation efforts (GAO). Newtok, located along the Ninglick River near the Bering Sea, experienced six extreme weather events between 1989 and 2006 accelerating flooding and erosion and severely damaging or destroying public infrastructure including the village landfill, barge ramp, sewage treatment facility, and fuel storage facilities (Bronen; Bronen and Chapin; Maldonado et al). The community began a relocation planning process in 1994. Almost 10 years later, after identifying a potential site nine miles from the original settlement, they obtained title through a land exchange agreement with the U.S. Fish and Wildlife Service (Bronen and Chapin). The Newtok Traditional Council, the local tribal government, has led the relocation effort, and in 2006, it established the Newtok Planning Group, a collaboration of federal, state, tribal governmental, and nongovernmental agencies to facilitate the process. Construction at the relocation site began in 2009, but funding constraints have slowed the process (Bronen; Bronen and Chapin; Maldonado et al).

The community of Isle de Jean Charles located in Terrebonne Parish, Louisiana, is also seeking to relocate. Terrebonne Parish has been losing land due to coastal erosion and salt-water intrusion since the 1800s, and since 1965, the parish has experienced 18 presidentially declared disasters (Terrebonne Parish). Land loss has been extreme in Isle de Jean Charles. In 1950 the community was 5 miles by 12 miles; by 2013, it had shrunk to approximately ¼ mile by 2 miles (Maldonado et al). Isle de Jean Charles and other communities in the southern reaches of Terrebonne Parish have lost population as residents have moved further inland to areas less susceptible to storm-related flooding (Hobor, Plyer and Horowitz; Maldonado et al). For the Isle de Jean Charles residents, the goal of resettlement is to reunite dispersed tribal members in a safe location and restore local culture (Maldonado et al). The community is currently looking for funding to support relocation efforts.

Concentrated Relocation and Redevelopment

While the relocation of entire communities might be feasible for small, closely-knit communities, different resettlement policies are needed for urban areas where community ties are diffuse. An alternative between community resettlement and the relocation

---

1   The Uniform Relocation Assistance and Real Property Acquisition Policies Act of 1970, or Uniform Relocation Act (URA), requires that individuals and families displaced as a result of a federally funded project or program be given the opportunity to secure comparable replacement housing that is within their financial means. If such housing is not available, the act provides for "housing of last resort." Housing of last resort may involve the use of replacement housing payments that exceed the URA maximum amounts and allows payments to be used to rehabilitate existing structures or building new ones (Rohe and Mouw; US HUD, n.d. (c)).

of individuals is to enable residents who take a buy-out to acquire properties in safer, strategically located relocation areas through mechanisms such as land swaps or relocation incentives. The challenge in these cases is twofold: (1) providing enough housing for all relocating residents and (2) finding a suitable relocation and redevelopment site can be difficult in a highly urbanized area. Nevertheless, they offer examples of programs that can help shift urban settlement to safer areas.

In April of 1997, the Red River topped the dikes in Grand Forks, North Dakota, and damaged 83 percent of homes and destroyed entire neighborhoods (US HUD, n.d. (b)). To reduce future flood risk and make way for the new dike construction, the city used disaster CDBG and Hazard Mitigation Grant Program (HMGP) funds to acquire over 800 properties, making the buy-out one of the largest land acquisition programs in the nation's history (FEMA). To address the housing shortage that resulted from the flood, the city simultaneously partnered with a nonprofit housing organization to develop 180 homes in two subdivisions. The city financed the housing with additional disaster CDBG funds and tax-exempt bonds (US HUD, n.d. (b)).

Though not a one-to-one land exchange as in the community resettlement examples discussed above, the initiative created options for households that participated in the buy-out to relocate within the city. Initially, the city-financed subdivisions did not sell as well as anticipated largely because the home prices exceeded the value of many of the homes lost in the flood. The new homes ranged in price from $105,000 to $147,000 while many of the homes lost in older neighborhoods were only valued at $50,000 to $80,000 (US HUD, n.d. (b)). The new houses eventually sold when housing prices were reduced.

After the 2005 hurricanes flooded New Orleans, a non-profit housing organization, Project Home Again (PHA), developed a home-building and land-swap program that concentrated redevelopment for nearly 100 low- and moderate-income households displaced by Hurricane Katrina's floodwaters (Nelson). PHA, funded by the Riggio Foundation, worked closely with the New Orleans Redevelopment Authority (NORA) to implement the land swap program. Through the program, displaced homeowners swapped their storm-damaged properties to PHA in exchange for a newly built house elevated above FEMA-designated flood levels in one of three target neighborhoods. If the parcels that participants turned in were located in one of PHA's target neighborhoods, PHA constructed homes on them for others. PHA purchased additional lots from NORA, and traded scattered site properties they received through swaps to NORA. By concentrating redevelopment, PHA sought to support the return of entire neighborhoods, reducing the risk that residents would return to neighborhood devalued by depopulation.

In the New Orleans case, residents resisted relocation proposals in the rebuilding period after the hurricanes (Nelson, Ehrenfeucht and Laska; Ehrenfeucht and Nelson). Despite this, PHA's land-swap program facilitated resettlement of almost 100 households. While most PHA participants had attempted to return and rebuild, logically and financially it was difficult, and this relocation option became the best way for the participants to return to the city.

Four Challenges for Equitable Resettlement

Relocation will become necessary at previously untested scales. Designing adaptive and responsive policies will ensure that they can be both cost effective and socially beneficial. There are four major challenges facing equitable relocation and resettlement policy development: financing, adapting existent programs, governance, and finding suitable sites. In this section, we briefly discuss each.

Challenge One: Financing Relocation.

A primary challenge for all relocation initiatives is adequate funding to facilitate an equitable resettlement process. Relocation assistance has come primarily from FEMA's disaster preparedness and recovery programs, various housing programs including Housing of Last Resort funds under the Uniform Relocation Act, and HUD's CDBG and Community Development Block Grant Disaster Recovery (CDBG-DR) programs.

   All these programs have limitations and none have been designed to address the process of long-term or gradual environmental or economic change. Housing of Last Resort funds provided under the Uniform Relocation Act may be used only when people are displaced due to federal action. Communities that are not incorporated, including the threatened Native villages in Alaska, are ineligible to receive CDBG funds under the federal law governing the program (GAO). Even when funds are accessible to support relocation and resettlement, funding allocations often fall short of the actual costs of replacing housing and community infrastructure (Hugo 2011). In the case of community resettlement, foremost among these challenges is the need for sufficient funding to reconstruct infrastructure and non-residential structures.

Challenge Two: Adapting Existing Programs.

The existing programs used for relocation and resettlement were designed to address different conditions than communities are now facing, requiring that current programs be adapted. For instance, FEMA recovery funds require a disaster declaration to trigger the availability of funds, but ongoing climate-induced environmental changes, such as increasing sea level rise and steady erosion, are not included as disasters under the Robert T. Stafford Disaster Relief and Emergency Assistance Act (Bronen; Bronen and Chapin), nor are the impacts of chronic economic decline. CDBG-DR monies, an important source of recovery funding, can also be used to fund relocation and resettlement but, here again, these require a disaster declaration.

   Amendments to the Stafford Act to expand the definition of disasters to include gradual climate induced changes would enable the release of federal funds to support relocation and resettlement initiatives in communities such as Newtok (Bronen and Chapin). A concentrated investment of federal and state resources is also needed in shrinking cities suffering economic decline. Given the inadequacy of funding from a single source to support relocation and resettlement efforts, the coordination of federal, state, philanthropic and other private resources is essential. With this in mind, policies and programs must be designed to be flexible, permit financial layering, and minimize conflicting requirements among programs.

Challenge Three: Governing Relocation.

Because relocation has been a policy of last resort, no institutional framework or agency has the authority to oversee and coordinate relocation and resettlement assistance, and therefore a governance structure that can coordinate comprehensive relocation and resettlement programs is necessary. The Newtok Planning Group, for instance, brings together dozens of federal, state, and tribal governmental and nongovernmental agencies to facilitate relocation, but none have a funded mandate to relocate the community, and there is no lead agency to coordinate efforts (Bronen; Bronen and Chapin). The local tribal government, the Newtok Traditional Council, has limited capacity to coordinate the relocation work of federal and state agencies and obtain and administer funding for the relocation process (Bronen). A lead agency could provide much needed assistance to communities and individuals to help them deal with conflicting regulations among programs and speed relocation and resettlement efforts in a coordinated fashion.

Challenge Four: Finding and Developing Suitable Resettlement Locations.

A final challenge is ensuring that people have access to suitable locations to resettle. In the case of individual buy-outs, unaffordable housing in nearby areas or different neighborhood character can create barriers for residents who are moving. If someone receives fair market value for a parcel, it will often not allow the property owner to buy an equivalent in the region, particularly in situations where the parcel is devalued by increased risk or decline. Property value is also a challenge in shrinking cities, where the process of depopulation reduces property values. In these situations, it is important to create mechanisms to enable residents to move to comparable and desirable neighborhoods.

In the case of community relocation, the challenge is finding a suitable resettlement site that is adequate in size and reduces vulnerability, while continuing to be located near work or other forms of livelihood. Proximity to employment is essential to reconstructing livelihoods (Oliver-Smith 1991; Hugo 2011). For resource-dependent communities, loss of land can result in loss of livelihood if that distance from resources increases. Given this, community relocations tend to involve resettlement close to the original site (Black et al). Yet suitable land near to the original site may not be available.

The adaptation strategy of the threatened bayou community of Jean Lafitte, in Jefferson Parish, Louisiana, illustrates the importance of a suitable site. The community considered relocation but has sought a localized structural solution instead in part due to the lack of available land nearby for resettlement (Bailey, Gramling, and Laska,). This historic fishing village is located at the edge of the greater New Orleans region. Suburban sprawl has transformed the once undeveloped land near Jean Lafitte into single-family residential subdivisions. Moving further inland would make it difficult for residents to continue to fish given the expense of commuting and the potential loss or destruction of networks of exchange and support (Bailey, Grambling, Laska).

The need for a suitable relocation site is also important for concentrated relocation. In the New Orleans case, PHA used land swaps to help guide redevelopment while providing resettlement options for displaced residents that enabled them to return to New Orleans. Collaboration with the city's redevelopment authority was critical to facilitate land swaps and coordinate resettlement with neighborhood and citywide recovery plans (Nelson). Concentrated relocation strategies may be better suited to places like New Orleans with large amounts of vacant property.

Conclusions

Designing effective relocation policy has ongoing challenges, but the stakes are high. Relocation is costly. In 2006, the Army Corps of Engineers estimated the cost to relocate

Newtok's 350 residents to be between $80 and $130 million (U.S. Army Corps of Engineers). Costs of repeated disaster response will be immeasurably higher. Although resettlement is often seen as a last resort solution that residents often resist, and the structure of private property complicates relocation efforts, there is an imminent need to design programs that will lead to long-term resettlement as communities grow more vulnerable.

Works Cited

Altman, Irwin, and Setha M. Low. *Place Attachment*. New York: Springer. 1992. Print.

Bailey, Conner, Robert Gramling, and Shirley B. Laska. "Complexities of Resilience: Adaptation and Change Within Human Communities of Coastal Louisiana." *Perspectives on the Restoration of the Mississippi Delta.* Eds John W. Day, G. Paul Kemp, Angelina M. Freeman and David P. Muth. New York: Springer, 2014. 125–40. Print.

Binder, Sherri Brokopp. *Resilience and Postdisaster Relocation: A Study of New York's Home Buy-out Plan in the Wake of Hurricane Sandy*. Boulder: Natural Hazards Center. 2013. Print.

Black, Richard, et al. "The Effect of Environmental Change on Human Migration." *Global Environmental Change* 21 (2011): S3-S11. Print.

Brash, Julian. *Bloomberg's New York: Class and Governance in the Luxury City*. Vol. 6. U of Georgia P, 2011. Print.

Bronen, Robin. "Alaskan Communities' Rights and Resilience." *Forced Migration Review* 31 (2008): 30–32. Print.

Bronen, Robin and F. Stuart Chapin. "Adaptive Governance and Institutional Strategies for Climate-Induced Community Relocations in Alaska." *Proceedings of the National Academy of Sciences* 110.23 (2013): 9320–9325. Print.

Campbell, John R. "Climate-Change Migration in the Pacific." *The Contemporary Pacific* 26.1 (2014): 1–28. Print.

Carling, Jørgen. "Migration in the Age of Involuntary Immobility: Theoretical Reflections and Cape Verdean Experiences." *Journal of Ethnic and Migration Studies* 28.1 (2002): 5–42. Print.

Cohn, D'Vera and Rich Morin. *"American Mobility: Who Moves? And Who Stays Put? Where's Home?"* Pew Research Center. Web. 29 Dec. 2008 <http://www.pewsocialtrends.org/files/2011/04/American-Mobility-Report-updated-12-29-08.pdf>.

Comerio, Mary C. "Disaster Recovery and Community Renewal: Housing Approaches." *Cityscape: A Journal of Policy Development and Research* 16.2 (2014): 51–64. Print.

De Sherbinin, Alex, Andrew Schiller, and Alex Pulsipher. "The Vulnerability of Global Cities to Climate Hazards." *Environment and Urbanization* 19.1 (2007): 39–64. Print.

de Vries, Daniel H., and James C. Fraser. "Citizenship Rights and Voluntary Decision Making in Post-Disaster US Floodplain Buy-out Mitigation Programs." *International Journal of Mass Emergencies and Disasters* 30.1 (2012): 1–33. Print.

Ehrenfeucht, Renia, and Marla Nelson. "Recovery in a Shrinking City: Challenges to 'Rightsizing' Post-Katrina New Orleans." *The City after Abandonment*. Eds. Margaret Dewar and June Manning Thomas. Philadelphia: U of Pennsylvania P. 2013. 133–50. Print.

Federal Emergency Management Agency. *Implementing Floodplain Land Acquisition Programs in Urban Localities.* Web. 26 May 2013. <http://jamescfraser.com/storage/publications/Floddplain%20Project%20Report.Final.pdf>.

Fried, Mark. "Grieving for a Lost Home: Psychological Costs of Relocation." *Urban Renewal: The Record and the Controversy.* Ed. James Q. Wilson. Cambridge, MA: MIT P, 1966. 359–79. Print.

Gans, Herbert J. *Urban Villagers: Group and Class in the Life of Italian-Americans.* New York: Free Press, 1982. Print.

Greenberg, Michael R., Michael Lahr, and Nancy Mantell. "Understanding the Economic Costs and Benefits of Catastrophes and Their Aftermath: A Review and Suggestions for the US Federal Government." *Risk Analysis* 27.1 (2007): 83–96. Print.

Hackworth, Jason. "The Limits to Market-Based Strategies for Addressing Land Abandonment in Shrinking American Cities." *Progress in Planning* 90 (2014): 1–37. Print.

Hanson, Susan. "Perspectives on the Geographic Stability and Mobility of People in Cities." *Proceedings of the National Academy of Sciences of the United States of America* 102.43 (2005): 15301–15306. Print.

Hobor, George, Allison Plyer and Ben Horowitz. *The Coastal Index: The Problem and Possibility of Our Coast.* New Orleans, LA: The Data Center, 2014. Print.

Hugo, Graeme. "Environmental Concerns and International Migration." *International Migration Review* (1996): 105–31. Print.

——— "Lessons from Past Forced Resettlement for Climate Change Migration." Eds Piguet, Etienne, Antoine Pécoud, and P. F. A. de Guchteneire. *Migration and Climate Change.* Paris: UNESCO Pub. 2011. Print.

Hunter, Lori M. "Migration and Environmental Hazards." *Population and Environment* 26.4 (2005): 273–302. Print.

Huq, Saleemul, et al., "Editorial: Reducing Risks to Cities from Disasters and Climate Change." *Environment and Urbanization* 19.1 (2007): 3–15. Print.

Joffe-Walt, Chana. "A Shrinking City Knocks Down Neighborhoods." *Planet Money, NPR,* 15 Mar. 2011. Web. 1 Oct. 2013 <http://www.npr.org/blogs/money/2011/03/15/134432054/a-shrinking-city-knocks-down-neighborhoods>.

Kefalas, Maria. *Working-Class Heroes: Protecting Home, Community, and Nation in a Chicago Neighborhood.* U of California P, 2003. Print.

Kick, Edward L. et al, "Repetitive Flood Victims and Acceptance of FEMA Mitigation Offers: An Analysis with Community–System Policy Implications." *Disasters* 35.3 (2011): 510–539. Print.

Kinder, Kimberley. "Guerrilla-style Defensive Architecture in Detroit: A Self-Provisioned Security Strategy in a Neoliberal Space of Disinvestment." *International Journal of Urban and Regional Research* 38.5 (2014): 1767–1784. Print.

Lambert, Bruce. "Roger Starr, New York Planning Official, Author and Editorial Writer, is Dead at 83." *New York Times* 11 Sept. 2001. Web. < //www.nytimes.com/2001/09/11/nyregion/roger-starr-new-york-planning-official-author-and-editorial-writer-is-dead-at-83.html>.

Leighninger, Robert D. *Long-Range Public Investment: The Forgotten Legacy of the New Deal.* Columbia, S.C: U of South Carolina P, 2007. Print.

Levine, Joyce N., Ann-Margaret Esnard, and Alka Sapat. "Population Displacement and Housing Dilemmas Due to Catastrophic Disasters." *Journal of Planning Literature* 22.1 (2007): 3–15. Print.

Maldonado, Julie Koppel, et al. "The Impact of Climate Change on Tribal Communities in the US: Displacement, Relocation, and Human Rights." *Climate Change and Indigenous Peoples in the United States*. Eds Julie Koppel Maldonado, Benedict Colombi, and Rajul Pandya. Springer International Publishing, 2014. 93–106. Print.

Marris, Peter. *Loss and Change*. New York: Pantheon Books, 1974. Print.

McGovern, Stephen J. "Philadelphia's Neighborhood Transformation Initiative: A Case Study of Mayoral Leadership, Bold Planning, and Conflict." *Housing Policy Debate* 17.3 (2006): 529–70. Print.

McLeman, Robert A., and Lori M. Hunter. "Migration in the Context of Vulnerability and Adaptation to Climate Change: Insights from Analogues." *Wiley Interdisciplinary Reviews: Climate Change* 1.3 (2010): 450–61. Print.

Mohl, Raymond A. "Ike and the Interstates: Creeping Toward Comprehensive Planning." *Journal of Planning History* 2.3 (2003): 237–62. Print.

National Center for Appropriate Technology. *Operation Fresh Start Case Studies*. Web. <http://www.freshstart.ncat.org/case.htm>.

National Research Council (NRC), Committee on U.S. Army Corps of Engineers Water Resources Science, Engineering, and Planning: Coastal Risk Reduction; Water Science and Technology Board; Ocean Studies Board; Division on Earth and Life Studies. *Reducing Coastal Risk on the East and Gulf Coasts*. 2014. Web. <http://dels.nas.edu/resources/static-assets/materials-based-on-reports/reports-in-brief/coastal-risk-brief-final.pdf>.

Nelson, Marla. "Using Land Swaps to Concentrate Redevelopment and Expand Resettlement Options in Post-Hurricane Katrina New Orleans." *Journal of the American Planning Association* (forthcoming).

Nelson, Marla, Renia Ehrenfeucht, and Shirley Laska. "Planning, Plans, and People: Professional Expertise, Local Knowledge, and Governmental Action in Post-Hurricane Katrina New Orleans." *Cityscape* (2007): 23–52. Print.

Oliver-Smith, Anthony. "Successes and Failures in Post-Disaster Resettlement." *Disasters* 15.1 (1991): 12–23. Print.

———. "Communities After Catastrophe: Reconstructing the Material, Reconstituting the Social." *Community Building in the 21st Century*. Ed. Stanley Hyland. Santa Fe, NM: School of American Research Press, 2005. 45–70. Print.

———. "Sea Level Rise and the Vulnerability of Coastal Peoples. Responding to the Local Challenges of Global Climate Change in the 21st Century." *InterSecTions* No. 7. Bonn: UNU-EHS, (2009). Print.

Olshansky, Robert B., Laurie A. Johnson, and Kenneth C. Topping. "Rebuilding Communities Following Disaster: Lessons from Kobe and Los Angeles." *Built Environment* 32.4 (2006): 354–74. Print.

Oosting, Jonathan. "With City Hurting, Mayor Dave Bing says Detroit Works Project Demonstration is Working." *Detroit News*. 16 Mar. 2012. Web. 18 Mar. 2012 <http://www.mlive.com/news/detroit/index.ssf/2012/03/with_city_reeling_mayor_dave_b.html>.

O'Sullivan, Feargus. "The City That is Moving Down the Road: How to Reassemble a Place in 100 Years or Less." *Next City*. 8 Dec. 2014. Web. 8 Dec. 2014 <http://nextcity.org/features/view/the-city-that-is-moving-9-kilometers-down-the-road>.

Peacock, Walter Gillis, Nicole Dash, and Yang Zhang. "Sheltering and Housing Recovery Following Disaster." *Handbook of Disaster Research*. New York: Springer, 2007. 258–74. Print.

Perry, Ronald W., and Michael K. Lindell. "Principles for Managing Community Relocation as a Hazard Mitigation Measure." *Journal of Contingencies and Crisis Management* 5.1 (1997): 49–59. Print.

Rohe, William M., and Scott Mouw. "The Politics of Relocation: The Moving of the Crest Street Community." *Journal of the American Planning Association* 57.1 (1991): 57–68. Print.

Rybczynski, Witold, and Peter D. Linneman. "How to Save our Shrinking Cities." *Public Interest* (1999): 30–44. Print.

Sassen, Saskia J. *Cities in a World Economy*. Thousand Oaks: Sage Publications, 2012. Print.

Schatz, Laura. "Decline-Oriented Urban Governance in Youngstown, Ohio." *The City After Abandonment*. Eds Margaret Dewar and June Manning Thomas. Philadelphia: U of Pennsylvania P, 2013. 155–86. Print.

Smith, Adam B., and Richard W. Katz. "US billion-dollar weather and climate disasters: data sources, trends, accuracy and biases." *Natural Hazards* 67.2 (2013): 387–410. Print.

Strauss, Ben, Claudia Tebaldi, and Remik Ziemlinski. *Surging Seas: Sea Level Rise, Storms & Global Warming's Threat to the US Coast*. Climate Central. 2012. Print.

Schmidt-Soltau, Kai, and Dan Brockington. "Protected Areas and Resettlement: What Scope for Voluntary Relocation?" *World Development* 35.12 (2007): 2182–2202. Print.

Schwab, James C., et al. *Planning for Post-disaster Recovery and Reconstruction*. PAS 483/483. Chicago, IL: APA Planning Advisory Service, 1998. Print.

Schwartz, Alex F. *Housing Policy in the United States*. New York: Routledge, 2010. Print.

Terrebonne Parish. *Terrebonne Parish Hazard Mitigation Plan Update*. 2014. Web. 8 Oct 2014 <http://www.tpcg.org/files/flooding/Terrebonne_Parish_HMPU_2010.pdf>.

U.S. Army Corps of Engineers. *Alaska Baseline Erosion Assessment: Study Findings and Technical Report*. March 2009. Web. 24 May 2014 <http://www.climatechange.alaska. gov/docs/iaw_USACE_erosion_rpt.pdf>.

U.S. Department of Housing and Urban Development (HUD). "About HOPE VI." n.d.(a). Web. <http://portal.hud.gov/hudportal/HUD?src=/program_offices/public_indian_ housing/programs/ph/hope6/about>.

U.S. Department of Housing and Urban Development (HUD). "Grand Forks Residential Buyout Program." n.d.(b). Web. <http://portal.hud.gov/hudportal/documents /huddoc?id=DOC_22577.pdf>.

U.S. Department of Housing and Urban Development (HUD). "Housing of Last Resort." n.d.(c). Web. <http://portal.hud.gov/hudportal/HUD?src=/program_offices/comm_ planning/affordablehousing/training/web/relocation/lastresortt>.

United States. Government Accountability Office (GAO). *Alaska Native Villages: Limited Progress Has Been Made on Relocating Villages Threatened by Flooding and Erosion*. Government Accountability Office Report GAO-09–551. 2009. Print.

Vale, Lawrence J. *Purging the Poorest: Public Housing and the Design Politics of Twice-Cleared Communities*. Chicago: U of Chicago P, 2013. Print.

Warner, Koko, et al. "Climate Change, Environmental Degradation and Migration." *Natural Hazards* 55.3 (2010): 689–715. Print.

Weber, Matthew. "Grief and the Shrinking City." Unpublished article. 2009. Print.

Waldorf, Brigitte. "The Location of Foreign Human Capital in the United States." *Economic Development Quarterly* 25.4 (2011): 330–40. Print.

Zhang, Yang, and Walter Gillis Peacock. "Planning for housing recovery? Lessons learned from Hurricane Andrew." *Journal of the American Planning Association* 76.1 (2009): 5–24. Print.

Chapter 5
Developing the Sustainable City: Curitiba, Brazil, as a Case Study

Hanna-Ruth Gustafsson and Elizabeth A. Kelly

I. Introduction

Across the globe, urban growth is occurring at an unprecedented rate and scale (Urbanization). In 2008, the percentage of the world population residing in cities reached 50 percent. By 2050, an estimated 70 percent of the global population will live in urban centers (Human Population: Urbanization). Much of this growth is concentrated in "mega cities," metropolises of over 10 million people (Human Population; Urbanization). These mega cities are increasingly merging to form "mega regions," defined as regional centers of interlinked economic and urban growth and often home to as many as 100 million people (Vidal).

Mega regions' economic and cultural benefits are well documented (Glaeser 8, 70). The world's 40 largest mega regions cover only a fraction of the earth's surface and contain just 18 percent of its population but produce 66 percent of all economic activity, and 85 percent of technological and scientific innovation (Vidal). Unfortunately, explosive growth also causes significant problems. Mega regions and mega cities face unprecedented urban sprawl, carbon emissions, and economic inequality. These problems exist both in the developed world, where a significant majority of the population already lives in urban areas, and the developing world, where the greatest urban growth is currently occurring (UN–HABITAT. State of the World's Cities). Each of these problems compounds the others. For instance, urban sprawl increases transport costs, raises energy consumption, and spurs resource use, thereby adding to carbon emissions and economic inequality. Indeed, cities account for just two percent of the earth's landed surface but produce 70 percent of all greenhouse gas emissions, a number likely to increase as cities and populations expand (UN–HABITAT. Cites and Climate Change).

The readily apparent challenges of urban growth have attracted the attention and problem-solving efforts of academics, planners and policymakers. However, while these are global issues, policymakers tend to seek solutions from a limited number of sources. In the United States, when we do look abroad for solutions, we often turn to European neighbors for insights and rarely look beyond the high-income membership of the Organisation for Economic Co–Operation and Development (OECD) for policy ideas.[1]

U.S. policymakers' narrow focus on OECD best practices is troublesome: it ignores swaths of potentially valuable lessons, many of which may actually be more relevant to the

---

1 The following countries are members of the OECD: Australia, Austria, Belgium, Canada, Chile, Czech Republic, Denmark, Estonia, Finland, France, Germany, Greece, Hungary, Iceland, Ireland, Italy, Japan, South Korea, Luxembourg, Mexico, Netherlands, New Zealand, Norway, Poland, Portugal, Slovakia, Slovenia, Spain, Sweden, Switzerland, Turkey, United Kingdom, and the United States.

U.S. than those currently studied. For example, Sweden is often heralded for its community development innovations and focus on environmentally sustainable development. However, Sweden's generous social spending and minimal economic inequality, coupled with the fact that its annual urban growth rate is far below that of the United States, renders many of its lessons inapplicable in the U.S. context (Urban Population Growth).

This chapter seeks to turn the reader's attention from the oft–cited practices of European cities to an example further south. It highlights innovations in Curitiba, Brazil, that have been successful in alleviating the interrelated problems of urban growth: sprawl, environmental degradation, and economic inequality. While perhaps less familiar than the examples of our OECD counterparts, the planning and development strategies employed by this Brazilian city over the past decades are equally noteworthy.[2] At its core, Curitiba exemplifies how the integration of land use planning, transportation infrastructure, and environmental sustainability efforts can enable a city to meet the needs of its expanding population and mitigate the negative effects of urban growth. While recent criticisms, discussed later in this chapter, highlight the obstacles still facing Curitiba, its innovations and the foundational principles underlying them remain deeply relevant to expanding metropolises worldwide.

In sum, this text aims to highlight Curitiba's innovations and illustrate their relevance to other growing cities. It proceeds as follows. The chapter begins with an overview of Curitiba's history and the basic framework developed to address the challenges attendant with tremendous growth. It then discusses specific practices employed by Curitiba to mitigate the negative impacts of growth while addressing the challenges Curitiba still faces. The chapter concludes by exploring how Curitiba's innovations can be applied in other urban contexts, both in the United States and abroad.

II. Urban Growth in Curitiba

Curitiba, the capital of the southern state of Paraná, is one of the fastest growing cities in South America. Its explosive growth over the past half century has required the city to demonstrate tremendous resourcefulness in tackling the accompanying problems of sprawl, congestion, environmental impacts, and social inequality. The strategies it has employed in combating these challenges are instructive for other growing cities.

The city's metropolitan population currently numbers 3.7 million, ten times the city's population in the mid-twentieth century, and approximately the same size as Los Angeles (IPPUC). The city experienced an average annual growth rate of 4.6 percent for almost 50 years, transforming Curitiba from a small city to a metropolis (Lindau). Its annual growth rate stabilized somewhat in past years, averaging 3.8 percent since the 1990s. Still, the cumulative effects of sustained growth required Curitiba to adapt and adjust rapidly to changing conditions, despite a municipal budget that did not expand nearly as quickly as the demands placed upon it (Power).

In 1943, shortly after its population growth began, Curitiba adopted a comprehensive plan designed by French urbanist, Alfred Agache, which focused on downtown development with long Parisian-style boulevards radiating from the city center (Irazábal). The plan was intended to concentrate commercial and cultural activity in the downtown area, but failed

---

2   Though less known in the United States, Curitiba has received international recognition for its efforts. Curitiba won the prestigious Globe Sustainable City Award in 2010 for its excellent sustainable urban development (Press Release).

to address the rapid population growth that was beginning to take place on the outskirts of the city. Nor did the plan account for the increased traffic congestion that this growth would likely cause. These factors coupled with limited city funds meant that most aspects of the plan were never fully implemented, and by the 1960s, with the city's population growth in full flux, it became clear that an alternate plan was needed (Nexus). The municipal government held a design competition in 1965, soliciting contributions for a new plan that would account for the city's rapid population growth (Lundqvist). The selected entry was the design of a team of architects led by Jamie Lerner, who would go on to serve three terms as mayor of Curitiba. This second plan, known as the "Curitiba Master Plan," provided the planning principles and framework that have guided the city's development over the past 40 years and on which the city still relies today (Nexus 2).

As further discussed in this chapter, the Curitiba Master Plan stresses the link between integrated urban transportation, appropriate land uses, and environmental preservation (IPPUC 7). The boulevards of the earlier plan were reenvisioned as linear commercial corridors to be serviced by public transportation; each corridor spreading outward from the city center in an axial design in order to distribute growth throughout the metropolitan area (Goodman 75–6).

In conjunction with the development of the Master Plan, a regional planning association, *Instituto de Pesquisa e Planejamento Urbano de Curitiba*, was formed to facilitate coordination across city agencies and to support implementation of the goals identified in the Master Plan. Lerner was appointed the first head of IPPUC, or the Institute of Urban Research and Urban Planning, and the organization was given freedom to operate largely independently of the city's formal agencies. IPPUC remains a semi-autonomous institution in order to ensure continuity across election cycles, yet it works closely with the agencies to coordinate planning, housing, and environmental policies throughout the city.

Under the direction of the Master Plan and IPPUC, Curitiba has pioneered a number of initiatives to provide services to its growing population, with the goal of integrating land use and transportation planning determining much of the scope and form of these initiatives. The next section examines these practices in greater detail.

III. Curitiba's Urban Planning Innovations

A. Land Use Planning

As first outlined in Curitiba's Master Plan, the city revolves around five primary corridors of development that form an axis around the downtown area. While commercial uses are most abundant in the city center, they are also encouraged along the length of each commercial corridor, such that the city center serves as a hub or terminus, rather than merely a final travel destination (Goodman 75). This allows for a variety of uses to be available to residents throughout the city including residents of less affluent areas at a distance from the city center, and alleviates traffic congestion in the downtown area.

The city employs the following zoning strategy to support this design: dense commercial development and high-rise buildings are encouraged along each side of the primary commercial corridors extending from the city center. Along these roads, buildings are permitted at a floor area ratio of six, meaning that a building may have floor space six times the area of the plot of land that it sits upon (Rabinovitch 65). As one moves further away from a central corridor, the permitted floor area ratio decreases to four, and development thereby loses height and density (Rabinovitch 65). Commercial uses give way to urban apartment buildings, which are then replaced by smaller scale residential homes with an even lower floor area ratio. In this way, the development corridors provide the framework for the city's pattern of growth and ensure that development is not solely concentrated in the downtown core. These corridors also provide the framework for the city's transportation network, facilitating coordination between density and transportation services.

B. Bus Rapid Transit

A keystone of Curitiba's Master Plan is the promotion of public transportation systems, which Curitiba provides primarily through a high–capacity bus system known as Bus Rapid Transit (BRT). BRT has been described as:

> [A] high-quality bus based transit system that delivers fast, comfortable, and cost-effective urban mobility through the provision of segregated right-of-way infrastructure, rapid and frequent operations, and excellence in marketing and customer service. BRT essentially emulates the performance and amenity characteristics of a modern rail-based transit system but at a fraction of the cost (Wright).

Or in other words, "think rail, use buses" (Barton). Unable to afford a rail system and struggling to deal with rapid population growth, Curitiba developed the BRT system in order to offer its citizens high–quality transportation services at a fraction of the cost of rail-based systems. Its example illustrates how municipal transportation systems can provide reliable and convenient transportation services on a limited budget. Perhaps the best evidence of the system's success is its continued popularity. The bus system transports approximately 2.3 million people daily (Lubow). Around 70 percent of the city's population commutes by bus, including 28 percent of users who previously commuted by personal automobile (Goodman 75). Due in part to BRT (as well as Curitiba's long–standing ban on polluting industries), Curitiba boasts one of the country's lowest rates of ambient pollution (Lubow; U.S. Federal Transit Administration).

i. bus routes and zoning

As outlined in Curitiba's Master Plan and as discussed above, the city revolves around five primary corridors of development that form an axis around the downtown area. Each commercial corridor forms the backbone of a "trinary system" of three roads, along which the BRT system operates. The central route along each commercial corridor consists of a designated busway, which is dedicated to exclusive use by buses and has a lane of traffic operating in each direction. On either side of the busway are traffic lanes that are open to all vehicles and that allow for access to the businesses and services fronting onto the main commercial corridors. Parallel to the high-density commercial corridors, usually a block away, are one–way roads that are also open to all types of vehicles and that allow for rapid movement in a single direction (IPPUC 10).

   The BRT system is adapted to this hierarchy of roads. Along the dedicated busways at the center of the commercial corridors are the *Expresso* (Express) bus routes. High–capacity bi–articulated buses run along these routes, serving the greatest number of passengers. Since they do not have to contend with other forms of traffic, the Express buses average much higher speeds than regular buses. Operating parallel to the Express line are the *Ligeirinho* (Direct) buses, which run along the one–way streets about a block from the central corridor.

Both articulated and conventional buses run along these routes, making more frequent stops than do the Express buses along the dedicated busways.

Feeding into these two rapid bus lines are the *Interbarrios* (Inter–neighborhood) and *Alimentador* (Feeder) services. The buses on these routes circle the city, connecting residents in lower-density areas to the primary bus routes along the commercial thoroughfares. The city also operates a dedicated hospital bus line, connecting local health care facilities, as well as a system of school buses and a system of tourist buses.

The BRT system is fully integrated to ensure that all city residents have access to public transportation, with the bus lines that serve lower density areas feeding into the routes that serve higher density areas. Linking density to the type of bus service offered creates further efficiency in the system, as the high-density corridors generate the greatest demand and support the use of high-capacity buses that run frequently along segregated busways. Lower density areas are also served but generally do not warrant the need for dedicated busways, and therefore the emphasis in these areas is on comprehensive coverage, ensuring that all residents live within walking distance of a bus stop.

ii. Transfer terminals and bus stops

Transfer terminals serve to connect the feeder services with the rapid bus lines. Most of the transfer terminals include convenience stores, post offices, and other commercial services. Each city district, of which there are 12, also has a transfer terminal called a "Citizenship Street." The Citizenship Street terminals provide a range of municipal services in addition to the usual commercial operations including health centers, vocational training centers, legal assistance offices, and social service centers.

Each bus route is also served by a number of smaller bus stops. These appear at 500m. intervals along the Express route and somewhat less frequently in lower-density areas. Most of the bus stops along the Express and Direct routes are equipped with GPS displays that indicate in real time when the next bus will arrive. The bus stops share a characteristic tubular design and offer protection from outdoor elements. Furthermore, bus stops are raised, so that when boarding a bus, passengers are already at the appropriate level to climb onboard. This tweak reduces the time a bus needs to wait at each stop and aids mothers with prams, the elderly and wheelchair users.

iii. Fares and fare collection

Users of the BRT system purchase their tickets prior to boarding for maximum efficiency. Riders pay one uniform fare, regardless of the distance being traveled or the number of transfers. This is called the "social fare," alluding to the fact that shorter journeys subsidize the cost of longer journeys disproportionately taken by low-income residents. This was a deliberate choice on the part of the municipal government in implementing the BRT system, and it is in keeping with the principle that the system should be accessible to everyone, regardless of physical location or socioeconomic status. Fares are regularly reviewed to ensure that the "average worker" pays no more than 10 percent of his or her income on transportation costs, significantly lower than Americans' average transportation costs: U.S. residents in the lowest 20 percent income bracket spend roughly 42 percent of their annual income on transportation, and middle-income Americans spend roughly 22 percent (Goodman 76; The Leadership Conference Education Fund).

iv. Financing and operations

The Curitiba BRT system receives no government subsidy, and is completely self–supporting. The services are provided by 16 private bus companies, which are contracted and regulated by the *Urbanizacao de Curitiba SA* (URBS), a government company. URBS is also responsible for regulating taxi services and public parking in the city. URBS collects all fares and distributes payment to the bus companies based on the distances they travel. Previously, the companies were paid based on the number of passengers they carried, but this led the companies to all focus their services on the busiest commercial areas, where the greatest passenger demand existed. By paying the companies based on the distances they travel, URBS is able to ensure that the companies have an incentive to provide services in less dense areas as well.

The bus companies are responsible for purchasing all of their own vehicles for which URBS reimburses their capital costs at a rate of 1 percent per month (Demery). The guarantee of a 12 percent annual rate of return incentivizes the bus companies to invest in new vehicles, and generally, buses in Curitiba are only used for 3–4 years, in an effort to make sure that the fleet remains clean, safe, and comfortable. The city buys back the buses at the end of this period for a nominal amount, and generally repurposes the buses. "Retired" buses are used as mobile libraries, classrooms, soup kitchens and health centers, in keeping with Curitiba's emphasis on sustainability.

v. BRT usage

The Curitiba BRT system carries about 2 million passengers a weekday, compared to the 5.5 million passengers who use the New York subway system each weekday (Nexus 12; Facts and Figures). The Express buses travel at an average speed of 20 kilometers per hour, transporting about 11,000 passengers per hour per direction (Nexus 12–3). Approximately 70 percent of the population relies on the system for their daily commuting needs, including many car owners (Goodman 75). Perhaps the best evidence of the system's success is that while Curitiba's population has more than doubled since the system's introduction in the 1970s, traffic has declined by 30 percent (ICLEI).

The success of the Curitiba BRT system can largely be attributed to the integration of land use and transportation planning, which ensures that supply and demand are balanced. In the commercial corridors with significant demand, one finds fast moving high capacity buses traveling along dedicated busways. In lower-density areas, there is also bus coverage, but the emphasis is on linking feeder systems to the rapid BRT lines that run along the high density corridors. This combination promotes system efficiency, discourages sprawl, and ensures affordable public transportation access for Curitibans of all socioeconomic classes.

C. Parks System

In addition to its Bus Rapid Transit system, Curitiba is known for its innovative use of parks and green space to improve the quality of life of its citizens and proactively address the effects of global warming. Nearly one-fifth of the city is parkland (ICLEI 4). By comparison, only 8.1 percent of the average American city is parkland (Trust for Public Land). As of March 2012, Curitiba boasted 35 parks, 1004 conservation areas (including woods, gardens, squares, mini-gardens, and activity axes), and 78,000 square meters of natural forest, for a total of 64 square meters of green space per inhabitant. These numbers represent a remarkable increase: 50 years earlier, Curitiba had less than one square meter of green space per person (Trust for Public Land 12).

Besides its obvious aesthetic and recreational value, the park system is vital for controlling increased flooding, protecting Curitiba's biodiversity and water quality, and limiting carbon emissions. Curitiba began experiencing frequent flooding in the 1950s and 1960s, due to the construction of houses and other buildings along the streams and river basins and the coverage of streams to allow for additional development (Roman). Starting in the 1970s, while other municipalities used Brazilian federal flood control funds to construct dams, Curitiba devoted its moneys to creating a park system that preserves valley floors, river basins, and protection strips along streams in order to avoid floods (Taniguchi 15). This initiative stood in stark contrast to the deforestation occurring across Brazil during the 1980s. Curitiba designated parkland strategically so as to divert floodwaters into the parks, which are all outfitted with numerous deep lakes, weatherproof playground equipment, and picnic benches, given the expectation of flooding. Each watershed area in the city is flanked by a line of parks, whose diameter has expanded by a factor of six since the Master Plan's introduction in 1968 (Guillem). The latest addition is the Linear Barigui Park, which links existing parks and creates new ones in order to revitalize the Barigui river basin. The strategic designation of parks also enables the preservation of Curitiba's tremendous biodiversity. A mix of temperate forest and Atlantic rainforest, the Curitiba metropolitan area is home to almost 4,000 species (Guillem). Finally, through the Reservas Particulares do Patrimônio Natural Municipal program (RPPNM or Municipal Natural Heritages Private Reserves), Curitiba protects CO_2 sink holes without dispossessing landowners of their real estate by subsidizing the creation of privately owned parks. The Environmental Secretariat has identified 1,000 possible RPPNM areas with the potential to protect 14,000,000 square meters of indigenous forest (Guillem).

Creating this extensive and carefully situated park system required the deployment of creative zoning and development tactics. Following the passage of Curitiba's new Zoning and Land Use Law in 2000, the city jumpstarted the preservation of the valley bottoms by institutionalizing legal instruments such as transfer rights (Taniguchi 18). Curitiba subsequently designated four environmental preservation areas (akin to the United States' historic preservation zones), restricted or prohibited development in those areas, then allowed owners to transfer development rights (both in Curitiba and outside) and gave them significant zoning concessions on construction in the areas to which they transferred. The Curitiba Environmental Secretariat plans to designate another 10 environmental preservation areas in the next decade (Guillem).

D. Recycling

Like its transportation and park systems, the third prong of Curitiba's environmental preservation efforts—its much heralded recycling program—combines sustainability, social inclusion, and good fiscal stewardship. Over 70 percent of the Curitiba's trash is recycled through its recycling programs (PBS Frontline).[3] Indeed, the volume of recycled paper alone saves nearly 1200 trees a day (ICLEI 4). Curitiba's recycling program employs both carrots and sticks. The city does not incinerate garbage, and residents must pay for garbage pickup (based on volume) as they would for electricity or water (Guillem). But

---

3   By comparison, New York City recycles just 34 percent of its waste, only slightly better than the 33 percent American average. Houston recycles a measly 2.6 percent of its trash (Ellick).

the city also encourages participation by ensuring ease of use. Each household separates organic waste and trash, plastic, glass, and metal to allow for easy pickup and processing. As opposed to charging for garbage pick up, the city picks up recycling curbside from most residences, at no cost to the household, between one and three times a week, the program is funded through sale of salvage. The pickup trucks transport recycling to one of 13 privately run recyclable sorting parks, each of which employs homeless individuals and individuals in substance abuse recovery programs.[4] Additional proceeds from the sale of salvaged and recycled materials go towards social programs.

In the *favelas*, or shantytowns, inaccessible to recycling pickup trucks, Curitiba operates the Ecocitizens program. Originally started by the Catholic Church as an employment initiative (Guillem), the Ecocitizens program encourages homeless and low–income persons to collect and separate recycling from inaccessible neighborhoods. In exchange for bringing recycling to one of 92 sites, "ecocitizens" receive bus tokens, fresh fruits, and vegetables, and children's school supplies (Guillem). Ecocitizens remove 500 tons of recyclables a day (Guillem), for a total of 11,000 tons of garbage since the program's inception (Roman). The program has benefited 60 impoverished neighborhoods with 31,000 families by improving neighborhood sanitation conditions and by providing an influx of resources: nearly a million bus tokens and 1200 tons of surplus food as in-kind payment (Roman). Ecocitizens' success has prompted other Western cities to consult with the Environmental Secretariat about the possibility of adopting the program (Guillem).

IV. Criticisms of Curitiba

Unsurprisingly, given the awards and recognition bestowed on it,[5] Curitiba's innovative approach to urban planning has produced macro returns in addition to the more micro successes of greater bus usage, green space, and tons recycled (discussed above). Although Curitiba is the eighth largest city in Brazil, it has the fourth largest GDP, with 66 percent of its GDP produced by the commerce and service sectors given Curitiba's resistance to the expansion of heavy industry (World Development book case study). Curitiba is, however, the second-largest manufacturer of cars in Brazil and home to Nissan, Volkswagen, Audi, Siemens, and Electrolux, among other transnational companies. The city's 30 year economic growth rate is 7.1 percent (compared to a national average of 4.2 percent), and the trend continues: the metropolitan area had a growth rate of 3.11 percent in the 1990s and 1.36 percent in the 2000s (Scruggs). And despite this booming growth, Curitiba still boasts one of Brazil's lowest rates of ambient pollution (Lubow) and its second highest human development ranking (Scruggs). Per capita income is 66 percent higher than the Brazilian average, and Curitiba has the lowest illiteracy rate and highest educational attainment of any Brazilian city (World Development book case study).

Still, the city and its approach to urban planning have faced criticism in recent years. Skeptics note that, despite the BRT, Curitiba has a higher per capita ownership of private cars than any city in Brazil (0.63 cars per resident vs. national average of 0.27) (Scruggs),

---

4   Another 13 recycling plants are slated for construction by 2013 (Guillem).

5   In 2010 alone, Curitiba won ITDP's Sustainable Transport Ward, won the Swedish Globe Sustainable City Award, and was called a "long-term sustainability pioneer in the region" and the "clear leader in the index" by the Economist Intelligence Unit's Latin American Green City Index (Scruggs).

with the licensing of cars 2.5 times higher than babies being born in Curitiba, and efforts to discourage purchases unsuccessful because cars are considered status symbols (Lubow). Between 2008 and 2012, BRT use declined by 14 million rides, or 4.3 percent, with riders complaining that the system had become dirty and overcrowded (Halais).[6] Bike paths have remained largely unused (Halais).

More troublingly, while the formal city largely continues to thrive, the broader metropolitan area faces problems. Suburban residents are overwhelmingly poorer than their urban counterparts: 83 percent of the metropolitan population earns a low to moderate income (less than five monthly minimum wages). Housing has lagged behind population growth: while the region grew by 42,000 families between 2008 and 2009, it generated only 20,000 new housing units (Scruggs). An estimated 10 to 15 percent of the metropolitan population lives in substandard housing (Halais). There are now 13,000 households living in "invasion settlements," 6,000 of them in ecologically fragile areas. Since fewer than 70 percent of metropolitan households have sewer connections, raw sewage from these settlements flows directly into the previously pristine rivers (Lubow). And crime is high, with an average of 4.7 murders each day or 56.8 murders per 100,000 inhabitants, which is unremarkable by Brazilian standards but on par with high crime American cities like New Orleans (Halais).

Several explanations exist for these recent problems. First of all, while the Master Plan was designed to address population growth, it still envisioned a far smaller city and metropolitan region than currently exists. When the Master Plan was drafted in 1965, urban planners anticipated that the city might grow from 350,000 to 500,000, but they never thought the city itself would grow to 1.8 million people (with a metropolitan population twice that) (Lubow). This explosion was due in part to the famously livable environment offered by Curitiba, but also by the state of Parana's shift from a labor-intensive coffee economy to a mechanized agriculture of soybeans, which resulted in job losses experienced by hundreds of thousands (Lubow). The plan envisioned a dense urban core, and sought to limit suburban sprawl, but today's urban core cannot house the current mass of people and suburban sprawl has become inevitable, if unexpected. And even the famous BRT system cannot fully cope with the influx of population, driving affluent riders away from crowded conditions towards their cars. Ironically, given Lerner's trumpeting of the BRT as a more efficient and cost-effective alternative to light rail, the president of the rapid bus system now says Curitiba needs a light rail system to complement its existing bus system and provide quicker transit for longer suburban routes (Lubow).

Curitiba's later difficulty in responding to its new population realities stems in part from its planning model's focus on the formal city, rather than the broader metropolitan area, and hence the poor integration of the city and suburbs from both a governance and planning perspective. BRT terminates at city limits rather than crossing over into poorer suburbs, meaning residents of these poorer suburbs must cope with suburban bus lines that often charge higher fares with slower transit times (Scruggs). Consequently, hundreds of thousands of typically low-income suburban residents cannot access the planning interventions for which Curitiba is known (public parks, green spaces, pedestrian streets, creative repurposing of land, historic preservation) (Halais) or the social services made available (municipal health, education, and daycare networks, job training) (World

---

6 Halais, Flavie. "Has South America's Most Sustainable City Lost Its Edge?" *Atlantic Monthly* 6 Jun. 2012. Web. 31 Dec. 2014 <http://www.citylab.com/commute/2012/06/has-south-americas-most-sustainable-city-lost-its-edge/2195/>.

Development book case study). Many farther out areas offer inadequate sewage, and cannot be reached by trash or recycling trucks. And there exists minimal governmental infrastructure to address the problems in outlying areas. While Curitiba is sometimes called the Portland of the Southern Hemisphere, it is distinguished from that American city by its difficulty in operating across the wider metropolitan area, while Portland is known for its strong regional government (Scruggs).

While concerning, Curitiba's more recent problems do not detract from the continuing success of its formal city or the broader environmental and economic improvement brought about by its plan. Rather, it suggests the need to adapt the model to the 21st century reality of even greater urbanization than Lerner could have imagined.

V. Applicable Lessons

As discussed, the Curitiba Master Plan outlined certain general principles that have informed the city's development over the past 40 years and do so to this day. To even a causal reader of this report, these principles are obvious: they include integrated land use and transportation planning, an unwillingness to accept congestion and sprawl, environmental sustainability and preservation, close attention to the health and well-being of all its citizens, social inclusion, ease of use, and cost efficiency. Fulfilling these goals requires close collaboration between the often siloed government agencies entrusted with developing and administering the component programs, and Curitiba has been largely successful, thanks in part to the efforts of IPPUC. One or more of these principles is readily apparent in the design and implementation of each of the innovations discussed herein, including but not limited to: the BRT system's construction and placement, the social fare, high density zoning along the major corridors, the strategic designation of parkland to address the effects of global warming, and its recycling and related employment programs. Other cities can learn from Curitiba's example and apply these principles in their own planning and policymaking initiatives, provided the necessary collaboration exists between municipal departments.

Though recognizing the primacy of these general principles in combatting the problems of urban growth, the sections that follow explore how specific policy innovations outlined here could be implemented abroad. We also begin to identify cultural, legal, and other obstacles that might impede the export of these initiatives and in some cases suggest potential solutions.

A. Develop a BRT System

The creation of a BRT system would address many of the problems facing car-clogged and cash-strapped foreign cities burdened by rapid population growth. Curitiba's experience illustrates the BRT system's potential efficiency and ability, when well-placed and integrated with appropriate land uses, to replace cars as the preferred method of commuting. Indeed, a well-developed BRT system in an American city could move between 20,000–40,000 passengers per hour per direction, rivaling the numbers associated with many subway systems (Henscher 4). Moreover, a BRT system can be constructed for a fraction of the cost of subway or light rail systems—a boon for municipal budgets still recovering from the recession. Whereas constructing a subway system costs between 30–160 million dollars per kilometer in the United States, a complete BRT system costs only 5–20 million dollars per kilometer (Henscher 30). The shorter construction times for BRT systems also increase their political feasibility. Many elements of a BRT system can be implemented within a single election cycle, while subway and light rail construction depend on continued financial and political support across administrations.

In some ways, a Curitiba–like BRT system better lends itself to less dense or still developing municipalities. As evidenced by some American mayors' failed attempts to designate bike lines or implement congestion pricing, many vested interests are vying for limited space on American roads. This logjam could make it quite difficult to solicit support for dedicated busways, particularly given citizens' experience-driven belief that buses are slow and inconvenient and the social stigma sometimes associated with riding the bus. Nor do other countries necessarily share the Brazilian belief in the social function of land, that is, the idea that property rights may be limited by social and collective interests (dos Santos Cunha). Even in countries with generous eminent domain authority, acquiring the land necessary to implement an effective BRT system—be it from existing public roadways or private property—could prove difficult in certain political climates.

Despite the limitations inherent to introducing a BRT system to an existing transportation network, successful examples do exist in the United States, albeit on a relatively small scale. Most notably, Los Angeles, California, a city more often associated with traffic congestion than transportation innovation, has adopted elements of the BRT system developed in Curitiba. In 1999, a delegation from Los Angeles, including the then mayor, county supervisor and Metro's chief planning officer, traveled to Curitiba to observe the workings of the BRT system, and although their subsequent efforts to implement a dedicated bus line in Los Angeles were not realized until 2005, the metropolitan area now boasts multiple BRT lines, comprising approximately 400 miles of roadway. These bus lines have seen promising results: less than a year after the first such bus line opened in 2005, the Orange Line, ridership tripled its projected number, and on average the BRT lines run about 36 percent faster than the regular bus lines (Bloomekatz). Plans to expand the use of the dedicated busways in Los Angeles are ongoing, especially due to the substantial cost savings compared to rail options.

In New York City, another metropolitan area experiencing rapid population growth, where there are multiple claims to the roadways and the widening of streets is seldom a feasible option for dealing with traffic congestion, implementation of aspects of a BRT system has also begun. While the city is already served by an extensive subway network, buses are used to supplement this network, providing access along routes not covered by the subway system. The city has begun the use of off-board fare collection and transit signal priority to provide cost-effective and efficient transportation links (About Bus Rapid Transit). Dedicated bus lanes are also being added where possible, and the New York City Department of Transportation and the Metropolitan Transit Authority are currently working on infrastructure improvements to allow for dedicated bus lanes in some of the busiest areas of Manhattan, including along 34th Street and en route to LaGuardia airport (34th Street Select Bus Service).

As the experiences of Los Angeles and New York City suggest, for those municipalities where implementation of a citywide BRT system is neither feasible nor desirable, there are potential tweaks that can be made to improve the operation and equity of existing bus systems. Implementing a single dedicated bus lane along a central thoroughfare is certainly the primary example of this, but other elements of the BRT system also deserve consideration. As discussed, Curitiba's BRT network includes both linear and circumferential elements that avoid transportation dead zones where access is not available. Such dead zones are all too common elsewhere. Cities sometimes rely on bus networks that radiate from the city center in a series of straight lines, leaving the areas between the lines without access, or requiring passengers to take a bus into the city center and transfer in order to reach an area directly adjacent to them but not served by a circumferential bus route. These gaps in service can be particularly detrimental to low–income communities, where car use tends to be more limited, isolating neighborhoods and making the task of getting to work, or simply the grocery store, a time–consuming endeavor. Mimicking Curitiba's integration of linear and circumferential bus lines and careful attention to eliminating dead zones could improve the ease and efficiency of existing bus systems and promote social inclusion.

Individual design elements found in Curitiba's BRT system could also be implemented at minimal cost. For example, cities could create raised bus stops, to allow for quick boarding and unloading, and/or design buses with low floors, making it easier for children, the disabled, and those with prams or packages to board. Off–board fare collection and transit signal priority are also inexpensive ways of improving transit times along existing bus routes.

B. Integrate Land Use, Transportation, and Environmental Planning

As discussed, Curitiba strategically uses zoning both to enable densities that create demand for public transportation and to concentrate development in less environmentally precarious areas. The BRT system's high-take up rates and the city's reduced flooding demonstrate the efficacy of these efforts. Municipalities encountering similar problems could easily employ both strategies prospectively through modifications to their zoning codes. While the grandfathering of existing buildings would be a practical and political necessity in heavily developed areas, such zoning strategies have the potential to make a great impact in areas that are in the process of being developed. The idea of connecting public transportation to areas of high density is well-established, but Curitiba's integrated planning model shows us the benefits of pro-actively concentrating development and transportation along an axis of commercial activity rather than simply focusing on creating access links to a downtown area.

It should be noted that any prospective land use planning does require a certain degree of political cohesion, which can be difficult to achieve across city agencies and election cycles. The introduction of the Curitiba Master Plan occurred in the 1960's, when neither the municipal government of Curitiba nor the national government of Brazil could claim to be democratic bodies (Lubow),[7] and Curitiba had its plan questioned in court for failure to comply with the rules on popular participation (Rodrigues; Lubow). The origination of the Curitiba Master Plan came under Jamie Lerner, who was able to direct much of the city's development as head of IPPUC and then as the three term mayor of Curitiba. The degree of planning control exercised by Lerner and his colleagues in the mid–1960's was greater than can be achieved in most democratic cities today and for good reason, as we strive to represent a greater variety of interests in the decision–making process than were included in the development of the Curitiba Master Plan. Nonetheless, despite the limitations on efficacy sometimes posed by democratic processes, coordination among agencies is certainly possible, particularly on the municipal level. The development of cross–agency institutions, such as IPPUC, is one such means of facilitating comprehensive planning. A comparable municipal institution in the United States is that of an economic

---

7   A military regime was in power during Lerner's first two terms (1971–5, 1979–83). Lerner was only elected to his third term (1989–92) after the restoration of democracy.

development corporation (EDC): a not-for-profit organization that provides advice to municipal governments and coordinates major investment programs. While in the United States these often focus on real estate investments, there is also room for these entities to take on greater rolls in integrated transportation, land use and environmental planning.

C. Strategically Designate Park Land

The creation and expansion of urban parks is already a cause du jour of many urbanists and environmentalists, who cite high-cost cities like New York City (19.5 percent park land), Washington, DC (19 percent), and San Francisco (18 percent) as examples of the ability of green space to increase property values, decrease carbon emissions, and improve quality of life (Trust for Public Land 10). Indeed, city parks experienced something of a renaissance beginning in the 1970s and continuing into the new millennium (Sherer). Designating park land with an eye towards mitigating the impacts of global warming and the accompanying flooding by diverting floodwaters and preserving river basins, as Curitiba has done, would be an easy strategic tweak to ensure existing efforts maximize the environmental bang for their buck. And many cities already possess the legal tools necessary to achieve these goals. New York City, for instance, includes numerous historic preservation zones where property owners within such zones are given transfer development rights to use elsewhere (often with significant zoning concessions) if historic covenants prohibit their proposed construction. Similarly, cities could create environmental preservation districts in precarious areas, restrict or forbid development there, and allow owners to transfer their development rights. Though the designation of such zones is likely to encounter opposition from current landowners, the prevalence of historic preservation districts and covenants demonstrates the feasibility of environmental analogues. Tax credits or other subsidies for private environmental preservation, as for historic preservation, are another option, albeit ones that are more costly.

Examples of such programs do exist, although they have yet to become commonplace in the United States. In King County, Washington, for example, development rights may be transferred from environmentally sensitive areas to support high-density development in urban areas. Transfers are most often made via a TDR (Transfer of Development Rights) land bank, enabling landowners to receive financial compensation in exchange for the right to develop their property (Higgins). A conservation easement is placed on the land being protected while the development rights are held by the land bank for purchase by developers of approved projects in high density areas. The program concentrates development in designated areas while preserving sensitive areas from development. To date, the program has been used to encourage private property owners to preserve over 140,000 acres of resource land, comprised of open space, wildlife habitat, agricultural land and parkland (TDR).

D. Implement Comprehensive Recycling Programs

Municipalities worldwide already employ recycling programs. But Curitiba's example suggests potential tweaks to increase existing programs' take-up rates, minimize costs, and promote social inclusion. As discussed, Curitiba's recycling program uses both carrots and sticks to maximize participation: residents must pay for garbage collection but receive free curbside recycling with pickup multiple times per week. Other municipalities could price their services accordingly, in addition to increasing recycling; the switch might also increase municipal revenues. Likewise, in many cities, homeless individuals already gather recyclables in order to earn pennies from the manufacturing company. Other cities could adopt Curitiba's model and subsidize these efforts in order to collect recyclables in low-density or inaccessible areas in a cost–effective way. Such initiatives would have the added benefit of providing much needed employment for hard-to-serve populations. For these reasons, among others, similar programs have been successful elsewhere in Latin America, initiated both by private companies and municipal governments.

In Pueblo, Mexico, a private company called *Monedero Ecológico* (Green Wallet) introduced a rewards based waste disposal system that was inspired by the Ecocitizens recycling program in Curitiba. City residents who participate in the program are given a debit card that can be loaded with credits. Credits are earned for depositing solid waste at designated drop-off centers located at convenient locations, such as schools and grocery stores (André). One credit is obtained for every kilogram of waste, and the credits can be redeemed for cell phone minutes, clothes, food, and school supplies at participating local retailers. In the first 19 months of the program's operation, 3,000 cards were distributed and 19 tons of waste collected (André).

Municipal governments have also been influenced by the recycling programs developed in Curitiba. In Belo Horizonte, Brazil, the city supported the growth of a local waste–pickers association, *Associação dos Catadores de Papel, Papelão e Material Reaproveitável* (Association of Paper, Carton and Recyclable Material Pickers, ASMARE), and formed an official partnership with the Association (André). ASMARE members are now integral to the city's waste management program, collecting garbage on a daily basis and delivering it to central collection facilities, where they sort it into reusable recyclable materials, including paper, plastic and cardboard. While the Association was originally formed to protect the interests of waste collectors, it is now run as a commercial enterprise, producing recycled paper and selling recycled materials (ASMARE).

VI. Conclusion

By prioritizing smart transit, strategic land use, environmental sustainability, and social inclusion, Curitiba has effectively addressed the problems attendant in its 50 year growth and paved the way for a less congested, ultra-sustainable, and more equitable next half century. Its lessons and policies—particularly its transportation network, parks system, and recycling program—are primed for export and increasingly relevant in this age of emerging mega cities and mega regions. Hopefully, other municipalities will be inspired by its example and adopt its principles and/or innovations for more sustainable, efficient, and socially inclusive development.

Works Cited

"34th Street Select Bus Service." *New York City Government.* 2014. Web. 31 Dec. 2014 <http://www.nyc.gov/html/brt/html/routes/34th-street.shtml#updates>.

"About Bus Rapid Transit." *New York City Government.* 2014. Web. 31 Dec. 2014 <http://www.nyc.gov/html/brt/html/about/about.shtml>.

André, Richard. "Turning Garbage into Gold." *Americas Quarterly.* Web. 31 Dec. 2014 <http://www.americasquarterly.org/waste-and-recycling>.

"ASMARE (Brazil)." *Solid Waste Management Association of the Philippines*. 2012. Web. 31 Dec. 2014 <http://www.swapp.org.ph/iws-global-best-practices/137-smare-brazil>.

Barton, Scott, and Joseph P. Kubala, *Bus-Rapid Transit Is Better Than Rail*. Colorado: Center for the American Dream, 2003. Web. 31 Dec. 2014 <http://www.jtafla.com/JTAFuturePlans/Media/PDF/BRT-LRT%20Comparison.pdf>.

Bloomekatz, Ari. "Orange Line Busway is Metro's Quiet Success Story." *Los Angeles Times* 27 Jun. 2012. Web. 31 Dec. 2014 <http://articles.latimes.com/2012/jun/27/local/la-me-orange-line-20120628J>.

Demery, Leroy. *Bus Rapid Transit in Curitiba, Brazil—An Information Summary*. Vallejo, CA: publictransit.us, 2004. Web. 31 Dec. 2014 <http://www.publictransit.us/ptlibrary/specialreports/sr1.curitibaBRT.pdf>.

dos Santos Cunha, Alexandre. "The Social Function of Property in Brazilian Law." *Fordham Law Review* 80 (2011): 1171–81.

Ellick, Adam C. "Houston Resists Recycling, and Independent Streak is Cited." *New York Times* July 29, 2008. Web. 31 Dec. 2014 <http://www.nytimes.com/2008/07/29/us/29recycle.html?pagewanted=all>.

"Facts and Figures," *New York Metropolitan Transportation Authority*. Web. 31 Dec. 2014 <http://web.mta.info/nyct/facts/ffsubway.htm>.

Glaeser, Edward. *Triumph of the City: How Our Greatest Invention Makes Us Richer, Smarter, Greener, Healthier, and Happier*. New York: Penguin Books, 2011. Print.

Goodman, Joseph, Melissa Laube & Judith Schwenk. "Curitiba's Bus System is Model for Rapid Transit." *Race, Poverty & Environment* 11.3 (Winter 2005/2006). Web. 31 Dec. 2014 <http://urbanhabitat.org/node/344>.

Henscher. David. "Frequency and Connectivity—Key Drivers of Reform in Urban Public Transport Provision." *Journeys* 1 (2008): 25–30.

Higgins, Noelle. *Transfer Development Rights*. Web. 31 Dec. 2014 <http://depts.washington.edu/open2100/pdf/3_OpenSpaceImplement/Implementation_Mechanisms/transfer_development_rights.pdf>.

"Human Population: Urbanization." *Population Reference Bureau*, 2014. Web. 30 Dec. 2014 <http://www.prb.org/educators/teachersguides/humanpopulation/urbanization.aspx>.

Institute for Research and Urban Planning of Curitiba [IPPUC]. *Urban Planning Process in Curitiba*. 2012. Print.

International Council for Local Environmental Initiatives (ICLEI), *Curitiba: Orienting Urban Planning to Sustainability*. 2002. Web. 31 Dec. 2014.

Interview with Carlos Guillem, Curitiba Enviornment Secretariat, in Curitiba, Brazil. Mar. 16, 2012.

Irazábal, Clara. "Urban Design, Planning, and the Politics of Development in Curitiba." *Contemporary Urbanism in Brazil: Beyond Brasília*. Ed. Vicente Del Rio and William Siembieda. Gainseville, FL: University Press of Florida, 2009. 203. Print.

The Leadership Conference Education Fund, *Where We Need to Go: A Civil Rights Roadmap for Transportation Equity*. 2011. Print.

Lindau, Luis, Dario Hidalgo, and Daniela Facchini. "Curitiba: The Cradle of Bus Rapid Transit." *Built Environment* 36 (2010): 274. Print.

Lubow, Arthur. "The Road to Curitiba." *New York Times* 2007 May 20. Web. 31 Dec. 2014 <http://www.nytimes.com/2007/05/20/magazine/20Curitiba-t.html?pagewanted=all>.

Lundqvist, Marie. *Sustainable Cities in Theory and Practice A Comparative Study of Curitiba and Portland*. Sweden: Karlstad University, 2007. Print.

Nexus Research Group [Nexus]. *Curitiba, Brazil: BRT Case Study*. Web. 30 Dec. 2014 <http://nexus.umn.edu/Courses/ce5212/Case3/Curitiba.pdf>.

PBS Frontline and World Fellows Project, *Brazil—Curitiba's Urban Experiment*. Dec. 2003. Web. 31 Dec. 2014 <http://www.pbs.org/frontlineworld/fellows/brazil1203/master-plan.html>.

Power, Mike. "Common Sense and the City: Jaime Lerner, Brazil's Green Revolutionary." *The Guardian* 2009 Nov. 5. Web. 2014 Dec. 30 <http://www.guardian.co.uk/environment/blog/2009/nov/05/jaime-lerner-brazil-green>.

Press Release, *The Brazilian city Curitiba awarded the Globe Sustainable City.* 2009. Web. 30 Dec. 2014 <http://globeaward.org/winner-city-2010>.

Rabinovitch, Jonas. "Curitiba: Towards Sustainable Development." *Environment and Urbanization* 4 (1992). Print.

Rodrigues, Evaniza, and Benedito Roberto Barbosa. "Popular movements and the City Statute." *The City Statute: A Commentary*. Sao Paulo: Cities Alliance and Ministry of Cities, 2010. 23–34. Web. 30 Dec. 2014 <http://www.citiesalliance.org/sites/citiesalliance.org/files/CA_Images/CityStatuteofBrazil_English_fulltext.pdf>.

Roman, Alejandro. "Curitiba, Brazil." *Encyclopedia of the Earth.* Boston: Boston University, 2013. Web. 31 Dec. 2014 <http://www.eoearth.org/article/Curitiba,_Brazil>.

Scruggs, Greg. "Cracks in the Curitiba Myth." *Next City* 2013 Nov. 1. Web. 31 Dec. 2014 <http://nextcity.org/daily/entry/cracks-in-the-curitiba-myth>.

Sherer, Paul, *The Benefits of Parks: Why America Needs More Open Space*. San Francisco: Trust for Public Land, 2006. Web. 31 Dec. 2014 <http://www.eastshorepark.org/benefits_of_parks%20tpl.pdf>.

Taniguchi, Cassio. *Transport and Urban Planning in Curitiba.* 2001. Print.

"Transfer of Development Rights (TDR) Program." *King County Government.* 7 May 2014. Web. 31 Dec. 2014 <http://www.kingcounty.gov/environment/stewardship/sustainable-building/transfer-development-rights.aspx>.

Trust for Public Land, *City Park Facts*. San Francisco: Trust for Public Land, 2011. Web. 31 Dec. 2014 <http://cloud.tpl.org/pubs/ccpe-city-park-facts-2011.pdf>.

U.S. Federal Transit Administration, Office of Research, Demonstration and Innovation. *Issues in Bus Rapid Transit*. Web. 31 Dec. 2014 <http://www.fta.dot.gov/documents/issues.pdf>.

UN-HABITAT. *Cities and Climate Change—Global Report on Human Settlements*. UN-HABITAT, 2011. Print.

UN-HABITAT. *State of the World's Cities 2008/2009*. Sterling, VA: Earthscan, 2008. Web. 2014 Dec. 30 <http://www.unhabitat.org/pmss/listItemDetails.aspx?publicationID=2562>.

"Urban Population Growth." *The World Bank*, 2014. 30 Dec. 2014 <http://search.worldbank.org/data?qterm=urban%20growth&language=EN>.

"Urbanization." *United Nations Population Fund*, 2014. Web. 30 Dec. 2014 <http://www.unfpa.org/pds/urbanization.htm>.

Vidal, John. "UN Report: World's Biggest Cities Merging into Mega-Regions." *The Guardian.* 2010 Mar. 22. Web. 2014 Dec. 30 <http://www.guardian.co.uk/world/2010/mar/22/un-cities-mega-regions>.

"World Development book case study: sustainable urban development in Curitiba." *New Internationalist.* 2014. Web. 31 Dec. 2014 <http://newint.org/books/reference/world-development/case-studies/sustainable-urban-development-curitiba/>.

Wright, Lloyd, and Walter Hook. *Bus Transit Planning Guide.* New York: Institute for Transportation and Development (2007). Print.

Chapter 6

Exploring Options for Urban Sustainability in an Era of Scarce Water Resources: A Possible Ban on Lawns

Sarah B. Schindler[1]

Introduction

Much of the United States is still feeling the effects of the worst drought in 50 years. At the same time, lawns, which front many suburban American homes, are the largest irrigated crop in the country. Lawns occupy approximately three times more space than corn (Lindsey) and twice as much as cotton (Steinberg 4), and consume up to 60 percent of potable municipal water supplies in Western cities and up to 30 percent in the East (Rappaport 898 n.114). As cities and towns confront water shortages and other concerns associated with climate change, many are beginning to adopt sustainability plans and ordinances that impose environmentally beneficial measures upon citizens and corporations—for example, mandatory green building ordinances, recycling requirements, plastic bag bans, and limits on what can be burned (Kuh 1148). As climate change adaptation measures become more common, it is likely that more municipalities will pass ordinances that aim to control individual actions that have a cumulatively significant impact on the environment. At the same time, existing regulation across much of the United States actively encourages and arguably requires the maintenance of lawns (Smith 216–18).

In this chapter, I consider how municipalities can use the law to reduce lawns and the harms they cause. Because lawns are so prevalent and use such a large percentage of potable municipal water, yet offer limited benefits, they are a logical point of attack for future sustainability ordinances. This is not far fetched: a small number of southwestern localities have begun to prohibit or limit new turf installation (Scottsdale; Tucson);[2] others have incentivized the removal of existing lawns (Chandler; Scottsdale);[3] and watering

---

1 This chapter was derived from a longer law review article that appeared previously in the George Washington Law Review: Sarah Schindler, Banning Lawns, 82 GEORGE WASHINGTON L. REV. 394 (2014) (lead article)

2 Scottsdale, Ariz., Rev. Code § 49–247 limits new model home landscaping by prohibiting new turf installation in front yards and limiting new turf installation to 10 percent of lots less than 9000 square feet and to five percent of the remainder of larger lots up to one acre. Tucson, Ariz., Land Use Code ch. 23, § 3.7.2.2 limits new turf installation by multifamily residential developments to five percent of the site, 100 square feet, or eigh percent of the required open space, whichever is greater. Las Vegas, Nev., Code § 14.11.150 prohibits new turf installation in residential front yards and limiting new turf installation in rear and side yards to the greater of 50 percent of gross area or 100 square feet, whichever is greater.

3 Scottsdale, Ariz., Rev. Code § 49–243 provides single–family residential customers up to $1500 in rebates and commercial and multifamily customers up to $3000 in rebates for removing existing turf and replanting with low-water-use landscaping. Chandler Arizonia rebates residents for removing at least 1000 square feet of turf and replacing it with at least 50 percent non-grass plants.

and fertilizer limitations are fairly widespread (Dothan; Garden Grove; Tucson).[4] To date, however, there has been little scholarly discussion of limits on lawns.

Part I of this chapter provides a brief history of the lawn and discusses the reasons for its predominance in the United States. It recognizes that, although lawns offer some benefits, they appear to be outweighed by the substantial number of harms that lawns create, including extreme water use in a time of water shortages; emissions tied to gas–powered lawn mowers and leaf blowers; pollution and runoff from petrochemical–based fertilizers; fire hazards in dry climates; and propagation of monocultures and the loss of biodiversity. When examined cumulatively, these sorts of "environmentally significant individual behaviors" may warrant prohibition (Kuh 112).

Part II outlines regulatory techniques that might be applied to target and correct the harms associated with lawns, and suggests that some local governments might consider legal mandates as a potentially powerful regulatory option, especially in the face of increasingly extreme climate conditions. Although a large-scale movement to ban lawns may currently be politically implausible in many parts of the country, such bans might be desirable or even necessary in the future as the effects of climate change—including water shortages—become more common. Part III considers the contours of a potential mandate against lawns. It addresses the sources of municipal authority to regulate lawns and the probable defeat of any takings challenge to a lawn ban. It then turns to the structure of a potential ban on lawns, discussing when to impose the ordinance and whether the ordinance should outlaw all turfgrass or just front yards. The chapter concludes by recognizing that many people would dislike the idea of a lawn ban. At least upon first impression, they may think it sounds like an unlikely, untenable, and possibly impermissible use of the police power. This chapter demonstrates, however, that it is in fact well within a municipality's police power to reduce or eliminate lawns, even by retroactively banning them. Moreover, these regulatory techniques are likely to become more common as climate conditions worsen and water becomes increasingly scarce. Thus, what might at first seem like an implausible proposal may turn out to be more likely than most would suspect.

I. Lawns

A. History

Most suburban neighborhoods in the United States have a few common aesthetic qualities. One of these qualities is that most homes are fronted by an expanse of green, non-native turfgrass. Although it is now hard to imagine neighborhoods without lawns, prior to the Civil War, turf cultivation was an uncommon use of property (Steinberg 11; Schroeder 5). Rather, it was common to see houses fronted with productive vegetable gardens or native vegetation mixed with dirt (Steinberg 11).

Some commentators suggest that the creation and maintenance of lawns stems from the human desire to dominate and impose order over nature (Pollan 41; Kolbert 82). The

---

4   Dothan, Ala., Code § 102–165 restricts the watering of lawns and gardens by Dothan Utilities customers to three days per week from April 1 to October 1 each year. Tucson, Ariz., Code § 27–95 prohibits all outdoor irrigation during a "water emergency," except with reclaimed water. Garden Grove, Cal., Code § 14.40.025 prohibits the watering of lawns between 10am and 6pm and limiting watering that is not continuously attended to no more than 15 minutes per day per station.

aesthetic has its roots in the English manor, where the lord of the estate maintained a neat, green expanse by employing a "band[ ] of scythe-wielding servants" or a shepherd and his flock (Kolbert 82). This suggests that the lawn has built–in class significance as well. The American lawn is a "democratized" form of the aristocratic manor lawn, which was more of a "setting for lawn games and ... a backdrop for flowerbeds and trees" than an aesthetic masterpiece in and of itself (Pollan 41).

In recent years, U.S. lawns have expanded at a rapid rate, such that the lawn is now "the single largest irrigated crop in America in terms of surface area, covering about 128,000 square kilometers in all" (Lindsey). Many theories have been advanced to explain why lawns occupy the dominant position that they do. People have long appreciated the lawn as an essential, beautiful component of the home, and lawn dominance has likely continued due to status quo bias and preference (Biber 1321–2). The lawn norm is deeply embedded.

Existing public and private laws also encourage and often effectively require a neat, short, turfgrass yard. From a public law perspective, after the Supreme Court declared zoning to be a valid exercise of the police power, suburban development flourished (Richmond 548). Many of the first suburban municipal zoning ordinances included setback regulations, which required buildings to be constructed a certain distance from the street or sidewalk and thus created an area of space between the building and the street (Williams 177–9). For commercial structures, this setback space is often filled in by parking lots; in residential neighborhoods, it is filled in with lawns. Many localities also have "weed ordinances" that effectively require lawns, both by mandating that ground cover be kept short, and by prohibiting certain native plantings or vegetable gardens in front yards.[5] From a private, contractual perspective, one fifth of Americans live in residential common interest communities that are governed by covenants, conditions, and restrictions ("CC&Rs") (Jouvenal A1). CC&Rs regularly require setbacks, limit fences, and may even require front lawns. Some CC&Rs also prohibit the cultivation of vegetables, fruits, or native plants (Schindler 289). Thus, both existing laws and agreed to property rules tend to reinforce the lawn as a staple of American landscape design.

B. Benefits

Whether the prevalence of lawns is a product of their entrenched legal status, or whether existing law simply reflects long held practice, many feel an attachment to their lawns and believe that lawns offer benefits to them and their communities. Lawns provide a consistent, unifying aesthetic when one looks down a street. Because they are what people expect, lawns tend to "keep[ ] the neighbors happy and add[ ] to their property value" (Steinberg 7). By maintaining a neat front yard, homeowners suggest that they have a relationship, and shared values with their neighbors (Pollan 23, 41). Thus, not mowing, tending, or maintaining a lawn could be viewed as a dereliction of one's civic responsibility and duty as a member of the community.

There are also some health and safety justifications for lawns: grass can help prevent soil erosion and runoff (Tekle 226); trap dust and particulate matter (Duvall); lower temperatures (Gibeault 1); and reduce glare and noise (ibid.). A lawn also provides a better carbon sink than a parking lot (Milesi 426; Bittman). Further, many people derive a

---

5   For example, Annapolis, MD., Code § 10.20.010 states "the height limit of grass, weeds and 'other rank vegetation' is 12 inches."

psychological benefit from having a buffer between their homes and the outside world, and the law protects that buffer (Wells and Evans, 321).

Some also see lawns as providing a community centered benefit—a space where neighbors can gather (Bormann 23). They are also more user friendly than, for example, a rocky desert landscape; lawns provide a soft place for children and dogs to play. Lawns can also enhance social capital in a given neighborhood by providing an area that facilitates such interactive behavior (Farr 147). This idea ties into the idea of the lawn as deeply seated in the ethos of the sanctity of the single family home and of home ownership itself. Justice Douglas, in *Village of Belle Terre v. Boraas*, famously exclaimed, "[a] quiet place where yards are wide, people few, and motor vehicles restricted are legitimate guidelines in a land use project addressed to family needs" (9). The police power is broad enough to accomplish these goals: it can be used to mandate wide yards under the guise of furthering the public welfare.

C. Harms

Although many individuals have a strong psychological attachment to their lawns, that attachment comes at a significant cost. In many localities, lawns may be inefficient and may cause harms that outweigh their benefits. Those harms include dramatic potable water consumption, high energy costs from water use, increased water and air pollution, and loss of biodiversity. Because lawns cover such a large percentage of our built environment, we must account for these harms cumulatively.

When considering large scale environmental harms, one often imagines commercial manufacturing facilities with polluting smokestacks. However, a growing area of legal scholarship focuses on "the environmental significance of individual behaviors and lifestyles"—actions that scholars term "environmentally significant individual behaviors" (Kuh 1116–17). Many existing environmental laws, especially comprehensive federal laws, fail to regulate individual actions that, cumulatively, result in significant harm to the environment. Though the actions taken by a single individual to keep her lawn neat and green might be environmentally insignificant, on a nationwide scale, or even one based on the local watershed, lawn care may have a substantial impact and thus warrants close consideration. Further, while the government can control some sources of environmental harm by regulating manufacturers, there is no upstream source through which to regulate the harms associated with lawns. Thus, the most logical place to impose regulation is on the individual's behavior.

Although regulation of lawns is technically a property restriction, it is also inherently a limitation on individual actions. If a property restriction is put in place that retroactively bans all lawns, an individual may not plant a new lawn, may not continue to water or mow an existing lawn, and may even be forced to tear up an existing lawn. Individual actions like these are environmentally significant because every person is a polluter; our individual actions "lie at the core of both the climate change problem and its potential solutions" (Ibid. 1114). As one commentator notes, "[w]e pollute when we drive our cars, *fertilize and mow our yards*, pour household chemicals on the ground or down the drain, and engage in myriad other common activities" (Vandenbergh 518). Thus, the law should find a way to capture these individual, cumulatively significant harms.

Lawns often require substantial quantities of water to maintain color, health, and appearance. For the last hundred years, Americans have come to see water as "abundant, safe, and cheap," living in what one commentator has referred to as "the golden age of water" (Fishman 9). Those days are waning, however, many parts of the United States are facing one of the worst water shortages in recent history, and climate change is altering weather patterns such that these droughts will become more common (Doremus 1104–5, 1115. 2011). Thus, climate change adaptation is inherently linked to water concerns, and therefore to lawns.

People like green lawns, and in many parts of the country green lawns mean heavily watered and fertilized lawns. Because most turfgrass is a non-native species, it often needs assistance to thrive; some people water their lawn twice per day (Lindsey). A large percentage of the potable municipal water supply is used for this purpose; studies suggest that approximately 60 percent in the West and 30 percent in the East is being used for lawn irrigation (Rappaport 898, Smaus). In real numbers, the EPA estimates that residential landscape irrigation accounts for approximately nine billion gallons of water per day and one–third of all residential water use in the United States (EPA, Reduce Your Outdoor Water Use).

The amount of water used on lawns is even more problematic when one considers the resultant energy costs. In the United States, water is typically collected, treated, and delivered to consumers before it is used to water lawns, consuming large amounts of energy at each step in the process (Cohen 2). Much of the water used for lawn care has to be transported from elsewhere, which contributes to emissions and thus global climate change (Jervey). Specifically, most municipal water is either surface water that must be extracted from rivers or streams or groundwater that must be pumped from aquifers (Cohen 2). The utilities that then treat and distribute the water must use energy to do so, and because many of the pipe distribution systems in the United States are old, a substantial amount of this already treated potable water is lost during transport (Ibid.). As for the water that does reach end users, it is often further heated or cooled, requiring the expenditure of additional energy. The high energy cost of water is also connected to the water subsidies that are prevalent in the United States. Although people require drinking water for survival and certain agricultural pursuits warrant subsidized water costs, there is no valid reason for allowing individuals to avoid paying the true cost of water, including its energy costs, when it is merely used for growing grass.

Beyond energy production, emissions are associated with lawns in other ways as well. Specifically, almost all people with yards mow them (or hire others to do so), typically with a gas–powered lawn mower, and many use leaf blowers to rid their lawns of debris. Thirty minutes of leaf blower usage creates the same amount of "polluting hydrocarbon emissions as driving a car seventy-seven hundred miles at a speed of thirty miles per hour" (Steinberg 8). Cumulatively, these individual actions substantially increase not only emissions, but smog and particulate matter. Lawns are also expensive and time consuming to maintain. Estimates suggest that people in the United States spend approximately 40 billion dollars each year on lawn care, and mowing and tending a lawn may occupy hours every week (Ibid.).

Another lawn related harm is tied to the petrochemical–based fertilizers with which many lawns are treated (Rappaport 901). Although front yards may look identical in Ohio, Arizona, and Georgia, their local geography, weather, and growing conditions are not. Homeowners require "the tools of 20th century industrial civilization—its chemical fertilizers, pesticides, herbicides, and machinery" to keep lawns green and growing in many parts of the United States (Pollan). These chemicals pollute stormwater runoff that often flows into local bodies of water (Tekle 215–16). The United States is not meeting water quality standards in large part because of urban runoff. Although some point source pollution—from defined sources like factories and wastewater treatment plants—has been substantially reduced through regulations, urban stormwater runoff is still a major source of environmental harm. It is within the purview of local governments to regulate much of the land use that results in that form of pollution. Further, although runoff is much more pronounced from truly impervious surfaces, such as pavement, "compacted soils mono-turf landscapes" like lawns can be "near impervious," and thus result in much greater amounts of runoff than would a natural landscape with a greater variety of topography (Rappaport 901). Runoff from lawns also contributes to the prevalence of pesticides in urban waterways (U.S.G.S. Fact Sheet). In addition to environmental harm, there is also some evidence that lawn chemicals and weed killers can increase cancer in pets and humans, respectively (Rappaport 923).

The non-native nature of turfgrass also results in harms associated with ecological principles and loss of species biodiversity. This landscape reduces the amount of habitat that might otherwise be available for native plants, thus "hasten[ing] the process of plant extinction" (Ibid. 885). For example, the prairie is an extremely endangered ecosystem that provides an important habitat for birds and butterflies, and requires little water (O'Connor A23). Additionally, in many dry climates, lawns are a potential fire hazard; fire hazards maybe be reduced if native plants or xeriscaping is used instead of lawns. Furthermore, lawns also tend to create more allergy-producing pollen than native plantings. They also fail to provide the same level of ecosystem services as native plants. All of these factors combine to lead some to view the lawn as "the most obvious example of humankind's disregard for Nature" (Rappaport 886).

Although in many ways the current legal structure mandates lawns, they are often environmentally and financially inefficient, for the reasons discussed above. Yet property rules and laws are typically organized in such a way as to incentivize or encourage the productive use of property and to avoid waste (Serkin 1275). These rules are informed by numerous strands of property theory. For example, Locke's labor theory of property suggests that people have ownership interests in property in which they invest their labor (Locke 15). Law and economics theorists have described the way that private property ownership serves to internalize externalities, thus fostering more efficient use of property (Epstein 762). Such theories form the basis of property doctrines such as adverse possession, which seek to decrease the inefficient use of land and increase its efficient use. In contrast, the law in many communities currently requires, and certainly allows, lawns that are inefficient and that affirmatively cause harm to those communities. Alternative productive uses of property could provide ecosystem services benefits and thus would be more efficient. For example, instead of lawns, landowners could use xeriscaping, which is often thought of as desert or dry landscaping, but can be used more generally to describe any landscaping that uses native plants and is thus sustained primarily by natural rainfall. Native plants are adapted to local climates, and thus typically require less maintenance and watering than non-native turfgrass. Such a result would be more in line with standard views of the purpose of property law.

II. Regulating Lawns

Change in environmental law and policy requires intentional action. Such action can take the form of law (including mandates and bans), norms, market–based mechanisms (including economic incentives), or architecture (Lessig 662–3). Although scholars have shown an increased interest in determining the appropriate regulatory scope for individual behavior that impacts the environment and have debated which techniques would be most appropriate for different types of harms, few have focused on bans (Czarnezki; Vandenbergh 554). Although some scholars and regulators view bans as too harsh and broad stroked for the harms that they target, they are currently used in some instances; for example, health concerns have led to bans on using asbestos and smoking. Further, worsening climate change might alter the physical and regulatory landscape, necessitating a more stringent approach to regulation in the future. And because "[c]limate change is a private property problem," it will likely lead to greater restrictions on individual behavior and the use of private property (Babie 19). Markets and architecture should play a role in reducing the predominance of lawns in the United States. However, in this chapter, I bracket those approaches, and focus on the role that norms play in maintaining lawns, and the role that legal regulation through mandates might play in reducing their harms.

A. Norm Change

The preeminent role of front lawns in the United States is due in large part to a pervasive norm (McAdams 359). Author Michael Pollan believes that this norm involves "a deep distrust of individualistic approaches to the landscape. The land is too important to our identity as Americans to simply allow everyone to have his own way with it." The strength of the norm, evinced by the fear of social sanctions for failing to maintain a neat front lawn, results in entrenchment despite the many harms associated with lawns. The norm could be what Robert Ellickson terms "welfare maximizing"—one that seeks to solve collective action problems (167. 1991.). Under this theory, the goal of the lawn norm might be maximization of aggregate property value in a neighborhood. But this norm appears to be self–reinforcing and circular—property value is tied to lawns due, in part, to the historic expectation of lawns. But this is not because the lawn norm is inherently good or valuable; it is because no one wants to defect from the norm for fear of social sanctions (and because the norm has likely resulted in some having a true preference for lawns) (Sunstein 929).

In some instances, informational campaigns combined with other tools can work to change norms, which can in turn result in more environmentally responsible behavior. For example, a locality could work to promote information about how much money a household could save by not watering its lawn; a similar approach has been used in the context of energy efficiency (Schindler 13–14). However, norms are often slow to change, even with the aid of informational campaigns (Posner 1712–13). One commentator suggests that the current lawn norm will only fade when more sustainable front yard norms rapidly attract broad public interest (Tekle 230). Perhaps this is beginning to happen independently—members of the popular press have begun to write about the wastefulness of lawns, and certain thought leader communities are adopting policies to promote alternatives to standard lawns in response to citizen demands (Kurutz D1). Although current norms might suggest that homeowners would prefer to retain their existing lawns, a few pioneering communities could lead to an avalanche of changing preferences. Further, movement away from an entrenched norm might occur more naturally when the historic norm is shown to be harmful in contemporary settings.

It is also possible, however, that something stronger, like mandates, might be necessary to force more rapid change surrounding an entrenched norm. Because climate change problems are intensifying, and the recent drought has worsened the water scarcity problems in much of the country, immediate action is necessary. Further, while the harms from norm defection are internalized in the first person in the neighborhood to replace a lawn with xeriscaping—she risks damaging her property value and angering her neighbors for little (cumulative) environmental benefit—harms from maintaining lawns are broadly dispersed. Law is useful at coordinating behavior in the face of these sorts of collective action problems and it helps internalize externalities (Demsetz 348).

B. Law: Mandates and Bans

As lawn–related harms become more pressing, localities will likely turn to a variety of regulatory methods to foster sustainability and ensure harm reduction. Although tools like incentives may be sufficient to reduce water consumption and alleviate other lawn–related harms at the present time, mandated reductions in water consumption might be a more important policy tool in the future. Further, legal regulation encompasses varying levels of strictness—from mandated reductions to outright bans.[6] This section addresses some concerns with legal regulation, but also discusses why these solutions may be appropriate to address the harms associated with lawns in the future.

As a starting point, it is important to acknowledge that, at the present time, pervasive lawn regulation is unlikely in all but the most drought–ridden areas. This is in part because mandates, and especially bans, are often seen as unlikely or politically untenable for a number of reasons. First, they are generally disfavored and unpopular (Cheng 659–62: Corrosion Proof Fittings 1215–16). Public choice theory suggests that mandates—those that would provide amorphous benefits to the community at large, but would substantially burden individual homeowners—would not garner enough organized support to persuade local politicians to implement them (Michelman 148). Further, some commentators suggest that mandates are politically unlikely because laws will not change until the norms underlying those laws change (Kuh 1117–18; Tekle 239). But this argument is specious, because it ignores the fact that norm change often follows legal change, and that the police power is broad enough to lead despite opposing norms.

Although widespread lawn bans are currently unlikely, climate change may put many of these issues on the table; "[c]risis can lower political barriers to legal change" (Doremus 1115. 2011). Some would call the recent drought in many parts of the United States a crisis, and looking forward, climate scientists almost universally predict that radical, catastrophic changes in the natural environment will soon occur as a direct result of climate change (Org. for Econ. Cooperation and Dev. 71, 87). In times of national crisis, policies that were previously politically untenable, or even viewed as illegal or unconstitutional, may become the controlling policies to address the problems that are causing the crisis. As more individuals in more parts of the country feel the effects of severe water shortage, pollution, and climate change, they may become more likely to support policies previously thought to be radical, as well as the politicians who support those policies, in order to target crisis–related harms.

Another concern is that mandates directed at individuals sometimes suffer from an "intrusion objection," which involves opposition to a perceived invasion of "privacy or other civil liberties in a manner unpalatable to the public" (Kuh 1119–20). Many believe that banning the actions of private citizens impinges too substantially on private rights,

---

6 A ban entirely outlawing all turfgrass is an extremely restrictive form of mandate—"the most burdensome regulatory option" (Corrosion Proof Fittings v. EPA 1215–16). However, a municipality could instead simply limit the amount of ground area that turfgrass can occupy.

including private property rights. Challenges to land use ordinances are often founded in general libertarian property rights theory, the idea being that the fewer regulations on the use of the property there are, the better, as this will foster more efficient use of property (Epstein 761–3). These views relate to lawns because "[a] strong view of private property empowers the landowner to do what she wishes with her yard" (Smith 215). Indeed, property rights proponents believe that having a lawn is a right—a form of democracy (Colomina 149). Even proponents of natural landscapes sometimes take this view, assuming that there is a right to environmentally unfriendly landscapes and ignoring the strength of the police powers (Rappaport 927). One commentator suggests that "it would stretch our customary understanding of the appropriate role of regulation to attempt to mandate that an owner ... systematically remove invasive species" (Echeverria 14). Although the police power is broad, local governments are still quite deferential to property rights, and thus often fail to pass otherwise legally permissible ordinances that would support principles of sustainability and biodiversity.

However, "the right to use one's real property as desired, historically cherished as it is, was never conceived as absolute" (Smith 215). And, as previously discussed, local governments already regulate lawns in the United States via weed ordinances, frontyard garden bans, and setbacks. This effectively eliminates the libertarian argument because the government already interferes with individual lawn choice. Because individuals own property subject to the government's police power, governments have the ability to "redefine the content of property rights" (Serkin 1259). The key is striking the appropriate balance between regulatory control and honoring the autonomy interest in using one's property as one wishes.

The collective fear of imposing mandates—both from the perspective of academics and policymakers—might also be unfounded and less formidable than imagined. That this fear is overstated is evidenced by the fact that "sustainability mandates"[7] are becoming more common. Courts have long recognized that the exercise of the police power "must become wider, more varied, and frequent, with the progress of society," and local government trends toward sustainable policies are an example of that progress (Boston & Me. R.R. Co. v. Cnty. Comm'rs of York 114). For example, green building ordinances, which require private developers to construct their private development projects to meet certain levels of energy efficiency or sustainable design, are now quite common. Some communities restrict leaf blower usage on "ozone action" days; others limit whether and when people can wash their cars or water their lawns; and in some areas, residents are required to separate their recyclables from their trash. Some municipalities have placed restrictions on watering golf courses and at least one prohibits restrictive covenants that require turfgrass.[8] Even affirmative removal mandates are not unheard of. For example, many cities require homeowners to remove snow from the sidewalks in front of their homes (State v. McMahon). From a private law perspective, some intentional communities are beginning to incorporate sustainability mandates, including bans on gas–powered mowers, leaf-blowers, and industrial fertilizers, into their CC&Rs.

---

7   I use this term to mean mandates or bans, imposed via public law, that aim to require or curtail an action that can lead to a more sustainable environment, and perhaps reduce some negative impacts of climate change.

8   Dothan, Ala., Code § 102–166 prohibits watering of golf course fairways during a water emergency. Denver, Colo., Code § 57–100 prospectively prohibits restrictive covenants that require turfgrass.

Further, a lawn ban should suffer less risk of an intrusion objection than other mandates on individual behaviors because a lawn ban is primarily a *property* restriction, not a direct restriction on individual behaviors. Of course, a ban on lawns de facto regulates individual behavior, because individual action is inherently limited by property restrictions, but that is not a lawn ban's primary purpose. Further, while many environmentally significant individual behaviors are conducted in private inside the home, this is not so with respect to lawns, on which all activities are conducted outside the home. Thus, the intrusion concerns related to privacy and civil liberties should be less pronounced in the face of a property restriction that only indirectly limits publicly visible behaviors.

Finally, as the effects of climate change make themselves more evident to citizens and policymakers, the externalities associated with environmentally significant individual behaviors should emerge as a natural target of regulation. Regulation serves as a means of forcing internalization of externalities, which is often necessary in the context of environmental law. Indeed, "[m]ost economic theorists recognize that some level of environmental regulation is necessary because environmental problems frequently involve significant externalities, require solutions that carry high transaction costs, and concern threats to a public good, all factors that may contribute to market failures" (Circo 749). Thus, mandates should be less troubling in the context of a lawn ban than they would be if private individual behaviors were being directly targeted, and they may even be necessary to target and alleviate the harms associated with lawns specifically, and climate change more broadly.

III. The Anatomy of a Lawn Ban

In the event that lawn bans become more necessary and accepted, this section considers the appropriate scale of regulation, the sources of governmental power to enact such a ban, and ways that it might be implemented.

A. Regulatory Structure

1. scope: state versus local
Because climate, resource scarcity, and environmental priorities are so dependent upon location, a ban on lawns would not make sense for all states or municipalities in the country. Certainly, a federal ban on lawns would not be appropriate, at least not given the current disparate U.S. climate and water usage patterns. State regulation might be useful in certain states that face similar water usage patterns and demands across their jurisdictions. Perhaps those states could establish standards that would trigger targeted incentives or disincentives, and eventually, as water shortages became more acute, bans on lawns. However, local governments are likely more adept at addressing climate change in larger states with many different climates. Further, because the real power to effect change lies in the police power, which is delegated to local governments, they are the ones who are already taking action on lawns (Las Vegas). A benefit of regulating at the municipal level is that municipalities can be innovative and local ordinances can be specifically tailored to the needs, concerns, and geographically related harms of each individual community (Schindler, LEED 289–90). Thus, pragmatically, local regulation seems to make the most sense. However, there are also theoretical justifications for local action to alleviate lawn harms.

Because a lawn ban is a controversial proposal, the political capital necessary to pass such an ordinance may be lacking in many (or all) states. However, localities are different. First, in many major thought–leader cities in the United States, lawn norms are starting to change. There already exists a market–driven desire in these places for sustainable policies and efforts to reduce greenhouse gas emissions. Effective regulation at the local level can harness that market desire, and when these cities adopt cutting–edge policies, others tend to follow. Further, according to Charles Tiebout, different communities provide different services and benefits—and adopt different policies—to attract different types of residents (420). So if lawn bans are enacted at the local instead of the state level, those with a preference for lawns can (theoretically) move to a jurisdiction without a ban, thus allaying some concerns over property rights and free choice that might otherwise be associated with lawn bans.

The Matching Principle provides some support for the idea that certain local harms tied to climate change are ripe for local regulation (Butler and Macey 36). Specifically, the Principle posits that the regulating jurisdiction should not be larger than the regulated activity (Ibid. 25). At base, the costs of lawns go beyond each individual municipality. Water crosses jurisdictional boundaries. Its availability and the harms that lawns impose manifest themselves on a regional scale—the watershed for water use impacts and runoff, and the grid–scale for energy issues. While regional or watershed–level governance might be ideal in this context, the United States generally lacks strong regional structures. Thus, the next smallest unit of government, local governments, would most appropriately address these problems. For example, most lawns exist in the suburbs, and greenhouse gas emissions in the suburbs are usually higher than in central cities (Claeser and Kahn 7–9). This suggests that the policies implemented in suburbs with respect to lawns might differ from those implemented in cities.

2. power: home rule, police powers, and zoning enabling acts
Although to some the idea of a local government banning lawns might seem draconian, local governments in fact have many sources of authority to enact lawn bans. First, a land use ordinance generally only requires a rational basis to be upheld (Lilburn v. Sanchez 355). So long as the municipality had "fairly debatable" reasons for the enactment, the ordinance will stand (Village of Euclid 388). This is due in part to broad police powers: local governments have the power to act in furtherance of the public health, safety, welfare, and morals of the community. This power flows from the state's plenary regulatory authority, coupled with municipal home rule authority, which now exists in most states (Hunter v. City of Pittsburg 178). When a land use ordinance is enacted pursuant to the locality's police power, it is presumed to be valid (Serkin 1257–8). Police powers are broad and may change to encompass the times and the context (Village of Euclid 387). Thus, scholars recognize that these powers justify "development regulations intended to conserve natural resources and protect the environment," including regulations that "broadly seek to curb unsustainable land development, even when they impose significant burdens on the landowner" (Circo 745).

An additional source of local power flows from enabling legislation, which exists in all states, and expressly grants zoning powers to municipalities. Because every state has adopted a zoning enabling act, "the question of inherent power to zone is rarely litigated" (Kenneth Young 2.15). However, there is a question whether the power to regulate lawns would be considered within a locality's zoning power, especially if one is in a jurisdiction with a fairly specific zoning enabling act. Many states began the process of zoning by enacting the Standard State Zoning Enabling Act ("SZEA"), which was promulgated in 1922, but many have now adopted their own modified, state specific acts. Thus, the power to regulate lawns as a form of zoning would vary based on the specific language of the state statute. Historically, because most zoning laws did not mention yard vegetation, many communities adopted "special purpose controls"—the aforementioned weed ordinances—outside their normal zoning ordinances to regulate landscaping (Smith 217). Some zoning enabling acts, however, do specifically address these issues. For example, the Texas SZEA expressly refers to the "size of yards" as zoning that is covered by the act (§ 211.003(a)(3)). Thus, a locality seeking to ban lawns in Texas could likely rely upon its express zoning powers, instead of falling back on its broader home rule authority to do so. On the other hand, if a locality is situated in a Dillon's Rule state whose SZEA does not specifically delegate or mention the ability to ban or control lawns, the locality might not be able to do so.

A final municipal source of power to enact lawn bans could derive from a determination that lawns are a nuisance, or a public bad, due to their negative impact on the health and safety interests of the public. To facilitate this legal construction, a municipality could identify a lawn as a nuisance per se, for example by defining weeds or noxious vegetation in a nuisance vegetation ordinance to include lawns. By labeling lawns a nuisance per se, a local government could engage in a direct attack on the very existence of lawns, as many have done against funeral homes or houses of ill–repute in residential areas. The ban would seek to directly prevent the harm that lawns cause. For example, because turfgrass is often a non-native species, a locality could address it in the same way it addresses other exotic species. Because non-native flora and fauna have the ability to harmfully modify local ecosystems, local governments might take an aggressive position on removal and remediation to alleviate the threat. Indeed, local governments often regulate and eradicate invasive species, and do so pursuant to their police powers.[9]

The nuisance approach may be attractive to local governments because of the deference that courts afford to governments that act to protect their communities. Of course, many landowners will probably object to a ban on lawns, and are likely to assert that a newly enacted land use ordinance works a regulatory taking of their property. In *Lucas v. South Carolina Coastal Council*, the Supreme Court held that a state regulation depriving a landowner "of all economically beneficial use" of her land is a taking, unless the use prohibited by the regulation is already precluded by the principles of nuisance and property law (1027–9). On balance, a local government enacting a lawn ban has support against such a constitutional challenge. First, the idea of a "regulatory taking" is a recent creation. Previously, the Supreme Court held that government regulations that control nuisances are not subject to property protections afforded by the Fifth Amendment because the police power is broad enough to cover these situations (Hadacheck 410). This is because "[u]nder the police power, rights of property are impaired not because they become useful or necessary to the public, or because some public advantage can be gained by disregarding them, but because their free exercise is believed to be detrimental to public interests" (Freund 511). In addressing nuisances to protect the public, "the government can regulate away a hazardous or injurious activity without paying compensation" (Serkin 1240).

Even if a regulatory taking challenge were to proceed to the merits, the claim would be examined under the per se test of *Lucas* only if the landowner could show that the regulation deprives her of all economically viable use of her property (Lucas 1027). Because land

---

9   Palm Beach County, Fla., Unified Land Dev. Code, art. 14, ch. D, §§ 1–10 requires owners of property located near "natural areas" to remove invasive non-native species, and prohibiting their planting.

uses that involve a lawn are typically not dependent on the presence of the lawn itself, a landowner would likely have a difficult time establishing that being forced to remove a lawn was a deprivation of all use of the property, especially because the doctrine of conceptual severance suggests that a court must look at the parcel as a whole when considering what has been taken (Ibid. 1016). A lawn ban would result in the homeowner losing the value of the lawn, but retaining the value of the rest of the property. Only in the rarest of cases would a lawn ban be assessed under the *Lucas* jurisprudence. Instead, a court would likely apply the ad hoc analysis set forth by the Supreme Court in *Penn Central Transportation Co. v. New York City*, which considers the regulation's economic impact, its interference with reasonable investment–backed expectations, and the character of the governmental action. In practice, "landowners rarely win these cases," and a court applying these factors to a lawn ban would likely find in favor of the municipality because the extent of the loss the lawn owner incurs is not likely to be dramatic (Karkkainen 90).

The only remaining question is whether a court might consider a retroactive lawn ban that includes an affirmative requirement that the landowner replace the lawn with something deemed more environmentally friendly, like xeriscaping, to be a permanent physical occupation of property constituting a per se taking (Loretto 440). Although there has been little scholarship addressing the government's authority to require people to take action on their private property, the Supreme Court has recognized that a regulation may force action without being deemed an impermissible taking. In *Loretto v. Teleprompter Manhattan CATV Corp.*, the Court explained that a physical occupation of land by a third party "is qualitatively more severe than a regulation of the *use* of property, even a regulation that imposes affirmative duties on the owner" (436). Thus, while there might be a taking if the government requires a landlord to allow third parties to enter his land and install something thereupon, there would not likely be a taking if the landlord were required to install the thing himself (440 n.19). Accordingly, it would seem that, so long as a lawn ban provides a homeowner with multiple replacement alternatives, and allows the homeowner to install those alternatives himself, the ban would not run afoul of *Loretto*.

B. Crafting the Ban

Because lawns are not only ubiquitous in most American residential communities, but also legally permissible and often required, banning them would be what some commentators have referred to as a "[r]egulatory transition []"— a movement away from one legal regime to another (Doremus 45. 2003). These transitions "are inevitable over the long run, and often represent socially adaptive responses to changed circumstances or increased information. They are difficult to achieve, however, because substantial psychological and political barriers stand in the way" (Ibid.). Times of legal transition also bring about the risk of legal challenges; whenever a local government considers adopting a new land use ordinance, especially one that is controversial, uncommon, or provocative, it must consider its likelihood of being sued. Municipalities face a "problem of how to be fair to landowners who acquired property under one set of rules, only to see the uses of the property drastically limited as morals, technology, or scientific understanding change" (Ellickson and Been 140. 2005). Sometimes, the threat of a lawsuit is enough to discourage a local government from enacting forward–thinking legislation. This section addresses the different temporal circumstances under which a lawn ban could be imposed, and the legal challenges that might accompany or inform that timing decision. It then considers who or what a ban could cover and control.

1. timing of ordinance imposition

In considering *when* to impose a ban on lawns, a municipality has three options. The mandate: (1) could apply only to new construction, thus allowing the continuation of existing lawns but prohibiting new ones; (2) it could apply only when the property at issue is sold, rented, or modified in some way, requiring, prior to a conveyance of the property, that any existing lawn be torn up and replaced with another acceptable form of ground cover; or (3) it could be imposed retroactively, requiring that all existing lawns be torn up, perhaps pursuant to an amortization schedule. This section addresses the arguments for, and legal consequences of, each option.

The most straightforward and least controversial approach to a ban on lawns would be to prohibit any new construction (commercial, residential, or both) from including a turfgrass lawn. This is the approach taken in Las Vegas, which generally prohibits new turf installation in front yards and limits it in rear and side yards (§ 11.14.150). Applying a ban prospectively avoids some of the political concerns that accompany retroactive ordinances that force existing community residents and homeowners to take actions that they might not support or desire. Existing residents often prefer land use patterns to freeze once they have moved in (Fischel 146–7 2001).

There are some substantial problems with only applying a new rule prospectively. Specifically, because a ban on lawns would be put in place to reduce harms, and hopefully alleviate some climate change–related concerns, applying it only prospectively would curtail its potential benefits. Especially in a community that is already substantially developed, it is unlikely that only banning new lawns would have much of a cumulative impact. In contrast, for a still–developing area, beginning with a baseline of existing lawns and banning new ones could still result in a substantial decrease in water and fertilizer usage.

A second approach that a municipality could take would be to prospectively ban all new lawns, but also to require the removal of existing lawns at the time that the owner of property covered by the ordinance sells the property, rents to someone new, or seeks discretionary permits to alter the property in a way that is related to the lawn. The imposition of ordinances at a point of change is a technique already used in some situations. For example, under the Clean Air Act, if a preexisting stationary source is "subsequently modified," it will then need to comply with new source performance standards (Serkin 1226). Similarly, some suggest that the only way that states will succeed in achieving energy efficiency will be to require homeowners to take certain actions such as retrofitting their homes with energy efficient appliances and fixtures prior to a sale (Schindler, Energy Efficiency 20). Further, the highest court in the state of New York upheld the facial validity of a local ordinance that banned mobile homes from certain areas of the Village of Valatie upon a transfer of ownership of the land containing the mobile home or the mobile home itself (Village of Valatie v. Smith).

Conditioning the point of ban implementation on change in ownership instead of applying it immediately is a form of amortization. In the context of land use law, some localities build in an amortization period to newly adopted ordinances that would otherwise immediately force lawful preexisting nonconforming uses to come into compliance with the new ordinance. The idea behind amortization is that property owners should be given enough time to realize on their investments before being forced to comply with the new law. Generally, an ordinance faces a greater risk of being deemed a taking or violation of substantive due process if it is immediately applied to a nonconforming use. Some jurisdictions, however, deem any amortization period to be per se unreasonable (Zitter 419–22). Further, there is a real question as to whether a lawn would be considered a lawful, preexisting, nonconforming "use" for purposes of this analysis, and thus subject to greater protections. This inquiry is relevant because heightened protections are often afforded to existing uses of land (Serkin 1244). At least one commentator argues, however, that the additional protection of existing uses is unjustified (Ibid. 1242–3). Thus, one could assert that lawn bans applied at the time of a property change should not be viewed differently and should not be more likely to result in a taking or due process violation than those applied only prospectively.

Assuming that amortization periods are permissible in a state, this approach—applying a ban at the time of a change in status of the property—would be beneficial in communities that are already built–out, in that it would cover more property than a purely prospective ban. There is, of course, an enforcement concern associated with this approach. For example, homeowners might attempt to get around the ban by failing to report new leases. It is unlikely, however, that a homeowner would forego the protections of a recording statute and not record the sale or transfer of an interest in her property merely to avoid having to remove a lawn. Further, because permit applications are matters of public record submitted directly to a municipality, the municipality would be on notice of such changes, and thus able to enforce the lawn ban.

There is also an efficiency concern with a ban applied at the time of sale in that there might be an incentive for people to hold onto their property. If there is a close community in a given neighborhood, there might be pressure from neighbors urging others not to sell or rent because they want to maintain the existing uniform lawn aesthetic.

Finally, a municipality could adopt a retroactive lawn ban: a ban imposed at the time the ordinance is adopted, which requires all covered property owners to tear up their existing lawns and replace them with something else, or let them die. At this time, there does not appear to have been a legal challenge to any existing lawn ban. However, a municipality considering implementing a retroactive ban pursuant to the discussion above might expect to be sued by landowners unhappy with the requirement that they remove their existing lawns. Indeed, some of these individuals might assert that they bought their homes, in part, because of the specific landscaping that fronts it, and that the ban interferes with their vested rights. But pursuant to the analysis in the previous section, retroactive bans would also likely withstand such challenges.

That said, a retroactive approach is bold and would certainly be viewed with disfavor by many. Scholars have described retroactive laws as "anathema to liberty and a well-ordered society" and "a monstrosity" (Serkin 1262, Fuller 53). Others view them as unfair, believing that individuals must be able to rely on existing laws in structuring their actions and behaviors (Doremus 14, 2003.). Further, they are somewhat rare; for example, this author is unaware of a recent situation in which a city adopted a residential zoning ordinance and forced all lawful, preexisting, non–nuisance commercial uses in the zone to immediately cease operation (Young 628).[10]

Despite its unpopularity, and thus political unlikelihood, there are a number of benefits that would derive from a retroactive lawn ban (assuming the lawns are replaced with environmentally friendly alternatives). Importantly, this approach would most immediately and directly target the harms caused by lawns because it would encompass the greatest amount of property. In addition to the broad societal and ecosystem services benefits that would flow from lawn removal, removal would also lead to individual savings: lawnless

---

10   It is likely that such action would impinge upon vested rights and constitute a taking.

homes would use less water on average, and their proprietors might gain free time that would otherwise be spent caring for the lawn.

2. What would be covered

A municipality considering a ban on lawns should think carefully about how much lawn to ban. Specifically, will all turfgrass be outlawed—that surrounding both residences and commercial properties—or just front yards? If all turfgrass were banned, golf courses, athletic fields, corporate campuses, and public parks would also be impacted. Such a broad ban would target and alleviate lawn harms most thoroughly, assuming that all turf, regardless of location or use, contributes to the same harms. However, such an extensive ban would intrude on some commercial uses of the lawn itself, and thus could raise additional considerations in a takings analysis. Further, if the replacement material for turfgrass in play fields and parks did not allow for play, their utility might be decreased, leading to a decrease in social capital in the neighborhood.

Finally, a ban on all turf would cover not just publicly visible private property in front of the house, but the backyard as well. At base, this should not raise additional legal concerns because the primary justifications for bans pertain to water consumption and environmental harm rather than aesthetics. A ban on backyard lawns might raise additional privacy concerns, however, which could lead a court to apply heightened scrutiny to the ban. The home itself, often surrounded by a lawn, has received exceptional levels of protection under the law (Payton v. New York 601). One reason for this heightened protection might be that expressed by Peggy Radin, who views certain property, including the home, as part of a person's identity (992). Generally, laws anger people when the laws are seen as "infringing upon personal autonomy … by preventing the home from providing a space for unfettered thinking, reflection, and the development of personhood" (Kuh 1173). Thus, the closer a regulation is to the self or its extensions, the greater the level of resistance to it. Front yards are visible to all passing by and thus lack a portion of the privacy, or the expectation of privacy, associated with the home's interior. However, the same cannot be said for backyards. Thus, courts might be more willing to protect backyards for the same reasons they protect homes. Further, the intrusion objections against a ban on backyards would likely be stronger than if only front yards were banned.

In contrast, a locality could decide only to ban front yards in residential areas or in front of commercial buildings or offices. Such an approach would raise fewer concerns with respect to issues surrounding privacy, intrusions, and takings, but it would also result in a smaller total benefit. Finally, a locality could simply ban "the industrial lawn," regardless of its location or use (Tekle 215). Thus, any playfield or green open space that, while perhaps non-native, did not need to be watered, fertilized, or mowed, could remain.

C. Affirmative Lawn Removal and Replacement Requirements

One might wonder why a municipality would choose to ban lawns rather than the practices that contribute to lawn–related harms: watering, mowing, and fertilizing. If these practices were banned, lawns would surely die out in the regions where they are causing the greatest harm, and most of the aforementioned takings concerns would be avoided. The problem is that if a locality only bans those behaviors, it misses out on the opportunity and benefits that might come from a ban that not only requires lawn removal, but also requires its replacement with landscaping that is more beneficial.

Lawn bans that are part of a broader sustainability plan can further ambitious community designs—the goal is not just to eliminate environmental externalities associated with lawns, but also to change the landscape into something more sustainable. Further, if a locality seeks to alleviate lawn–related harms but does not control what can be installed in their stead, the harms that the mandate seeks to eliminate may not actually be avoided. For example, if a locality forbids lawn maintenance, but does not require lawn replacement, a homeowner whose lawn died due to lack of water and fertilizer could simply install Astroturf, which may be less environmentally sound than a lawn. Thus, a lawn mandate that does not address subsequent replacement will not necessarily result in a net environmental benefit.

Although affirmative requirements do raise autonomy issues, municipalities have historically regulated what individuals may plant on their property. For example, statutes in colonial Virginia prohibited people from overplanting tobacco and required them to grow other crops (Ellickson and Been 135. 2005). Similarly, lawn maintenance is not always a personal choice. Weed ordinances, which require people to mow their lawn and remove native plants, have been upheld despite the fact that some view them as "irrational" (Rappaport 918 n.165).

Likewise, many historic preservation ordinances require landowners to maintain the historic features of their property, or to install new ones, often at great expense.[11] Although the affirmative requirement to install a certain type of landscaping might require the expenditure of money, that alone would not make it an unconstitutional taking of property (Maher v. City of New Orleans 1067). There are political concerns associated with requiring individuals to spend money, however, and because many municipalities are facing severe budgetary shortages, it is unlikely that they could contribute funding for lawn replacements. Thus, municipalities should think creatively about how to require replacement of lawns with sustainable alternatives that would not cost their citizens a great deal of money.

Finally, because the removal of lawns might result in a decrease in social capital or in spaces where children can play, a locality requiring removal of lawns could commit to constructing additional public parks in their communities. These could serve the role of third places, and would be more inclusive than front yards, as they would be truly public open space, a commons instead of private property.

Conclusion

Although lawns inflict numerous harms on the communities in which they are located, most localities have not banned or even limited them. However, water shortage is quickly becoming one of the most dire problems facing much of the country, and the world. Limiting lawns, especially in parts of the country where water shortages or water pollution are most acute, is a direct way to reduce their harms while simultaneously providing an opportunity to improve biodiversity. Although the hurdles to implementing lawn bans are currently more political than legal, the changing climate might lead to changing attitudes.

Fifty years ago, if a city told a developer that her new homes had to be green buildings, which incorporate certain features to make them more sustainable and efficient than

---

11 For example, homeowners in Portland, Maine's historic district who seek to replace gutters that are "a significant and integral feature of the structure" may be required to use historic but expensive materials, such as wood (Portland, Me., Code §§ 14–634(a)(2)(a), 14–650(b), (e), (f)).

standard, more cheaply constructed homes, she would likely have been incredulous. Today, however, green building ordinances are quite common in many localities. In part, this legal change followed a shift in norms as growing builder interest in green development was evidenced through the use of voluntary market mechanisms like the Leadership in Energy and Environmental Design ("LEED") standards. In sum, this chapter suggests nothing more than an expansion of the notion that local governments appropriately can regulate the sustainability of the built environment: that principle simply needs to be taken beyond buildings and into the yard.

Although the idea of a local government ordering its citizens to tear up their lawns and replace them with xeriscaping seems far–fetched, norms are already moving in that direction in some communities. Thus, as with green building, it is possible that as voluntary actions become more common, and as droughts lengthen and water and energy become more expensive, local politicians will become less wary of the concept of a lawn ban. And as far as trade–offs go, "[t]he lawn is an easy sacrifice, compared to trees and shrubs—or taking a shower" (Smaus).

Works Cited

11 Las Vegas, Nev., Code of Ordinances Sec. 14.150. 2009. Print.

14 Las Vegas, Nev., Code of Ordinances. Sec.11.150(a). 2 December 2009. Print.

35 Dothan, Ala., Code. Sec. 102–166 Supp. 2013. Web. 17 Feb. 2015. <http://library. municode.com/index.aspx?clientId=10658>.

64 Scottsdale, Ariz., Rev. Code. Sec. 27–95, 49–247. Supp. 2013. Print.

211 Texas Local Government Code Ann. Sec. 003(a)(3). 2006. Print.

Annapolis, MD., Code. Sec. 10.20.010. 2011. Print.

Babie, Paul. "Climate Change: Government, Private Property, and Individual Action." *Sustainable Development Law & Policy* 11.2 (2011): 19. Print.

Biber, Eric. "Climate Change and Backlash." *New York University Environmental Law Journal* 17 (2008): 1321–2. Print.

Bittman, Mark. "Lawns Into Gardens." *New York Times Opinionator*. New York Times, 29 Jan. 2013. Web. 10 Feb. 2015. <http://opinionator.blogs.nytimes.com/2013/01/29/ lawns-into-gardens/?_r=1>.

Bormann, F. Herbert, Diana Balmori and Gordon T. Geballe. *Redesigning the American Lawn: A Search for Environmental Harmony*. 2nd ed. 2001. 9 Print.

Boston & Me. R.R. Co. v. Cnty. Comm'rs of York. 10 A. 113, 114. Maine Supreme Judicial Court. 1887. *Atlantic Reporter*. WestlawNext. Web. 10 May 2015.

Butler, Henry N. and Jonathan R. Macey. "Externalities and the Matching Principle: The Case for Reallocating Environmental Regulatory Authority." *Yale Law & Policy Review* 23 (1996): 36. Print.

Chandler. *Rebate Programs*. Chandler, Arizona. Web. 7 Jan. 2014. <http://www.chandleraz. gov/default.aspx?pageid=746>.

Circo, Carl J. "Using Mandates and Incentives to Promote Sustainable Construction and Green Building Projects in the Private Sector: A Call for More State Land Use Policy Initiatives." *Penn State Law Review* 112 (2008): 749. Print.

City of Lilburn v. Sanchez. 491 S.E.2d 353, 355. Supreme Court of Georgia. 1997. *South Eastern Reporter, 2d Series*. WestlawNext. Web. 10 May 2015.

Cohen, Ronnie, Barry Nelson and Gary Wolff. "Energy Down the Drain: The Hidden Costs of California's Water Supply." *Natural Resource Defense Council.* 2014. Web. 15 Feb. 2015. <http://www.nrdc.org/water/conservation/edrain/edrain.pdf>.

Colomina, Beatriz. "The Lawn at War: 1941–1961." *The American Lawn.* Georges Teysott ed. New York City: Princeton Architectural, 1999. 135–49. Print.

Corrosion Proof Fittings v. EPA, 947 F.2d 1201, 1215–16. United States Court of Appeals, Fifth Circuit. 1991. *Federal Reporter, 2d Series.* WestlawNext. Web. 10 May 2015.

Czarnezki, Jason J. *Everyday Environmentalism: Law, Nature & Individual Behavior.* Washington, D.C.: ELI, Environmental Law Institute, 2011. Print.

Demsetz, Harold. "Toward a Theory of Property Rights." *American Economic Review* 57.2 (1967): 348. Print.

Denver, Colo., Code. Sec. 57–100. 2002. Print.

Doremus, Holly. "Climate Change and the Evolution of Property Rights." *University of California Irvine Law Review* 1 (2011): 1104–5, 1115. Print.

Doremus, Holly. "Takings and Transitions." *Journal of Land Use & Environmental Law* 19.1 (2003): 11, 45. Print.

Duvall, J.B. *Maryland Turfgrass Survey: An Economic Value Study.* Maryland Department of Agriculture, 1987. Print.

Echeverria, John D. "Regulating Versus Paying Land Owners to Protect the Environment." *Journal of Land Resources & Environmental Law* 26.1 (2005): 14. Print.

Ellickson, Robert C. *Order without Law: How Neighbors Settle Disputes.* Cambridge, M.A.: Harvard University Press, 1991. Print.

Ellickson, Robert C. and Vicki L. Been. *Land Use Controls: Cases and Materials.* 3rd ed. New York, N.Y.: Aspen, 2005. Print.

Epstein, Richard A. "How to Create—or Destroy—Wealth in Real Property." *Alabama Law Review* 58.4 (2007): 762. Print.

Farr, Douglas. *Sustainable Urbanism: Urban Design with Nature.* Hoboken, N.J.: Wiley, 2008. 147. Print.

Fischel, William A. "Why Are There NIMBYs?" *Land Economics* 77.1 (2001): 146–7. Print.

Fischel, William A. *The Homevoter Hypothesis: How Home Values Influence Local Government Taxation, School Finance, and Land-use Policies.* Cambridge, MA: Harvard University Press, 2001. Print.

Fischer Kuh, Katrina. "When Government Intrudes: Regulating Individual Behaviors that Harm the Environment." *Duke Law Journal* 61.6 (2012): 1127–8, 1132–3, 1148. Print.

Fishman, Charles. *The Big Thirst: The Secret Life and Turbulent Future of Water.* New York: Free Press, 2011. Print.

Freund, Ernst. *The Police Power, Public Policy and Constitutional Rights.* Chicago: Callaghan, 1904. 511. Print.

Fuller, Lon Luvois. *The Morality of Law.* Rev. ed. New Haven: Yale University Press, 1969. Print.

Garden Grove, Cal., Code Sec. 14.40.025. 2012. Print.

Gibeault, V.A., S. Cocker-ham, J.M. Henry, and J. Meyer. "California Turfgrass: It's Use, Water Requirement and Irrigation." *California Turfgrass Culture*, 39.3–4 (1989): 1–14. Print.

Glaeser, Edward L. and Matthew Kahn. "The Greenness of Cities 7–9." *Harvard University School of Government.* March 2008. Web. 15 Feb. 2015. <http://www.hks.harvard.edu/var/ezp_site/storage/fckeditor/file/pdfs/centers-programs/centers/taubman/policybriefs/greencities_final.pdf>.

Hadacheck v. Sebastian. 239 U.S. 394, 410. Supreme Court of the United States. 1915. *Supreme Court Reporter*. WestlawNext. Web. 10 May 2015.

Hunter v. City of Pittsburgh. 207 U.S. 161, 178. Supreme Court of the United States. 1907. *Supreme Court Reporter*. WestlawNext. Web. 10 May 2015.

Jervey, Ben. "The Waterless City." *GOOD*. 3 May 2011. Web. <www.good.is/post/the-waterless-city>.

Jouvenal, Justin. "The Spat That Laid Low Olde Belhaven." *Washington Post*, 10 February 2013: A1. Print.

Karkkainen, Bradley C. "Biodiversity and Land." *Cornell Law Review* 83.1 (1997): 90. Print.

Kolbert, Elizabeth. "Turf War." *New Yorker*, 21 July 2008. Print.

Kurutz, Steven. "The Battlefront in the Front Yard." *New York Times*, 20 December 2012: D1. Print.

Lessig, Lawrence. "The New Chicago School." *The Journal Legal Studies* 27.S2 (1998): 662. Print.

Lindsey, Rebecca. "Looking for Lawns." *Feature Articles*. NASA Earth Observatory. 8 November 2005. Web. 15 Feb. 2015. <http://earthobservatory.nasa.gov/Features/Lawn/>.

Locke, John, and J. W. Gough. "An Essay Concerning the True Original, Extent and End of Civil Government." 1948. *The Second Treatise of Civil Government and A Letter concerning Toleration*. Oxford: B. Blackwell, 1690. 15. Print.

Loretto v. Teleprompter Manhattan CATV Corp. 458 U.S. 419, 440. Supreme Court of the United States. 1982. *Supreme Court Reporter*. WestlawNext. Web. 10 May 2015.

Lucas v. S.C. Coastal Council. 505 U.S. 1003. United States Supreme Court. 1992. *Supreme Court Reporter*. WestlawNext. Web. 10 May 2015.

Maher v. City of New Orleans. 516 F.2d 1051, 1067. United States Court of Appeals, Fifth Circuit. 1975. *Federal Reporter, 2d Series*. WestlawNext. Web. 10 May 2015.

McAdams, Richard H. "The Origin, Development, and Regulation of Norms." *Michigan Law Review* 96.2 (1997): 359. Print.

Michelman, Frank I. "Political Markets and Community Self-Determination: Competing Judicial Models of Local Government Legitimacy." *Indiana Law Journal* 53.2 (1977): 148. Print.

Milesi, Cristina, Steven W. Running, Christopher D. Elvidge, John B. Dietz, Benjamin T. Tuttle, and Ramakrishna R. Nemani. "Mapping and Modeling the Biogeochemical Cycling of Turf Grasses in the United States." *Environmental Management* 36.3 (2005): 426–38. Print.

O'Connor, Hollie. "Saving the Prairie, and Planting Some New Ones." *New York Times*, 18 August 2012: A23. Print.

Organization for Economic Cooperation and Development. "Climate Change." *OECD Environmental Outlook to 2050* (2012). Print.

Palm Beach County, Fla., Unified Land Dev. Code, art. 14, ch. D. Sec. 1–10. Supp. 15. 2013. Web. 17 Feb. 2015. <http://www.pbcgov.com/pzb/ePZB/uldc.pdf>.

Payton v. New York. 445 U.S. 573, 601. Supreme Court of the United States. 1980. *Supreme Court Reporter*. WestlawNext. Web. 10 May 2015.

Penn Cent. Transp. Co. v. New York City. 438 U.S. 104. Supreme Court of the United States. 1978. *Supreme Court Reporter*. WestlawNext. Web. 10 May 2015.

Portland, Me., Code. Sec. 14–634(a)(2)(a), 14–650(b), (e), (f). 2013. Web. 17 Feb. 2015. <http://www.portlandmaine.gov/citycode.htm>.

Pollan, Michael. "Why Mow?" *N. Y. Times Mag.* 28 May 1989: 22–23. Print.

Posner, Eric A. "Law, Economics, and Inefficient Norms." *University of Pennsylvania Law Review* 144.5 (1996): 1712–13. Print.

Radin, Margaret Jane. "Property and Personhood." *Stanford Law Review* 34.5 (1982): 992. Print.

Rappaport, Bret. "As Natural Landscaping Takes Root We Must Weed out the Bad Laws—How Natural Landscaping and Leopold's Land Ethic Collide with Unenlightened Weed Laws and What Must Be Done About It." *John Marshall Law Review* 26.4 (1993): 865–898. Print.

Richmond, Henry R. "Sprawl and its Enemies: Why the Enemies are Losing." *Connecticut Law Review* 34 (2002): 539–48. Print.

Schindler, Sarah B. "Of Backyard Chickens and Front Yard Gardens: The Conflict Between Local Governments and Locavores." *Tulane Law Review* 87.2 (2012): 289. Print.

Schindler, Sarah. "Following Industry's LEED." *Florida Law Review* 62 (2009): 285–350. Print.

Schindler, Sarah. *Encouraging Private Investment in Energy Efficiency.* January 2011. TS. University of Connecticut School of Law, Center for Energy and Environmental Law Policy. Web. <http://papers.ssrn.com/sol3/papers.cfm?abstract_id=1805891>.

Serkin, Christopher. "Existing Uses and the Limits of Land Use Regulations." *New York University Law Review* 84 (2009): 1275. Print.

Smith, James C. "The Law of Yards," *Ecology Law Quarterly* 33 (2006): 203, 216–18. Print.

State v. McMahon, 55 A. 591, 593. Conn. 1903. Supreme Court of Errors of Connecticut. 1903. *Atlantic Reporter.* WestlawNext. Web. 10 May 2015.

Steinberg, Theodore. *American Green: The Obsessive Quest for the Perfect Lawn.* New York: W.W. Norton, 2006. Print.

Sunstein, Cass R. "Social Norms and Social Roles," *Columbia Law Review.* 96.4 (1996): 903–929. Print.

Tiebout, Charles M. "A Pure Theory of Local Expenditures." *Journal of Political Economy* 64.5 (1956): 420. Print.

Tucson, Ariz., Land Use Code. ch. 23, Sec. 3.7.2.2 .1995. Print.

U.S. Environmental Protection Agency. *Reduce Your Outdoor Water Use*, U.S. Environmental Protection Agency. May 2013. Web. 15 Feb. 2015. <http://www.epa.gov/WaterSense/docs/factsheet_outdoor_water_use_508.pdf>.

United States Geological Survey. *Pesticides in Stream Sediment and Aquatic Biota.* USGS Fact Sheet No. 092–00. U.S. Geological Survey, n.d. Web. 15 Feb. 2015. <http://water.usgs.gov/nawqa/pnsp/pubs/fs09200/fs09200.pdf>.

Vandenbergh, Michael P. "From Smokestack to SUV: The Individual as Regulated Entity in the New Era of Environmental Law." *Vanderbilt Law Review* 57.2 (2004): 515. Print.

Village of Belle Terre v. Boraas. 416 U.S. 1, 9. United States Court of Appeals for the Second Circuit. 1974. *Supreme Court Reporter.* WestlawNext. Web. 10 May 2015.

Village of Euclid v. Ambler Realty Co., 272 U.S. 365, 388. Supreme Court of the United States. 1926.

Village of Valatie v. Smith, 632 N.E.2d 1264, 1265. Court of Appeals of State of New York. 1994.

Wells, Nancy M. and Gary W. Evans, "Nearby Nature: A Buffer of Life Stress Among Rural Children," *Environment & Behavior.* 35.3 (2003): 311–321. Print.

Williams, Frank B. *The Law of City Planning and Zoning.* New York: Macmillan, 1922. 177–79. Print.

Young, J.S. "City Planning and Restrictions on the Use of Property." *Minnesota Law Review.* 9.7 (1925): 593, 628. Print.

Young, Kenneth H. *Anderson's American Law of Zoning.* Vol. 1. Deerfield, IL: Clark Boardman Callaghan, 1996. 2.15. Print.

Zitter, Jay M. "Validity of Provisions for Amortization of Nonconforming Uses." *American Law Reports.* 5th ed. Vol. 8. New York, NY: Thomson Reuters, 1992. 391, 419–22 § 3[b]. Print.

Chapter 7
Think Global, Pay Local

Ray Brescia

There is hardly an economic policy issue that has been studied more over the last 20 years than the impact of raising the minimum wage on a host of issues, for example its effect on the purchasing power of the working poor, whether it increases unemployment, and what type of effect it may have on economic growth. One aspect of the way that the minimum wage law has developed in the United States that has received somewhat less attention is the extent to which the overall minimum wage, that which is set by federal law, has evolved over time, through experimentation at the state and even the local level. This chapter explores the role that local governments can play in setting the local minimum wage. It identifies the questions that advocates must ask if they wish to press for local minimum wage increases in their cities and localities, and encourages aggressive local experimentation on the minimum wage question as a strategy for promoting a higher minimum wage at the national level and for reducing income inequality in the United States.

Before I begin to discuss the role of local experimentation in the minimum wage, some background on the way the minimum wage works in the US might be useful. In the US, the federal government sets a floor that is supposed to serve as the absolute minimum hourly wage employers can pay most workers. I say "most workers" because there are certain exemptions to the minimum wage requirements of the Fair Labor Standards Act, the law that sets the federal minimum wage. For example, employees who receive tips, like wait staff in many restaurants, can receive a wage that is lower than the federal minimum wage.

At the same time, while federal law sets a minimum hourly wage, a state government can choose to require employers within that state to pay an amount that is higher than the federal minimum wage, and many state legislatures have adopted laws that do just that. A more recent phenomenon that has taken hold over the last few years is that local governments are also looking for ways to improve the wages of their residents and those who work within city limits and have begun to explore passing local minimum wage ordinances. These laws work a lot like federal and even state requirements, but just apply to employers within the jurisdiction of that local government. Offering further variation, some cities have passed local ordinances that apply only to companies that have contracts to undertake work with or for the city, meaning that if a company has a contract with the local government to provide a good or service, the employees working under that contract must receive pay that meets the threshold set by the city.

Since the explicit goal of many of these local minimum wage laws is to improve the economic prospects of the working poor through a certain degree of economic redistribution—from employers and customers to employees—it is one of the tools available to local governments to address economic inequality. As such, it is one of the ways that cities, by addressing income inequality, can "save the world." It is thus an appropriate topic for consideration in this collection.

As an assessment of the value and viability of local minimum wage laws for addressing economic growth and income inequality, and the strategies necessary to achieve these goals, this chapter proceeds as follows. First, I will discuss the history of minimum wage laws and show that local experimentation on the subject has a long history in the US.

Second, I will discuss the state of the growing body of economic research on the impact of minimum wage laws on economic growth, wages overall, and employment. Because of this history of minimum wage laws, where states, and some localities, have experimented with raising the minimum wage over the last few decades, these experiments have served as societal petri dishes where one can not only observe the impact in one community of raising the minimum wage, but one also has "control" communities: neighboring states and localities that have not seen an increase in the minimum wage yet share similar demographic, industrial, economic, political, and historical profiles. Because some states and localities have raised the minimum wage, and, at the same time, those areas have neighboring communities that have not done so, one can compare the effects that the increase in minimum wage, and the lack of an increase in the minimum wage, have on those communities that are, but for the increase in the minimum wage, otherwise similarly situated. These experiments help bring much needed and critical information to bear on the debate around the value and impact of increases in the minimum wage.

Third, after a discussion of the research on the economic impacts of raising the minimum wage, this chapter will provide an overview of the different local governments that have taken it upon themselves to increase the minimum wage in their communities. This section will discuss the state of play in terms of local minimum wage ordinances: where have they succeeded and where have they failed. It will focus on one community in particular, Long Beach, CA, to explore some of the tactics used there to increase the minimum wage in that California city for certain service industry workers. Long Beach is a city that has transformed itself from a busy working port with a thriving manufacturing base to a tourist destination and has gone from a blue collar to a service economy town in just a few decades. A review of how proponents of an increase in the local minimum wage there went about achieving that for hotel workers helps provide insights into how local economies that have seen similar economic change can navigate the shoals of minimum wage advocacy.

Finally, I will conclude with a legal analysis: that is, a discussion of the legal requirements for local governments to adopt local minimum wage ordinances. While some cities have already passed local minimum wage ordinances, and many more are considering them, there are potential legal barriers that stand in the way of some cities should they wish to pursue raising the minimum wage for employees working within city limits. These barriers reside in the relationship between each city and its respective state government. The constitutional infrastructure of many states authorize local governments to take such actions as raising the local minimum wage. In some states, however, local governments are limited in the extent to which they can take action on a local minimum wage without express state government authorization permitting them to do so. For these cities, to adopt a local minimum wage law will require state legislative action authorizing these cities to raise their own minimum wage. Finally, state action on the minimum wage in some other states will displace local government power to take action in this area through the doctrine known as preemption.

This section will review the different ways the legal landscape may differ across states, identifying issues advocates must address when assessing the authority of local governments to take action on the minimum wage. This section will also address the potential legislative backlash some communities may face should they take action to raise the minimum wage by drawing on a recent example of state legislatures across the country stepping in to curb local government action in the area of handgun control and regulation.

At the federal level, where political inaction and intransigence has meant the federal minimum wage level has stagnated and even declined, on average, in constant dollars for decades, state legislatures may also suffer from the same political paralysis. State legislatures, where they must act to authorize local government action in terms of the minimum wage, may be unwilling or unable to adopt such authorizing legislation. While local legislative bodies may wish to pass legislation that increases employee wages, state lawmakers may present insurmountable barriers by barring local action. This chapter addresses these and a host of other issues that arise when local governments attempt to raise the minimum wage, at least where they might have the authority: i.e., at the local level. As it turns out, however, the question of whether those cities have the authority is a thorny legal issue, one which I will explore in greater detail here. Nevertheless, as the history of minimum wage advocacy reveals, aggressive experimentation at different levels of government sometimes creates the political space for broader action on the subject. It is to these issues that I now turn.

The Historic Roots of Minimum Wage Experimentation

The idea that the political reality in Washington, DC, may stand as a barrier to national minimum wage advocacy in the US is not new. In fact, it is as old as minimum wage advocacy itself. But, as the story of the decades-long fight for a national minimum wage reveals, local experimentation that pushes the outer boundaries of what may seem possible politically, and even pushes beyond those boundaries, was essential to the adoption of a federal minimum wage standard. It suggests that local experimentation today might blaze a similar path forward, despite what might appear as unwavering political opposition.

During the Progressive Era, as the general public became aware of oppressive conditions in the factories of the day and the pay received for the work there, particularly as both related to women and children, states began to experiment with setting a minimum hourly wage. The first of these, passed in Massachusetts, gave a state wage board the authority to set a minimum wage that employers were encouraged to pay women and children employees, but it was not mandatory. By 1920, 13 states had passed their own minimum wage laws: some that empowered wage boards, like in Massachusetts, to adopt voluntary wage guidelines; others that set wages based on the perceived ability of employers to pay them; and a third type that simply passed across-the-board minimum wage requirements (Quigley 517–18).

While this experimentation occurred at the state level, the ability of governmental bodies to pass minimum wage ordinances would face a legal challenge that went all the way to the US Supreme Court, but it was not these state level ordinances that were challenged, at least not yet. What employers challenged was Congress's creation of a wage board to consider setting *local* minimum wages. In the case *Adkins v. Children's Hospital*, employers challenged the authority of Congress to establish a wage board that would set minimum wages in the District of Columbia for women and minors. The statute authorized this board to "ascertain and declare … [s]tandards of minimum wages for women in any occupation within the District of Columbia, and what wages are inadequate to supply the necessary cost of living to any such women workers to maintain them in good health and to protect their morals." With respect to minors, the board would assess "standards of minimum wages for minors in any occupation within the District of Columbia, and what wages are unreasonably low for any such minor workers." In 1923, the Supreme Court struck down this legislation, holding that the so-called freedom to contract, upheld consistently in prior instances by the

Supreme Court, included the private ability to set wages between employer and employee, and this ability was beyond the reach of governmental regulation.

The Court also went on to question the ability of a wage board to assess what would, in fact, serve as a minimum wage that could "maintain" working women "in good health" and "morals." "What is sufficient to supply the necessary cost of living for a woman worker and maintain her in good health and protect her morals is obviously not a precise or unvarying sum—not even approximately so," the Court's majority would write. Furthermore, "[t]he amount will depend upon a variety of circumstances: The individual temperament, habits of thrift, care, ability to buy necessaries intelligently, and whether the woman lives alone or with her family. To those who practice economy, a given sum will afford comfort, while to those or [sic] contrary habit the same sum will be wholly inadequate." The Court also went on to find that "[t]he relation between earnings and morals is not capable of standardization" and that "[i]t cannot be shown that well-paid women safeguard their morals more carefully than those who are poorly paid." The Court's skepticism of the ability of a governmental body to assess, in the aggregate, some of these issues was what ultimately doomed this legislation. As the Court found "the inquiry in respect of the necessary cost of living and of the income necessary to preserve health and morals presents an individual and not a composite question, and must be answered for each individual considered by herself and not by a general formula prescribed by a statutory bureau."

Despite this ruling with respect to Congress's authority on the minimum wage question, states still pressed on with minimum wage legislation. The Great Depression, with mass unemployment, made minimum wage ordinances critical to sustain worker wages. By 1938, 25 states had some form of minimum wage ordinance, even though the Supreme Court, in 1936, held New York's minimum wage statute unconstitutional. The Court did this after having held, just one year earlier, that federal legislation that enabled the Roosevelt Administration to set industry-wide minimum wages was also unconstitutional. These decisions, including *Morehead v. New York ex rel. Tipaldo* that addressed New York's minimum wage, together with *Adkins*, mentioned previously, were just some of a series of decisions of the so-called *Lochner* court, named for an earlier decision of the Supreme Court that recognized freedom to contract as a way to invalidate first Progressive Era legislation, and then the efforts of President Roosevelt's New Deal. The decision in *Morehead*, and a case that followed soon thereafter, played an outsized role securing the ability of both the federal government, and state legislatures, to pass economic legislation, including minimum wage laws.

In 1937, just 10 months after reaching its decision in *Morehead* invalidating New York's minimum wage law, the Supreme Court revisited the issue, ruling on Washington State's effort to raise the minimum wage in that state. This time, however, in *West Coast Hotel v. Parrish*, the Court upheld the law, ultimately reversing prior decisions, placing little weight on the notion of freedom of contract, and paving the way for Roosevelt's New Deal legislation to stand constitutional muster. The Court in *West Coast Hotel* found as follows with respect to the freedom of contract:

> The Constitution does not speak of freedom of contract. It speaks of liberty and prohibits the deprivation of liberty without due process of law. In prohibiting that deprivation, the Constitution does not recognize an absolute and uncontrollable liberty. Liberty in each of its phases has its history and connotation. But the liberty safeguarded is liberty in a social organization which requires the protection of law against the evils which menace the health, safety, morals, and welfare of the people. Liberty under the Constitution is thus

necessarily subject to the restraints of due process, and regulation which is reasonable
in relation to its subject and is adopted in the interests of the community is due process.

While this decision signaled a significant shift in the Court's view of minimum wage
laws, one has to consider the larger political context in which this shift occurred. Just two
months' before the Court issued its decision in *West Coast Hotel*, President Roosevelt,
frustrated by the Court's stance towards his New Deal legislation, introduced what came to
be known as his "court packing" plan which was designed to increase the number of justices
on the Supreme Court from nine to fifteen. Associate Justice Owen Roberts's decision to, in
effect, switch sides, and rule in the majority in *West Coast Hotel*, ultimately paved the way
for other New Deal legislation to survive constitutional challenge. While Roosevelt's court
packing plan died in Congress, the main impetus for it, the seeming intransigence of the
Supreme Court with respect to his legislative agenda, also seemed to fade as well. Once the
Court backed state-based minimum wage laws, Congress passed the Fair Labor Standards
Act, described earlier, which set the federal minimum wage, and opened the door to further
state and local government experimentation in the minimum wage field.

The likely impact of minimum wage experimentation on state and even the federal
minimum wage, despite seeming political intransigence and judicial skepticism, cannot be
underestimated. It is to this issue that I will return later in this chapter.

The Impact of Increases in the Minimum Wage: What the Research Says

Before getting into the legal issues surrounding local experimentation with respect to
minimum wage levels, it is critical to ask whether increasing the minimum wage has a
net positive impact on the incomes of low-wage workers. If a minimum wage increase
results in the wages for some employees improving in the aggregate, but forces employers
to reduce their workforce, to close down, or to leave a jurisdiction, any of which would
result in worker unemployment, and, by extension, reduced wages for some workers, it is
hard to say that an increase in the minimum wage is a net positive, or that it would have a
significant impact on economic inequality. Before embarking on an effort to raise a local
minimum wage, city officials would clearly want to know, as best as they could in advance,
whether taking such action will end up helping low-wage workers and not hurting them.
As it turns out, this question—what are the economic impacts of raising the minimum
wage—is one that has been studied extensively over the last two decades. This body of
research gives elected officials and advocates a relatively clear sense of the likely impacts
of raising the minimum wage. It is to this research that I now turn.

One of the first studies to look at the impact of increasing the minimum wage on
employment was commissioned by the US Congress when it created the Minimum Wage
Study Commission in the 1970s. One of the tasks of the commission was to analyze
what impacts, if any, would come about as a result of creating automatic increases in the
minimum wage to correspond with inflation. The key findings of that study were that for
almost all age groups, there would be no negative impact on employment by increasing the
minimum wage (Minimum Wage Study Commission, 1981). According to this research, the
only age group that would experience a slight reduction in employment—one of just 1.5
percent—was teens (Brown, Gilroy and Kohen, 1994).

As mentioned, the federal minimum wage sets a floor for wages, and states, and even
localities in select states, can set their own minimum wage law that is higher than the

federal, or even a state, minimum wage. Because of this phenomenon, this state-based and local experimentation enables social scientists and economists to measure how raising the minimum wage in one state or community might have a disparate effect on the economic performance of neighboring states and communities that have not increased their minimum wage. This phenomenon has permitted researchers to assess the impacts of raising the minimum wage and develop side-by-side comparisons of the economic life of low-wage workers and the broader economy in those areas. Mostly what these studies find is that the harm to employment in these communities from an increase in the minimum wage is minimal, and the benefits, to employees and employers alike, are considerable.

One of the first of these side-by-side studies was one conducted by David Card and Alan Krueger in the early 1990s. This study compared the impacts of raising the minimum wage on fast food workers in New Jersey, where the legislature had raised the minimum wage, and neighboring Pennsylvania, which had not done so. The first of its kind, and seminal in the field, this study revealed that the increase in the minimum wage did not decrease employment in the fast food industry when compared to employment numbers in that same industry in Pennsylvania (Card and Krueger, 1994).

Since that 1994 study, it is hard to find an issue in the social sciences that has been studied more than the effects of minimum wage increases. There have been so many studies that researchers have been able to conduct meta-studies, which attempt to cull the most consistent findings from previous research. In fact, there have now been enough meta-studies that someone has even conducted a meta-analysis of the meta-studies themselves. John Schmitt of the Center for Economic and Policy Research conducted such research by compiling decades of research findings to reach the following conclusion: there is, at most, a negligible negative impact on employment as a result of increasing the minimum wage. While this impact on employment is negligible, there are other benefits to raising the minimum wage other than simply increasing the earnings of, and, in turn, expenditures by, low-wage workers. While Schmitt found that employers may reduce non-wage benefits, like health subsidies, other economic benefits accrue when there is an increase in the minimum wage: for example, worker productivity increases, there is less worker turnover, and employers have to expend fewer resources finding and training new employees (Schmitt, 2013).

While an increase in the minimum wage does not appear to have a significant negative impact on employment, what are its prospects for decreasing economic inequality? While that may be a difficult question to answer, we know one thing: the failure over the last 40 years on the part of the federal government to increase the minimum wage to keep pace with inflation has corresponded to a period of increased economic inequality in the United States. Through the 1970s, a relatively higher minimum wage than that which exists today helped elevate earnings for the lower end of the economic spectrum (Autor, Manning, and Smith, 2010). Using 2013 dollars, the minimum wage in 1968 stood at $10.69 an hour. By 1979, it dropped to $9.67. By 1990, it had fallen 40 percent, to $6.84 in constant dollars (Elwell, 2014).

Analyzing the connection between raising the minimum wage and reducing poverty, Arin Dube of the University of Massachusetts at Amherst found that just a 10 percent increase in the federal minimum wage would reduce the poverty rate by 2.4 percent. In 2012, economists at the Economic Policy Institute estimated that a phased-in increase in the federal minimum wage to $9.80 an hour would help 28 million workers. These workers would receive nearly $40 billion in additional income. Most of these workers would be in households earning less than $60,000/year (Hall and Cooper, 2012).

How does the minimum wage relate to economic inequality? As stated above, the last four decades have seen a dramatic erosion of the federal minimum wage in terms of constant dollars. This erosion in constant value over the last 40 years corresponds with an increase in economic inequality in the US over that same period. One analysis of the impact of the decline in the minimum wage from 1979 to 1988 concluded that this decline accounted for roughly 25 percent of the income inequality among men and 30 percent of the income inequality among women (DiNardo, Fortin, and Limieux, 1996). Increasing the federal minimum wage would, without doubt, help to reduce economic inequality. Is there the political will to accomplish this task, and at what level of government is it possible to take such action? An increase in the federal minimum wage would certainly be the easiest way to put more money in the hands of low-wage workers across all sectors and from all geographic areas. Is such an effort possible?

Public opinion is strongly on the side of raising the minimum wage. The American public, on the right and the left, generally believes the federal minimum wage should be higher. Yet, political gridlock in Washington, DC, means that an increase in the federal minimum wage does not appear possible at this time. But, as stated earlier, the federal minimum wage merely sets a floor, and not a ceiling, on the minimum wage level. Even though an effort to raise the minimum wage at the federal level would be the simplest and most complete way to accomplish this increase, states are always free to set their own minimum wage laws, and the laws in many states permit local minimum wage experimentation as well. But the question of the ability of cities and other local governments to pass their own minimum wage laws is a complicated one.

Sometimes the state passes legislation that enables minimum wage experimentation. Sometimes state legislative silence on the issue means local governments can step in and pass their own minimum wage legislation. At other times, local voters utilize ballot referenda to make such changes. There are even regulatory fixes to the minimum wage in some states that do not require legislative approval. What this means is that there are many tactics that advocates can deploy to raise the minimum wage, whether it is at the state or local levels, or in just certain industries.

Even where state legislatures may face some of the same gridlock that plagues Washington, advocates have other tools at their disposal to either circumvent state legislatures on the state level, or advocate locally, either through legislative initiatives, ballot initiatives, or other means. An overview of the different tactics advocates in states and many cities have used to promote higher minimum wage levels follows in the next section.

Local Success Stories

Many states, and some cities, have pursued a range of approaches designed to improve the wages of those on the bottom end of the pay scale. With political gridlock in the federal government, despite polls that show overwhelming popular support for raising the federal minimum wage, state legislatures and local governments have proven fertile fields for increases in minimum wage levels. And there are a number of tools available to states and cities to promote such increases.

Some states, like New York, have passed minimum wage laws through the state legislature that are higher than the level set by the federal government. All in all, 29 states and the District of Columbia have set their minimum wage level above that required by federal law (National Conference of State Legislatures). Cities have also gotten into the act. Seattle, WA, recently raised its minimum wage for all workers. It has begun to phase in slowly, with the current wage level set at $11 an hour. By the end of the decade, however, it will increase to $15.00 an hour for almost all workers in the city (there is a longer rollout for workers in small businesses). Presently, Oakland, CA's minimum wage is the highest in the nation: $12.25 an hour, which was recently matched by San Francisco, CA, which will also increase to $15 an hour before the decade is out.

Other cities have attempted to promote an increase in wages for workers employed by companies doing business with the city, and these increases are often higher than most minimum wage levels, which is why the legislation promoting such higher wages are often called "living wage" ordinances. Cities as diverse as Norwalk, CT; Lincoln, NE; and Bloomington, IN, have all passed living wage ordinances, which generally seek to insure that an individual working a full-time job covered by the ordinance will meet or exceed the federal poverty level. In other words, these ordinances attempt to ensure that a full-time worker and his or her family does not have to live below the poverty line.

In addition to passing legislation that raises the minimum wage in a state, there are other options available for advocates and elected officials alike to find ways to increase wages for low-wage workers. In some states, the minimum wage has been raised by ballot initiative. In November 2014, for example, four so-called "red states" (i.e., those that are dominated by a Republican-leaning electorate), Alaska, Arkansas, Nebraska, and South Dakota, passed ballot referenda that would increase the minimum wage in their respective states, sometimes by a roughly 2–1 margin.

In New York, after failing to get the state legislature to pass a new increase to that state's minimum wage—raising the wage to a higher level in New York City, but also raising it in other parts of the state—Governor Cuomo has turned to a regulatory mechanism for finding ways to raise the minimum wage, in at least one sector: fast food. In a letter-to-the-editor penned to the *New York Times*, the Governor described this process as follows: "State law empowers the labor commissioner to investigate whether wages paid in a specific industry or job classification are sufficient to provide for the life and health of those workers—and, if not, to impanel a Wage Board to recommend what adequate wages should be." Cuomo added that he was going to direct the commissioner "to impanel such a board, to examine the minimum wage in the fast food industry." If the wage board makes a recommendation to increase the wages in this sector, such a change will not require legislative approval.

In Cuomo's letter expressing his reasons for supporting this move, he pointed out that many fast food workers either qualify for financial support through welfare payments, or have a family member who does. In fact, he pointed out that New York ranked sixth in the nation in terms of the amount of public assistance spent on fast food workers, costing state taxpayers $700 million a year. This notion that the broader community suffers when low-wage workers' earnings are insufficient for them to achieve self-sufficiency, and that the community as a whole gains through an increase in the minimum wage for low-wage workers, are common arguments in minimum wage advocacy. And research consistently shows that an increase in the minimum wage means an increase in consumer spending in the communities where those increases have taken hold and even that job growth is stronger—not weaker—in those communities. Indeed, a recent study of the 13 states where minimum wage increases went into effect in January 2014 showed that in those states, job growth in the months following the increase in the minimum wage exceeded that which took place in the states that did not increase their minimum wage level. As the Center for Economic & Policy Research found when it analyzed Bureau of Labor Statistics data for this time period: "Of the 13 states that increased their minimum wage in early 2014, all but one (New Jersey) are seeing employment gains." The Center found further that "[t]he average change in employment for the 13 states that increased their minimum wage is +0.99 percent while the remaining states have an average employment change of +0.68 percent." (Center for Economic & Policy Research, 2014).

While many states and some cities have adopted across-the-board minimum wage hikes, in some other cities, higher minimum wage laws apply in only certain industries. One of these cities is the city of Long Beach, CA, a community that has seen a dramatic economic shift in just the last few decades. Advocates in Long Beach pursued an increase in the minimum wage for hotel workers in hotels with more than 100 beds. Long Beach's experience with increasing the minimum wage for these workers helps identify strategies that local governments and advocates can pursue should they wish to advance a minimum wage for all, or even some, workers in their communities.

Long Beach went from being a blue collar, union-rich economic environment, one dependent on its port traffic and manufacturing and with large plants operated by companies like Boeing, to one dominated by the service economy. Moreover, in the wake of the loss of its manufacturing base, the city government spent $750 million to attract and retain service industry companies, like hotels and restaurants, through both direct subsidies and infrastructure improvements in the city to make it more conducive to tourism. For the city's workers, this industrial shift meant a dramatic loss in earnings. The loss of its manufacturing base, coupled with restrictions on increases in the ability of local governments to raise taxes imposed in California's local governments through Proposition 13, led the federal government, in a 1978 report, to list Long Beach as one of the nation's most financially distressed cities. As a result of heavy subsidies to the hospitality industry, by 2007, the Urban Land Institute listed Long Beach as having one of the nation's top 10 revived downtowns. Despite this revival of the downtown area, spurred by government incentives and subsidies, a study by the Brookings Institution identified Long Beach as having a high concentration of low-income people; indeed, according to Brookings, it ranks sixth in the country for concentrated poverty.

Two groups, the Los Angeles Alliance for a New Economy (LAANE), and a local union that organizes hotel and restaurant workers, UNITE HERE, worked together to promote a minimum wage increase for hotel workers in larger hotels. Their strategy was crafty and smart. The first thing these groups did was design an approach that used a ballot initiative as the vehicle for passing the minimum wage law. Knowing that local elected officials were likely heavily influenced by the hotel industry, one of the city's largest employers, the leadership of the groups thought a ballot initiative—taking the issue directly to the voters—was the right strategy.

First, they drafted union workers, paid staff, and volunteers to canvas neighborhoods throughout the city to obtain the number of signatures needed to secure a place on the ballot for the initiative. California law is fairly liberal in both allowing local governments to raise the local minimum wage and empowering citizens to place voter referenda on the ballot. This combination made Long Beach ripe for this campaign.

Second, the campaign's leaders amassed data, both on the level of subsidies offered and spending by the city to bolster the tourism and hotel industry, as well as the earnings of the city's lowest paid workers. Their research found that 40 percent of the city's workers earned less than twice the poverty level. An analysis of just one hotel's workers found that 41 percent of them qualified for some form of public assistance because their earnings were so low. Finally, hotel workers earned just $19,000 a year, which was less than half the median wage in the city, and a third of the earnings of someone working in the city's remaining manufacturing plants.

Third, they stressed the notion that many of the hotel workers lived within the city of Long Beach and, because most families on the low end of the economic scale spend a lot more of their earnings on consumer products and services than higher wage workers, an increase in earnings for these workers meant more money would be spent in the community, which would bolster the local economy. This message resonated with local small businesses (not the typical allies in a minimum wage fight) so much so that over 200 local businesses supported the campaign to pass "Measure N," the minimum wage referendum.

Finally, they launched a civic engagement campaign that matched campaign workers and volunteers based on their backgrounds and demographics to different groups and neighborhoods throughout the city. A volunteer would speak to his or her block association about the effort; a union worker would walk through his or her neighborhood, going door-to-door to rally support. These conversations—not just to secure signatures to get the issue on the ballot, but also to garner support on election day—took place throughout the city, in poor neighborhoods as well as wealthy, in predominantly white communities and communities of color. Matching messenger with audience—demographically, economically—proved effective, as did changing the message that messenger delivered. Low-wage workers would hear about the importance of raising wages for hotel workers, some of whom might be relatives or friends. Wealthier communities heard about the value of boosting small business development in the community by putting more spending power in the hands of local workers (Pierce, Shelton and Masoumi).

This multi-pronged strategy worked. On election day in 2012, in a year with a closely fought presidential election and a divided electorate that yielded a clear winner in the electoral college, and relatively close victory in the popular vote, Measure N passed by a 2–1 margin, securing higher wages for hotel workers in a city that has seen a dramatic shift in its economic base, from manufacturing to hospitality and tourism. Furthermore, spurred on by their victory in Long Beach, the leaders of LAANE and UNITE HERE took their campaign to the much larger and neighboring City of Los Angeles, where, in the spring of 2015, they won an increase of its minimum wage to $15 an hour by 2020, although there they accomplished the increase by means of legislation passed by the local legislative body, not through voter referendum.

Long Beach's experience with raising the minimum wage offers advocates a glimpse into one method—the local ballot initiative—for raising the minimum wage in a community. As the previous discussion shows, that is just one avenue through which advocates can try to bring about such an increase. Legislators at the federal level are always free to raise the minimum wage at any time, but politics often get in the way, despite wide-spread popular support for such an effort. Minimum wage increases at the state level, through state legislatures, which have a rich history in the United States, is also possible, but politics may also serve as a barrier to an effort in state houses throughout the country. This leaves advocates for an increase in the minimum wage few options but to seek out opportunities at the level of local government: in cities, counties, towns and other smaller units of government.

When are local initiatives—like those in Long Beach—a possibility? Putting aside the question of local politics, where the local political landscape may be just as hostile to a change to local economic conditions as it is at the federal and at some state legislative arenas, what conditions in the local legal ecosystem must exist for local minimum wage experimentation to thrive? When are the local legal conditions ripe for local experimentation? While no discussion of legal conditions is complete without a discussion of political realities, identifying the political landscape in each community is a task for local advocates. When the legal landscape lends itself to local experimentation on the minimum wage is a question to which I now turn.

Legal Issues Related to Local Minimum Wage Experimentation

The question of whether local governments or voters have the ability to raise the minimum wage is a complex one, and the legal ecosystem in each state, and even every locality, varies to a certain degree. The powers of local governments are typically subject to not just state oversight, but also those very powers typically emanate from either state constitutions or state law. In the federal system envisioned by the framers of the US Constitution, state governments are critical actors in the system, and local governments are the product of state governmental authority. State governments create local governments and the powers of local governments, such as they have them, emanate from state authority. In some states, state constitutions or state law delineate the power of local governments in great detail. In others, local powers are not as clearly defined, and, over time, legislatures and courts flesh out those contours on a case-by-case basis. Still, there are several principles that typically govern the relationship between state and local governments.

First, local powers emanate from state authority. Local governments are generally considered the product of state authority and only where state permission has been granted, either explicitly or implicitly, is a local government permitted to act in a particular area. Second, where there may be a "gray area" between state and local authority, different states take different approaches for resolving the question of local authority in a particular field. There is a great deal of variation within how each state approaches some of these questions, so it is sometimes difficult to make sweeping statements about local authority, without addressing a particular state's laws and constitution. What follows is an overview of some of the key questions advocates should explore when approaching the issue of local authority to address the minimum wage. Answers to these questions vary by state, and advocates must consider the answer that applies in their respective states when embarking on a local minimum wage campaign. A state-by-state survey of answers to each of these questions is beyond the scope of this chapter; instead, I will flag some of the key questions and issues with which advocates must familiarize themselves in determining whether a local minimum wage campaign might be feasible in a particular local jurisdiction.

First, does the state constitution or state law permit local minimum wage experimentation? Local experimentation on the minimum wage question is easiest in communities where the relevant state constitution or state law explicitly grants local governments broad authority to pass laws affecting the health, well-being and safety of their constituents. In many states, this authority extends to local minimum wage laws. As is clear from the many communities in California that have raised their local minimum wage, California is one state that gives local governments a great deal of leeway in passing their own local laws, including minimum wage laws. In such states, local governments can move through local legislation, passed at the local level, to implement a higher minimum wage than both the federal government and the relevant state requires. Similarly, as is also the case in California, which has a history of local ballot activism, state law may also permit an increase in the minimum wage through local ballot initiative. Remember, this is how advocates in Long Beach were able to increase the minimum wage in the hotel industry there despite their fears that local elected officials, though generally favorably inclined to support initiatives that helped low-wage workers, might be less willing to oppose a powerful local lobby, like the hotel industry, that would fight such an increase had advocates pressed for it at through local legislative processes. Where a state's laws or constitution permits local experimentation on health and safety laws, and such authority is interpreted broadly to encompass the power to impose a local minimum wage on local employers, advocates are on stable legal ground when they pursue minimum wage variations at the local level.

Second, is local authority limited or narrow, such that it does not encompass the ability to engage in minimum wage experimentation? While some states appear to grant local governments broad authority to pass local minimum wage ordinances, the power of local governments to pass health and safety legislation in some other states is narrower such that local governments need an explicit grant of power to pass such legislation, including that respecting the local minimum wage. Where state laws or constitutions limit local experimentation, generally local governments only have that authority granted to them by the states. Thus, in states like Virginia, local government authority is limited to that which is explicitly permitted by the state legislature in that state.

This approach dates back over 100 years to a doctrine that emanated from John Foster Dillon, the Chief Justice of the Iowa Supreme Court, who was suspicious of local government power and authored several influential opinions and a legal treatise that stated that local government authority should be limited to that which is essential to the functions of local government or which is expressly permitted by state law. Since local governments are not expressly empowered in states that still follow Dillon's Rule to pass local minimum wage ordinances, they likely lack the authority to so. Advocates can obviously lobby their state legislatures in such states for the express power to increase the minimum wage at the local level, but until such time as the state legislatures grants local governments the authority to engage in minimum wage experimentation, they are prohibited from doing so.

Third, does state law preempt local experimentation? In many respects, the two scenarios described so far are the easy cases. Where a state's legal ecosystem in effect grants or denies local governments the authority to pass local minimum wage ordinances, local governments must follow the direction of the state's laws. But sometimes it is not so straightforward. State laws do not always give explicit direction to local governments as to whether they can pass local minimum wage ordinances. Whether local governments can pass their own minimum wage laws in such settings where the law is unclear will often hinge on the doctrine known as preemption.

When assessing whether preemption occurs, one asks whether a particular state's laws on a subject prevent and displace the authority of local governments to pass legislation in that area. Sometimes this preemption is explicit, where state legislatures have stated clearly that their word on the subject of the minimum wage is final and local experimentation is thus prohibited. Sometimes preemption appears implicitly, however, where a state's body of laws and regulations address the subject, and the courts in the state determine that the state has spoken on the subject, foreclosing related local action. In turn, any attempts by localities to address it are preempted; that is, the state law is the superior authority and any efforts by the local government to pass legislation related to the subject are void.

For example, a state system of laws and regulations may cover the practice of hydraulic fracturing, or fracking. That legal ecosystem may not state explicitly that local governments can or cannot impose additional requirements on companies that wish to engage in this practice within that locality's jurisdiction, but due to the fact that the state system of laws and regulations may cover the practice, courts might say that the state's intervention in this area of law "occupies the field," leaving no room for local governments to take legal action.

A majority of states have passed their own minimum wage laws. State courts in those states may determine that this entry into the field of minimum wage regulation preempts local governments from passing their own minimum wage laws within a particular state. For local governments to have the authority to pass minimum wage laws in such states, it is not just that the state's laws cannot bar them from passing such legislation or that they have the explicit authority to do so. Rather, advocates must ensure that state passage of minimum wage laws within a state do not displace attempts by local governments to pass their own minimum wage legislation through the doctrine of state preemption on the subject.

Of course, as is the case in states where local authority is narrow and does not extend to the power to pass a local minimum wage, advocates can lobby state legislatures to grant legislation that overcomes the issue of state preemption. Even where state courts might interpret some state legislation in the field of the minimum wage at the state level as displacing and preempting local efforts on the issue, if the state legislature steps in and grants local governments the authority to pass their own minimum wage laws, the preemption issue disappears.

A campaign to grant local governments minimum wage authority is underway in New York State, where preemption has blocked local minimum wage experimentation. Workers in that state face great variation in the standard and cost of living in different parts of the state. Rural communities in Upstate New York are more affordable, but the economic erosion in many communities, communities that have lost their manufacturing base, means jobs can be scarce. Yet, a worker earning the minimum wage in such localities might maintain a viable standard of living, and forcing employers to pay a higher minimum wage in such communities might just drive out employers (though the research on this question, discussed earlier, does not suggest this is too great a risk).

Downstate, in places like New York City, low-wage workers have greater job opportunities because of the robust economic activity in the region, but they also face an extremely high cost of living, making the minimum wage inadequate to sustain families without additional government aid. Because of the variation in the cost of living throughout the state and the varying economic opportunities for workers in the state, New York would seem ripe for local minimum wage experimentation that is sensitive to local conditions. Nevertheless, political realities in the state legislature, to date, have blocked local governments from receiving the authority to engage in local minimum wage experimentation. In such states, advocates must continue to press their state legislatures to authorize local authority over minimum wages should they hope to vest local governments with such power.

Fourth, do local or state politics threaten local experimentation? While advocates who wish to see a local government or local voters promote legislative initiatives or ballot efforts that will increase minimum wage levels for all workers in a city, or for certain workers within that city, those advocates will need to assess whether a local government or its voters can achieve such an increase given the legal constraints imposed on local governments and their voters in a particular state. As discussed above, California has fairly liberal rules that permit local legislatures and local voters to adopt higher minimum wage rules for local workers. It is no accident, then, that many localities in that state have raised the minimum wage for workers throughout California, in big and small cities alike. While advocates need to assess the legal backdrop in their respective states and the authority that cities have within them to pass their own minimum wage legislation, even if such authority exists, the political landscape in some states may threaten that authority, even where the present state of the law may permit local experimentation.

Even if cities have the authority to pass minimum wage legislation, the state-wide political climate in a state might mean that legislatures, when faced with cities taking aggressive steps related to raising the minimum wage, might step in to strip those cities of the authority to pass such legislation. In other words, cities taking aggressive steps towards raising their constituents' wages may face a backlash from state legislatures that withdraws the authority to pass such legislation, ultimately overturning these cities' efforts. Were state legislatures to take such action, it would not be the first time that state elected officials have stepped in to limit the power of city officials to take aggressive action to address a pressing social issue. The recent experience of cities in the realm of gun control policy may give some city officials pause should they consider action on the minimum wage.

In the early 1980s, individual victims of gun violence brought personal injury litigation against gun manufacturers and other entities connected to the sale and distribution of illegal handguns. After individual litigants in these cases achieved some initial successes, other litigants began to bring similar types of cases. In the late 1990s, seeking to apply traditional legal principles—like negligence and public nuisance—cities began to line up to bring handgun manufacturers and distributors to court, pursuing such remedies as compensation for the additional costs to those municipalities that were associated with policing areas plagued by the presence of illegal handguns, or injunctions to stop the proliferation of such handguns. All in all, across the span of nearly a decade, over 30 municipalities brought litigation against different entities involved in the sale and distribution of handguns, claiming, among other things, that those entities were responsible for the proliferation of illegal handguns within city limits as well as the harms caused by such handguns. Cities also claimed that these companies were liable to those governments for the costs borne by municipalities associated with dealing with those harms (An Overview of Lawsuits, 1).

Unlike their private counterparts, individual litigants who had some success in the courts, few of the cases brought by municipalities achieved much litigation success. What these cities did do was attract the attention of the gun lobby, mostly the National Rifle Association (NRA). And what the NRA did in response to the wave of municipal litigation was lobby state legislatures to, in effect, strip municipal governments of the ability to bring this litigation. They did this by creating legislatively crafted immunity provisions for gun manufacturers and others involved in the distribution of handguns so that they would not have to defend against municipal lawsuits related to handgun proliferation. All in all, 32 states passed such legislation, granting immunity to the gun industry from lawsuits by cities (The NRA, the Brady Campaign, 166).

When cities advanced litigation that sought to address the problems associated with the proliferation of illegal handguns, they faced not just a skeptical judiciary, but also a potent legal force: the NRA and a powerful anti-gun reform lobby. These well-organized and well-funded advocates for gun rights convinced a majority of state legislatures to insulate gun manufacturers and others involved in the distribution of handguns from litigation by municipalities seeking to recoup the costs of dealing with the proliferation of illegal handguns. Indeed, the fear of the power of the gun lobby was probably what made municipalities pursue their goals through the courts in the first place, rather than seek remedies through legislative channels, where they probably figured they would have little success. Those fears were realized when the gun lobby stepped in to take aggressive action to halt the reform sought by municipal governments through the courts.

The avenues for the types of reform that minimum wage advocates seek—local legislative reforms and ballot initiatives—are different from those pursued by municipal gun reform advocates, who attempted to work through the courts to advance their policy agenda. Nevertheless, the fear that anti-minimum wage reform forces might step in to take the sort of action the gun lobby did—advancing their cause at the state legislative level—is real, and one that minimum wage advocates might have to face. That is, those interested in pursuing minimum wage experimentation at the local level might find themselves up against stiff opposition in those state legislatures that might, on the whole, oppose raising the minimum wage anywhere in the their respective states. Even where the current state of the law in a particular jurisdiction is such, at present, that it might permit a city to set its own minimum wage, there is always the threat that an unsympathetic state legislature could step in—at the behest of some lobbying group or on its own out of a hostility to local minimum wage experimentation—to strip municipalities of that authority and invalidate any measures taken to increase the local minimum wage. As a result, it is quite possible that advocates in some states may face strong opposition in a particular state legislature to this power, even if it currently exists on the books in that state. Where aggressive local action on the minimum wage question might spur the legislature in a state to take steps to strip local governments of such authority, advocates might be hesitant to press the issue.

Conclusion

For local experimentation to take hold on the question of raising the minimum wage, the legal ecosystem in a particular state must permit it, or at least not prohibit it. Advocates must assess the landscape in their communities to determine the extent to which experimentation is possible. Furthermore, they must assess the political climate to ensure that their efforts will not encourage a legislative backlash that threatens such efforts. But even where advocacy faces an uncertain future, the history of minimum wage experimentation might give advocates hope that such experimentation can lead to surprising—and unforeseen—results.

The history of the passage of the first federal minimum wage recounted above offers at least two insights to advocates who might wish to see adoption of a higher minimum wage, whether at the federal, state or local level. First, local experimentation works, and might just push "higher" level adoption of the minimum wage. States were the first to adopt minimum wage laws at the beginning of the 20th century. Over time, political and economic pressure prompted action at the federal level, setting a national standard that would serve as a floor and a springboard for future experimentation by the states, as well as intermittent increases to that federal wage.

Second, political, social, and judicial winds may change, but minimum wage experimentation can help spur that change along. Few expected the *Lochner*-era Justice Roberts to switch his vote on the minimum wage issue, in particular, and New Deal legislation, in general, which ultimately took the steam out of Roosevelt's court-packing plan: what historians sometimes call "the switch in time that saved nine." And it was minimum wage experimentation that gave Roberts that opportunity to make the switch.

Today, support for a higher minimum wage has strong bi-partisan support among the American public. Elected officials and the powerful lobbies that support them can serve as barriers to passage of higher federal, state-based and local minimum wage laws. But one of the lessons of the effort to pass a federal minimum wage law, which started at the level of the states (and the District of Columbia, through Congressional action), is that experimentation helps push the political envelope and creates political opportunities, opportunities that might create pressure to align with pro-minimum wage forces.

Cities stand at the juncture of this opportunity. Action at the federal level on the minimum wage and in some state legislatures may not seem possible at this time. But local experimentation on the issue, which may create opportunities presently unforeseen and alliances unpredicated may hold the key to success. Advocates seeking to promote an increase in the minimum wage should turn their sights to cities, where thoughtful action, where it is most strategic, might help advance their cause, create opportunity, and ultimately advance social change.

Works Cited

"2014 Job Creation Faster in States That Raised the Minimum Wage." Web blog post. Center for Economic and Policy Research. N.p., 2014. Web.

"An Overview of Lawsuits Against the Gun Industry." *Introduction. Suing the Gun Industry: A Battle at the Crossroads of Gun Control and Mass Torts*. Ed. Timothy D. Lytton. Ann Arbor: U of Michigan, 2005. Print.

Autor, David H., Alan Manning, and Christopher L. Smith. The Contribution to the Minimum Wage to US Wage Inequality over Three Decades: A Reassessment. Working paper no. NBER 16533. Washington DC: Federal Reserve Board, 2010. Finance and Economics Discussion Ser., Divisions of Research and Statistics and Monetary Affairs. Web.

Brown, Charles, Curtis Gilroy, and Andrew Kohen. "The Effect of the Minimum Wage on Employment and Unemployment: A Survey." *Journal of Economic Literature* 20.2 (1982): 487–528. Print.

Card, David, and Alan Krueger. "Minimum Wages and Employment: A Case Study of the Fast Food Industry in New Jersey and Pennsylvania." *American Economic Review* 48.4 (1994): 772–93. Print.

Dinardo, John, Nicole M. Fortin, and Thomas Lemieux. "Labor Market Institutions and the Distribution of Wages, 1973–1992: A Semiparametric Approach." *Econometrica* 64.5 (1996): 1001–1044. Print.

Dube, Arindrajit. Minimum Wages and the Distribution of Family Incomes. Working paper. 2013. Web.

Elwell, Craig K. Inflation and the Real Minimum Wage: A Fact Sheet. Fact Sheet.: Congressional Research Service, 2014. Web.

Hall, Doug, and David Cooper. How Raising the Federal Minimum Wage Would Help Working Families and Give the Economy a Boost. Rep.: Economic Policy Institute, 2012. Web.

"The NRA, the Brady Campaign, and the Politics of Gun Litigation." *Suing the Gun Industry: A Battle at the Crossroads of Gun Control and Mass Torts*. Ed. Timothy D. Lytton. Ann Arbor: U of Michigan, 2005. Print.

Pierce, Jeanine. Personal interview. 29 May 2014. MP3. (on file with author).

Schmitt, John. Why Does the Minimum Wage Have No Discernible Effect on Employment. Rep.: Center for Economic and Policy Research, 2013. Feb. 2013. Web.

State Minimum Wages: 2015 Minimum Wage by State. Rep. National Conference of State Legislatures, 19 May 2015. Web.

Shelton, Leigh and Lopex Masoumi, Loraina. Personal interview. 8 May 2014. MP3 (on file with author).

United States. Minimum Wage Study Commission. Report of the Minimum Wage Study Commission. [Washington, D.C.]: The Commission, 1981. Print.

Chapter 8
Cities and Immigration Reform:
National Policy from the Bottom Up

John DeStefano, Jr.

'Please everyone get out and vote against the mayor. He is out of his mind. This is not a legal action … . You know they mostly get paid cash, so no taxes but plenty of services. So now we will pay for the elderly and the ILLEGAL. Soon nothing will be left in this vibrant city we hear so much about.'

(Bailey, "City Unveils a New ID." *New Haven Independent*)

Picking the Room

I am convinced of the importance of choosing just the right room. The room contributes to the mood and to the reaction of the audience. So for meetings on the budget, taxes, and property re-valuation? For those I always preferred a church. The sanctuary if possible. If not the sanctuary, then the church hall. People are always better behaved in church. For neighborhood meetings? I always found school cafeterias the best spot. Avoid the school auditorium, however. Inevitably in an auditorium you will end up with the city officials up on the stage at a long table looking down at the audience, sitting in row upon row of fixed seats. It invites and almost always results in an adversarial dialogue. So for neighborhood meetings, use the cafeteria with chairs moved into groups of seats around a table with a city official or two mixed into each group. Avoid fixed seats as well. People like to turn their chairs. Through long years of practice, I bear the scars of choosing the wrong rooms to do a meeting or the wrong location to do a press event. There is enough else that might and will go wrong without having the choice of the room contribute to a negative outcome. It is often said that the choice of the right room is a critical first step in community organizing. Choose something that is too big, even if one hopes for a large turnout, and it appears that there is little interest in the cause. Choose a room that is a little too small for the crowd, however, and participants feel a sense that a movement is afoot, and there is great community support for the effort.

But the choice of a community meeting room is really a metaphor for how one wants to approach an issue. A welcoming setting invites an open and constructive dialogue. It builds bridges. It promotes trust and fosters the growth of social capital, that community glue that helps make life a little easier, a little more productive, and a little more enjoyable.

This chapter is about making choices. The specific choice it describes is the way one city, the City of New Haven, where I had the privilege of serving as mayor for 20 years, chose to work with its community of recent immigrants. Like choosing a room, the choice of how to deal with any community is about how receptive one wants to be, the extent to which one wants to promote an open and co-equal dialogue, to engage in meaningful and constructive ways. While a national debate rages over how to deal with recent immigrants to the U.S., many engaged in a national discourse on the subject, including national politicians, have chosen a response that is jingoistic and close minded. This chapter describes some of those responses, including the national government's response to some of the choices made at the local level in dealing with New Haven's immigrant community. Unfortunately, some of the progressive policies we chose actually seemed to invite harsh and repressive tactics from the national government, such tactics being one choice for dealing with this community. Our city took a very different approach. This chapter is about the choices made by the City of New Haven in dealing with its community of recent immigrants. It helps situate a dialogue around how to deal with recent immigrants in the context of this volume; that is, it shows how cities can chart a way forward on immigration policy, even when the national dialogue has soured, and the national government seems paralyzed to make any real progress on immigration policies. What this chapter is also about is how city leadership can relate to and deal with grassroots organizing, here the effective organizing that had started in New Haven in 2004, as the burgeoning new immigrant community was finding its voice.

One meeting that became a key inflection point in community organizing in the city of New Haven around immigration reform took place in January 2007, and it was the first of its kind. The city's growing, largely Latino, largely undocumented immigrant community had been organizing since 2004. And it was past time that we in city government acknowledged what was so plainly occurring; that these immigrants, these undocumented immigrants, were our residents. They were our neighbors, the kids in our schools, the workers in our businesses. There were more and more of them coming to New Haven each year. And they were working in the community as employees and finding ways to start businesses in increasing numbers.

By early 2007, New Haven's residents, social justice advocates, and city government were in the midst of redefining how the immigrant community and the rest of the city viewed their respective rights and responsibilities to one another. That night in January 2007, city government was planning to explain the new way that city police would engage so many of our residents living in the shadows as undocumented residents. So we needed the right room. And we found it.

The meeting was scheduled for the Fair Haven Branch of the Public Library. The branch was opened on Grand Avenue in 1916 through a philanthropic campaign underwritten by steel magnate Andrew Carnegie. Generations of Fair Haven kids and residents have read, studied, attended programs, and taken out books there. Everyone liked the Fair Haven Library. It is a community institution that asks nothing of the neighborhood other than that they take advantage of it.

Neutral. Welcoming. Familiar. A place that serves the whole community. The perfect place for a meeting. Perfect for a frank and open discussion. It was at this meeting that I, after having served over 14 years as Mayor, finally understood the breadth of change immigration was bringing to our city and began to understand the tremendous opportunity it presented to all members of our community. And I realized then that it was the kind of change that ought to be embraced, change that was about choices. To trace the evolution of these choices and the campaign around the rights of recent immigrants in the City of New Haven, one must understand the historical context and the city's relationship to its immigrant population over the years. It is to these issues that I now turn.

Immigrants in New Haven

Over the preceding years, the city experienced the kind of gradual changes that, in isolation, might have passed unnoticed. For example, there were increasing numbers of children in the school district speaking English as a second language. In fact, they were speaking English as a second language in multiple dialects of Spanish. Similarly, a different kind of housing code violation began to present itself in the city. The housing code complaints began to turn up as a result of large numbers of adults and children living in one apartment. The giveaway was often the increased trash generated from some buildings.

We also noticed other changes. There were growing numbers of workers in service industry companies who came from outside the country. Employers hired large numbers of relatives or residents from a particular village or Mexican state. Hiring patterns emerged that the city had not seen since the pre–World War I immigration surge in the city. That generation of immigrants had built guns and hardware in city factories and the neo-gothic stone buildings of Yale University's Central Campus as well. This generation of immigrants was working in manufacturing too. But increasingly these new immigrants were baking bread, working in laundries, landscaping, working at restaurants, and toiling in food processing plants. They were filling voids in the workforce and laboring throughout the service industry in increasingly large numbers in New Haven. Demand for recreational opportunities was also changing. There was a larger demand for soccer and volleyball courts. For cricket courts. These were not the playing fields of choice when I was growing up on the east side of New Haven. This was different. And the traditional flag raisings on the central green celebrating Irish, Italian, and Polish independence days made room for Mexican, Guatemalan, and Ecuadorian independence celebrations as well.

The mushrooming attendance at some of the city's churches was also notable. The old Catholic churches of my youth that had been built by our previous generations of immigrants experienced dwindling attendance from their suburban-bound first- and second-generation sons, daughters, and grandchildren. Near empty 20 years ago, these churches were experiencing a resurgence in attendance. As a politician you go where people are. I started attending services for the traditional Mexican feast day of Our Lady of Guadalupe at one of these churches that had seen a demographic shift in its congregation, from older immigrants from European countries to more recent ones from Latin America.

The influx of recent immigrants also brought about negative consequences, like an increase in street robberies and home invasions that seemed to target the Mexican immigrant community. This phenomenon was described to me as criminals seeing certain immigrants as a new kind of "ATM." It was a walking ATM: a person one could identify because the individual had "Mexican" facial characteristics. Muggers presumed that the Mexican-looking targets did not have a bank account so they probably kept large sums of cash on their person or at home. And since they were "illegal" (mostly meaning that their visa had expired), the victims did not go to the police to report that they had been a victim of a robbery.

Incidents of domestic violence in the new immigrant community posed similar problems. The quickest way to avoid having a victim of domestic violence file a police report was for the attacker to compound the abuse with a threat to go to the police and report that the victim was undocumented. In such situations, the victim would often choose not to file a police report.

Compounding the problems facing new immigrants was that for many of them they had not experienced a trusting relationship with their local police in their country of origin. In New Haven, the immigrant community did not know the police and often had little understanding of its role with regard to immigration enforcement; that is, that a local police force like New Haven's did not play any role in enforcement of the immigration laws. In addition, our police officers did not know the new immigrant community. A lack of trust and infrequent positive and direct experiences with the police give way to ignorance, prejudice, and fear as a substitute for a relationship. That was a problem. And it is for many United States citizens as well. Think of the events of 2014 in Ferguson, Missouri, and the killing of two police officers in New York City, and you can see the consequences of the failure of community–police relationships.

Perhaps any one of these things alone might have passed unnoticed, but in city hall, where the schedule and your days are organized around the rhythms and events of the community, one would have to have had one's head buried in the sand not to see what was happening little by little, day by day. New Haven was returning to its immigrant city roots. Large immigrant populations were nothing new in New Haven. In 1910, one third of New Haven's population was foreign born and another third had at least one immigrant parent (Dahl 31). And in 2004, New Haven's economic and social resurgence was once more being fueled by the aspirations and values of these new immigrant residents.

Any one of these changes happening is no big deal. Taken together, clearly something much bigger was happening. And mostly it was happening in the City's Fair Haven Neighborhood, the new immigrant hometown in the city. Located on a peninsula between the Mill and Quinnipiac Rivers, it was originally known as Dragon. No one knows why (Mills Brown). By 2007, Fair Haven was over 50 percent Latino. Fair Haven was one of the city's first neighborhoods to be built off the central green laid down by the City's Congregational Church founders in 1638. As Fair Haven grew, it became home to successive waves of immigrant families seeking to resettle themselves in America. Their churches, grape vines, some of their neighborhood bars, and grandchildren remain behind. However, by 2007, Fair Haven was a neighborhood powerfully vibrant and alive, animated by the ambitions, hopes, possibilities, and challenges of the newest wave of immigrants to its streets from Mexico and Central and South America.

And this new generation of immigrants wanted to work. They wanted to own their own homes. And they wanted what every prior generation of immigrants worked for –the chance that their kids would have the opportunity to do better than they had. It is really not more complicated than that. That is the reason they came to the United States and New Haven. They came for the very same things for which so many prior generations of immigrants had come to New Haven. They came so that their children would have a better chance at life.

Self-improvement seems to be a theme in New Haven, as it is in many cities large and small across the country, and in immigrant and non-immigrant communities alike. Higher education is one of the means through which many seek to improve their opportunities in life, and there are few educational institutions in the world that enjoy the prestige of New Haven's most storied institutional resident, itself named for a benefactor who was also a recent immigrant to New World. Indeed, just 1.7 miles from Fair Haven sits one of New Haven's oldest "neighborhoods." In 1701 a collegiate school was organized in Saybrook, Connecticut, for the purpose "wherein youth may be instructed in the arts and sciences, who … may be fitted for public employment both in church and Civil State." (Osterweis 93). Incorporated through an act of the Colonial Assembly on October 9, 1701, the collegiate school remained in Saybrook through 1716, when it was relocated to New Haven at the urging of many of the descendants of the same Congregationalists who had founded New Haven in 1638. Acting once again in the fall of 1718, the Assembly endorsed the move of the newly named Yale College to New Haven. The new college was seated on the nine squares laid out by the city founders.

Today Yale University's Central Campus has expanded well beyond the city's original nine squares. Yale University stretches some 200 acres across Downtown New Haven.

Occupying some 340 buildings and 12.5 million gross square feet of space, the University comprises Yale College, 10 professional schools, and possesses the world's seventh largest library (Cooper, Robertson & Partners 3). One of the world's leading research institutions, the Yale Medical School's clinical practice is located at the separately owned and governed Yale New Haven Hospital which is, by bed count, the fifth largest hospital in the nation.

With a total enrollment of 12,336 students and full-time faculty and staff numbering some 13,463, Yale is a powerful presence in the city. It is the City's largest employer and fifth largest taxpayer. Adding in state payments in lieu of taxes, voluntary payments in lieu of taxes, and fees, Yale revenue accounts for some 13.3 percent of revenues in the city's $508.4 million annual budget. But there is more.

Over the last two decades, there has been a large transfer of knowledge and research from Yale's academic labs to commercial workbenches and production in the city. New Haven's new economy is not just academic: the eds and meds. Today it is also a large center of knowledge-based businesses, jobs, taxpayers, and entrepreneurs whose energy can only be fully expressed through large numbers of immigrant talent. And all that activity importing wealth into New Haven's households as income is recirculated through the service economy.

In the area of biotechnology alone, of the more than 50 companies in Greater New Haven, half are in the city. Since 2000, biotechnology companies have attracted over $2 billion in investments in Greater New Haven and employ over 1,000 people. The growth of knowledge-based companies is driven by students as well. Since the Yale Entrepreneurial Institute (YEI) was founded in 2006 to support students who wish to start companies, over 79 student-founded ventures have been fostered and grown. More than $104 million has been invested in YEI ventures and more than 380 full-time jobs have been created.

Critical to technology transfer out of the university to private markets and families is a robust talent platform. The attraction of talent, wherever it can be sourced from the city itself, from the United States or from around the world, and brought to New Haven is an essential strategy of and a distinct advantage to the growth of the city economy. Twenty years ago, international students made up just 12 percent of enrollment at Yale University. Today that enrollment stands at 2,477 or 20 percent of total student enrollment. Another 1,236 noncitizens and nonpermanent residents can be counted among Yale faculty, post–doctoral, and staff appointments. This is a good thing.

Nationally, the number of students studying at U.S. colleges and universities rose last year to nearly 900,000, leveraging some $27 billion into the U.S. economy (Belkin). Capturing as much of this talent as possible should be a core economic development strategy for U.S. cities.

Yet for all this, the federal government's approach to this potential talent pool and the economic activity it can generate is extremely short sighted. H–1B visas are issued annually by the U.S. Citizen and Immigration Services Agency for high skilled workers. H–1B's are three-year visas. In a recent year, Congress authorized the issuance of 85,000 of these visas to be made available on April 1, 2014. On that date, the U.S. received 172,500 applications for H–1B visas, meaning some 87,500 applications were turned down. Bright, hard–working, highly skilled workers were refused entry into the United States. Put aside the fact that with an average fee of $2,000 per application, the U.S. turned away some $177,000,000 in revenue to the federal treasury. Instead these workers went to countries such as Canada, where US companies set up subsidiaries to capture this talent and their tax dollars (Lawler and Stock). Economies are not a zero sum game. Time and experience has proven over and over again that growing the pool of talent geometrically increases wealth in our economy.

The fact is that just as immigrants have been fueling the growth of the city's service sector, so too is immigration critical to the growth of the knowledge-based, higher-wage sector of the city economy. In both instances, policies that are welcoming and promote robust immigration are in the self –interest of the community. Yale, like Fair Haven, is a bustling center of immigrant activity the production and benefits of which ripple throughout Connecticut and the nation. The growth of immigrant populations at both places is central to the prosperity of all.

Cities are locked in a competition for wages, and economic and social capital acquisition. A welcoming immigration policy would seem to have a more direct impact on economic development, much more direct than the construction of any baseball stadium (we have tried that), sports arena (we have tried that too), or program of urban renewal (where New Haven ranked number one in the nation in the 1960's for federal dollars awarded per capita in a self-defined mission to build America's first "slumless city"). More central to any of these things as a strategy for growth and well– being is the development of human capital. Ultimately, cities are about people, not places.

For New Haven and cities like it all across the nation, robust immigration to our shores will write the story of our nation's future through the development of our nation's greatest resource–our people. And if history is to be any guide, the nation has been renewed and remade generation after generation by successive waves of new Americans bound together not by birth but by a shared set of values. These are traditional values of hard work, persistence, community, and social tolerance. These are the kinds of values that bind communities not just in a moment of time, but over time as well. These are the kinds of values and behaviors that promote norms and cohesion in a community, values and behaviors that create stickiness, a kind of glue that cements and builds social capital.

Unfortunately for the people of New Haven and places like it, the jurisdiction responsible for regulating immigration, the national government, has in recent years been at best useless and, more often than not, antagonistic toward the kind of immigration policies that have grown both our economy and our human capital over our nation's history. Cities like New Haven and others like it across the country must lead where the national government seems to have failed.

By 2004 it was clear that New Haven was changing. The nature of change is that either it engages you or you engage it. The federal government was frozen in inaction. And cities like New Haven had a choice. Do nothing or embrace and shape the change. Change was happening in New Haven and our newest residents had, by 2004, decided to shape it.

Setting the Agenda

In 2004, the city's undocumented residents began to organize. The platform upon which they came together was in large measure being organized by three Fair Haven organizations. Two were secular, community-based social justice organizations and the third was St. Rose of Lima Catholic Church (St. Rose). The first secular organization was Junta for Progressive Action, Inc. (JUNTA). JUNTA was founded in the 1960's and is located on Grand Avenue in New Haven, just across from Fair Haven School and the Fair Haven Library. It is the oldest community-based organization in New Haven serving the Latino community. In 2004, JUNTA was led by Kica Matos. Kica, a lawyer, displayed tremendous organizing and strategic skills in this effort.

The second secular organization working with the undocumented immigrant community was Unidad Latina en Accion (ULA) founded in 2002 as a grassroots social organization made up of immigrants to address human rights abuses. ULA was led then and now by John Jairo Lugo. John was and remains a passionate advocate on behalf of the community.

JUNTA, ULA, community advocates, and immigrants living largely in Fair Haven began to advocate for change in 2004 by talking about the challenges they were facing in their lives in New Haven and around their hopes for themselves and their children. As these talks began to take shape, so too did an agenda. As the months proceeded, city hall was contacted and a dialogue began between my administration and these community members.

At this point, it was a dialogue occurring inside city hall and not a public discussion. We met in November 2004 and January 2005. At these early meetings, we covered lots of topics including the possibility of driver's licenses for the undocumented community. The group reached consensus that accomplishing this would be difficult. Any such change would have to occur at the level of the state legislature, which was unlikely to favor such an approach any time soon. One of those present at one of our meetings, Antonia Armas, spoke up and asked about a city government-issued identification document. Many of those present in the room embraced that idea. I had doubts about the legality of such a document, however. At this point the Yale Law School's Community Lawyering Clinic (CLC) was invited to join the effort to conduct research on this and other immigrant community legal concerns.

And then there was St. Rose. Now St. Rose's was familiar to me. My first interactions with the church occurred in the late 1960's when I attended parochial school across town and we played each other in basketball. Back then it was a mainstream Italian congregation and diversity meant that some of the Irish families who did not attend St. Francis some eight blocks away attended mass there. But it was not until 1995 that St. Rose really took me to school.

When I became Mayor in 1994, St. Rose's was part of an inter-faith social action church group called Elm City Churches Organized ("ECCO"). In 1995, I had a memorable night there with then-Police Chief Nick Pastore. The occasion was an issue that related to a Super KMart store that I had helped locate to the city. The fact that the store sold guns in a city that experienced then, and does to this day, an unacceptably high level of gun violence was offensive to some members of our community. So ECCO invited Chief Pastore and me to St. Rose to answer some questions regarding the sale of guns at the new store.

Upon arrival at St. Rose, the Chief and I were escorted to the church hall and given seats on the stage. In fact we were the only ones on the stage. The part of the hall where the audience was seated was dimly lit while spotlights were shining on the Chief and me. Our invitation stipulated that we would be each asked the same three questions—to which we could only respond yes or no. ECCO knew how to set up a room and a meeting.

Legendary community organizer Saul Alinsky once said that "No politician can sit on an issue if you make it hot enough." (xxiv). The Chief and I left the hall against the guns, having engaged a faith organization that over the next 20 years would always be a competent ally, a worthy adversary, and its members good friends.

By 2005 St. Rose had become the largest and most visible place of worship for the immigrant community in Fair Haven. Boasting congregation members from 18 countries and 15 Mexican states, St. Rose was stewarded by Fr. James "Jim" Manship. Fr. Jim had been assigned to St. Rose in 2005.

At this point in 2005 what was remarkable about the city's immigrant agenda was that it had been conceptualized, articulated, and pressed into action by the immigrant community itself. Organized by JUNTA, ULA, and St. Rose along with legal counsel from the CLC, it was a grass roots creation.

At the outset, it was fair to say that we in city hall were empathetic to immigrant issues in so far as they represented the real challenges being faced by some of our residents who happened to be immigrants. In 2004 and 2005, there was no sense that the community or city hall was organizing a challenge to the federal government or anyone else. There was no sense of fashioning immigration policy as a strategy toward economic growth. The connection between the undocumented in Fair Haven and those on H–1B visas at Yale was one of the furthest things from my mind at that time and would all come later. Instead this was an engagement with a constituent group similar to the hundreds such engagements held before and after, with the caveat about the reluctance and, in some instances, fear, of our undocumented residents to attend and participate directly with civil authorities.

A New Agenda

It is a truism to me that an effective politician wields power. For effectively engaging change only occurs through the exercise of power. Typically, as Mayor, I preferred that I was the one wielding the power, on an agenda that I set and on a time frame of my making. However, as much as one might like to set the agenda, sometimes others shape the agenda. It is thrust upon the public scene and that agenda is so very appropriate to the moment. It is an agenda that speaks directly and clearly to the values of the community and is so powerful in its aspiration that one simply surfs it like a perfect wave. Although one can feel the power and general direction of the wave, one does not know where it will lead and one needs to be careful to avoid being flipped off of it. As it turns out, New Haven's immigrant community itself had created a powerful agenda that set the stage for the perfect wave of reform.

This agenda did not come from city hall. It was a community-inspired and community-driven agenda. The agenda the city and I were presented with resulted from an incredibly well-researched and documented report titled *A City to Model* (the Report) prepared by JUNTA and ULA. The report was researched by the CLC under the direction of Professor Mike Wishnie.

The report identified the principal concerns of the New Haven immigrant community and organized them into six recommended actions for city government. The Report stated that:

> The major security concerns of the New Haven immigrant community include 1) the high rate of theft in their homes and on the way home from work because they cannot open bank accounts to protect their earnings and so must carry cash; 2) the under-reporting of crimes to police, due to immigrant's fears that they lack identification that the authorities will recognize or other worries based on the lack of trusting relationships between the residents and law enforcement; and 3) the underpayment or nonpayment of wages when unscrupulous employers take advantage of immigrant workers' vulnerabilities (JUNTA and ULA 1).

As Mayor of New Haven I sat through countless meetings with residents, businesses and constituency groups of every kind. From the outset, the dialogue around a local immigration reform agenda was different. It was well thought out, it was well researched and the immigrant community had been organized to make my seat hot. Meetings were direct without being antagonistic. The agenda was shaped to communicate a message of traditional values and social tolerance. Fashioned in this way, it was a powerful and effective message

In the meetings between the immigrant community's leaders and my administration, we covered the six recommendations from the Report, and one additional recommendation that came up in the course of the discussion. The actions from the Report called for:

1. Notifying New Haven financial institutions that customers can open banks accounts without Social Security numbers.
2. Creating a municipal identification card for the City of New Haven.
3. Developing a New Haven Police Department policy of non-enforcement of federal immigration laws.
4. Enforcing criminal wage laws through the New Haven Police Department.
5. Looking to best practices nationwide for improving police-community relations.
6. Creating an Office of Immigrant Affairs.

During the course of discussion a seventh action was identified:

7. Translating into and making available commonly used city documents in Spanish.

As we moved forward a couple of things appeared clear.

First, it was in all of New Haven's interest to see that the immigrant community, particularly the undocumented community, and the New Haven Police Department get to know each other directly and personally. And it was equally important that the City sort out its role with regard to federal immigration enforcement.

This new immigrant community was not going anywhere. Its members represented the best of our city's traditional values of hard work, personal responsibility, and the reciprocal responsibilities to the community that ensured everyone's rights and freedoms. The failure to engage this community would result in more than just its disenfranchisement. It would break down cohesion within the broader community itself. We are either each on our own or all in the city together. My view then and now remains that a city's most powerful interest will always be found in the broadest definitions of community membership rights and responsibilities that we can fashion. Networks of inclusion trump those of exclusion every time.

In order to promote inclusion, the first order of business was strengthening the relationship between the New Haven Police and the immigrant community. This was not just a matter of sorting out roles, it was a matter of legitimizing the reciprocal responsibility of civil government in New Haven among its residents–both immigrant and non-immigrant. And since many nuclear families were comprised of both documented and undocumented members, the idea that you could draw a line between the two just ignored reality.

Second, people who lived in New Haven and played by the rules, who respected themselves and their neighbors, deserved and were owed a city government that was responsive to their needs. I believed we should take those actions that we could take to accommodate all reasonable concerns, perspectives, and interests.

Simply put, New Haven City Government emerged from our meetings with a set of principles to guide our actions. They were the following:

- Clarifying civil and community rights and responsibilities;
- Making reasonable accommodations for community needs;

- Promoting social cohesion;
- Building legitimacy among residents and their government; and,
- Doing so in ways that sought to benefit all the members of our city. In other words, this would not just be an immigrant issue.

As a practical matter we agreed to proceed on three initiatives in the near term within the context of our seven principles:

- Making city documents available in Spanish: this would be easy;
- The drafting and issuance of a Police General Order spelling out that New Haven Police would not be an extension of federal immigration enforcement and would encourage undocumented residents to report crime: this would take some work; and
- Issuance of a municipal identification card: although I did not realize it at the time, this would become a bigger task and challenge than first imagined.

Leading from Behind

At times in conversations among fellow mayors, we have discussed leadership styles. My default view was always that we were elected to lead. So I always viewed my job as to have an opinion, to build support for an initiative and, to get out there and lead. The lesson I learned from this experience suggests that, at some moments, different leadership styles are appropriate. It is not just one way or another. The real genius of leadership is distinguishing between when to lead and when to lead by following.

The community, acting through JUNTA, ULA, St. Rose, and staffed by the CLC, had laid down a vision that not only reflected the needs of the immigrant community at a moment in time but did so in a fashion that reflected the long-standing traditions and values of the city. It was beautiful. City government's job was not going to be the conception of the direction. That job was done. City government was, however, going to be responsible for birthing the actions. And so we began.

Hablamos Español

My view about learning to swim is to start in the shallow end of the pool. Start in knee-deep water and work towards getting into the water up to your waist and then your chest. Then as you develop the confidence, skills, and the direction in which you want to go, by all means get in the water over your head. So it was to be with New Haven immigration initiatives.

"*Hablamos Español*" was the title given to the initiative to translate and make available in Spanish public documents that were frequently used by residents of New Haven. The City started with the basics: birth, and death certificates. And then we included portions of the City website. At the same time, great thought was given to including documents and access to city government that not only reflected the rights of citizens but also their responsibilities. So the City made sure forms and documents were made available in Spanish that explained how to pay taxes and parking tickets. It is a two way street after all.

We launched the language access program at a press conference at city hall on October 3, 2005. The launch went fine as far as the *Hablamos Español* initiative. But then the press had a question–what about identification cards? Is the city going to do those too? My response to these questions was a quick "identification cards? Sure!"

At that moment, I was on my way to becoming the Democratic Party's candidate for Governor of Connecticut in 2006, and the universe of press following New Haven affairs had grown as a result.

It only took 24 hours of my first round of state-wide and national press on immigration matters to back me down from immigrant identification cards. I went to school that day on the necessity and importance of how we talk about and organize support on behalf of immigration matters.

In fact, there was little public or bureaucratic reaction to *Hablamos Español*. The press attention generated by the discussion of an identification card made *Hablamos Español* seem like a rather small matter. Step number one, *Hablamos Español*, was done. Now the city turned to step number two, a Police General Order on immigration enforcement, with a pause put on immigration cards. And then Manuel Santiago was murdered.

Manuel Santiago

Manuel Santiago came to New Haven from Tlaxcala, Mexico, in 2001 to join his brother Basilo. Like many before him, he moved to the Fair Haven neighborhood of the city. He was a member of St. Rose of Lima Church. Manuel worked at a local bakery in Fair Haven. Many of his coworkers came from Tlaxcala like Manuel.

Manuel was murdered on October 18, 2006. He was cashing his paycheck in Fair Haven and a mugger was waiting for him. They got into a fight over the money. It was cash after all. Manuel was undocumented and he did not have a bank account. Manuel was stabbed to death. He was not in trouble in New Haven and seemed to keep to himself. He was a target because he was undocumented. Everyone at the Police Department, in city hall, and, most poignantly, in the community, understood that he was targeted just because he fit a profile, because of what he looked like. It chilled everyone and everything we were doing.

Manuel's wake was held at the Lupoli Funeral Home in the City's Wooster Square neighborhood. As Gene Lupoli, the funeral director, said in a news report at the time:

'Their lives are being taken because they are easy prey,' Lupoli said. 'These thieves know they won't go to the police. While (undocumented workers) may be guilty of a federal crime–and it's a nonviolent crime–or could be guilty of a crime, they're not reporting a federal crime. They're giving testimony of a robbery or an assault type issue. Italians, Jews,' Lupoli reflected, 'they have jumped ship (in the past to enter the country). Somehow they're all legal over time.' (Bass).

Two months after the murder of Manuel Santiago on December 21, 2006, New Haven took step number two. New Haven Police Chief Francesco Ortiz issued General Order #06–2. Concurrent with issuance of the General Order, Chief Ortiz and I organized a community meeting to explain the new direction that city police would pursue with the undocumented community. And so we went to the Fair Haven Branch of the New Haven Public Library on January 18, 2007, to the meeting I described to open this chapter. At that meeting I finally came to the view that this was a matter bigger than the immigrant community in New Haven: that how our city chose to act in immigration policy would meaningfully and positively impact the future of all of New Haven. Although we had taken our direction from the immigrant community to this point, it was time for city government to lead the way.

Fair Haven Public Library

Residents' views of local police, shaped by their experiences in their home countries, was not always positive. Additionally, the role of the police to enforce federal immigration law and ultimately the consequence of immigration enforcement, deportation, and family breakups was unclear. The truth be told, this was a matter of policy somewhat unarticulated by both city hall and police headquarters prior to issuance of the General Order.

City police had their concerns as well. Their ability to make cases was compromised when witnesses and victims refused to come forward due to immigration status concerns. Classic 'broken windows' policing strategies would suggest that failure to prosecute crime successfully in the undocumented community would result in its magnification across the city

Police staff had been at work on the General Order for months. Sgt. Luiz Casanova was then District Manager for the Fair Haven Police District. He was one of many working hard on structuring the language and intent of the Order. He provided the direct, hands-on experience that validated the component of community policing that involves engaging positively with the community in which the police work. Luiz Casanova is now Assistant Chief of the Department.

Internally, the members of the command staff of the Police Department were all on board with the new policy. There was some concern from Local 530, the union that represents all sworn personnel in the city police department except for the ranks of Chief and Assistant Chief. The Local raised some questions about the draft General Order having to do with seeking clarification regarding officer liability for actions not taken to enforce federal immigration laws. Once those questions were answered (there is no officer liability), the Local did not file any challenge to the draft Order.

In fact, immigration enforcement had rarely been a factor in the actions of the New Haven Police Department. Traditionally, these were policies and enforcement activities that the federal government would carry out on its own. Additionally, immigration violations were civil offenses, not criminal offenses. We would rarely judge going after civil offenses as the best use of local police resources.

Of course, many attitudes about immigrants changed after the September 11, 2001, attacks on the United States. As an example, the U.S. Department of Justice took the position that local law enforcement had the right to make arrests in the case of civil immigration offenses after the 9/11 attacks.

Justice Department priorities in the years after the 9/11 attacks appeared to be increasingly narrow from the point of view of places like New Haven. In 2007, our community experienced 13 homicides, 163 non-fatal shootings, and reported 9,672 so-called "Part I," or major, crimes to the FBI. Manuel Santiago was one of those statistics. His death and the other crimes experienced by our community bore little resemblance to the priorities and resources of the Justice Department at that time. New Haven was interested in addressing the actual behaviors that were harming our residents and putting our children at risk. Failing to do so would delegitimize the government and our police in the public's mind, and destabilize our neighborhoods, all of them, not just those made up of recent immigrants.

The General Order issued by Chief Ortiz was a straightforward articulation of the City's rapidly developing immigration policy. The Order stated:

> The City of New Haven is home to a diverse population. Many of its residents have emigrated here from other countries, and some are not citizens of the United States. The

City and the Police Department are committed to promoting the safety and providing proactive community policing services to *all* (italics added) who live here. Residents should know that they are encouraged to seek and obtain police assistance and protection regardless of their immigration status.

The department relies upon the cooperation of all persons, both documented citizens and those without documentation status, to achieve our goals of protecting life and property, preventing crime and resolving problems. Assistance from immigrant populations is especially important when an immigrant, whether documented or not, is the victim of or witness to a crime. These persons must feel comfortable in coming forward with information and filing reports. Their cooperation is needed to prevent and solve crimes and maintain public order, safety and security in the entire community. One of our most important goals is to enhance our relationship with the immigrant community, as well as to establish new and ongoing partnerships consistent with our community policing policy (New Haven Department of Police Service).

General Order 06–2 fit very comfortably in the city's immigration policies as well as within the culture and practice of the New Haven Police Department's community policing strategies. New Haven community policing is predicated on a core conception: to know the community. And to know the community means to spend time in the community: walking the beat, visiting with residents on playgrounds, sharing in their moments of both joy and loss. It is an all in approach. New Haven community policing rejects the role of the police as an occupying force in the neighborhood. Our neighborhoods are not enemy countries; if you think and act that way, you have lost.

The General Order went on to stipulate these key points:

- Police officers shall not inquire about a person's immigration status unless investigating criminal activity;
- It shall be the policy of the department not to inquire about the immigration status of crime victims, witnesses, or others who call or approach the police seeking assistance;
- No person would be detained solely on the belief that he or she is not present legally in the United States, or that he or she has committed a civil immigration violation. There is no general obligation for a police officer to contact U.S. Immigration and Customs Enforcement (ICE) regarding any person unless that person is arrested on a criminal charge; and
- Officers shall not make arrests based on administrative warrants for arrest or removal entered by ICE into the FBI's National Crime Information Center (NCIC) database, including administrative immigration warrants for persons with outstanding removal, deportation, or exclusion orders. Enforcement of the civil provisions of U.S. immigration law is the responsibility of federal immigration officials (New Haven Department of Police Service).

Enforcement criteria and actions were consciously articulated within the philosophical framework of the Police Department's community policing program. Officers were trained to the order and have consistently performed not only to the letter of its provisions but its spirit as well. No one wants civil relationships with the neighborhoods more than the city's beat cops. When residents fear the police, it is only a short step to police fearing the residents. This too is a prescription for disaster.

As would later turn out to be the case, less-than pleased with this action were national immigration officials, who began to see the city and its police department as increasingly antagonist to its goals of civil immigration enforcement. Meanwhile, in Washington federal policy makers continued to kick the immigration policy can down the road.

The City announced the General Order at a press conference on December 21, 2006. We also indicated at that time that step number three, the issuance of municipal identification cards, a first in the U.S., would be undertaken in 2007.

Setting the Table for Resident Identification Cards

As New Haven began planning for issuance of the identification card, we took practical steps to tamp down the sense among some in the city that undocumented immigrants were somehow being treated as privileged. While all of us in my administration recognized that the politics surrounding immigration issues were more welcoming in New Haven than in many other places around the U.S., care was nonetheless taken to treat the identification card as an initiative the benefits of which the community as a whole would enjoy.

The first issue was leadership. Throughout the time of engaging our immigrant community up through the release of General Order 06–2, the city was alternately shepherded, prodded, and poked by the Fair Haven-based community organization JUNTA, headed by Kica Matos. Kica was a lawyer who practiced defense law in death penalty cases and came to New Haven some years earlier. Through that time, I came to possess an incredible respect for her skills as a lawyer and social justice activist. In January 2007, as the city was moving to issue the nation's first resident identification card that would be made available for undocumented and documented residents alike, Kica joined the administration as Deputy Mayor for Community Services. This was the area of the government responsible for, among other things, immigrant communities and activities. There was no better general to direct this initiative than Kica Matos.

Second, the city consciously communicated that undocumented residents not only had rights, but responsibilities as well. And so on January 29, 2007, the city announced that its Volunteer Tax Assistance Program (VITA) would be expanded to the undocumented community. Up until this time, VITA was aimed at assisting low- and moderate-income residents in the city with filing their tax returns. In 2006, my administration had assisted some 773 filers pay their taxes and qualify for benefits such as Earned Income Tax Credits.

In 2007 we decided to expand the VITA program. While undocumented filers did not have a social security number, VITA would now assist them in getting Individual Tax Identification Numbers (ITINS) so that they could file their first tax returns. The goal was for all residents to pay their fair share of taxes and to build a record of tax receipts and residency for undocumented residents. It was also the city's goal to communicate that residency without legal status did not mean a free ride for anybody. The city was saying that our expectations for all residents were the same. At the last minute, the Internal Revenue Service cancelled its participation in the press event making this announcement. It did, however, accept the taxes due as calculated from the tax returns filed on behalf of our documented and undocumented residents alike.

Third, our goal was to make the card an initiative that did not come at the expense of other public services. I judged that the noise that would result from funding the initiative with taxpayer dollars would obscure the larger goal. Accordingly it was decided that the expense of the program would not be funded by tax dollars. First City Fund Corporation (FCFC), a nonprofit foundation founded in the city in 2004 to support economic development principally through the creation of a community development bank (Start Community Bank chartered in December 2010), agreed to fund the cost of the first year of the program.

Fourth, the city decided that the identification card should be as broadly designed in its utility as possible. This decision was made to avoid branding the card as the illegal immigrant card, available to and to benefit only one class of city residents. The card was named the Elm City Resident Card (the Card) and embraced a number of uses. Not only an identification card, the Card could also be used to access libraries, parks, and other city services. It was adopted by the public schools as a student identification card. The Card could be used as a debit card and for discounts at a number of city retail businesses. The Card was designed to load money on to it to pay the fare at city parking meters as well.

Finally the city engaged in broad due diligence. City staff met with the IRS to establish the standards by which personal identification and residence would be determined for each cardholder. The decision was made to use the same identification and resident standards as the IRS used to issue ITIN's for collection of federal taxes. City staff contacted and held meetings with the Connecticut Bankers Association, the Connecticut Attorney General, and the U.S. Attorney's Office. While the U.S. Attorney was noncommittal, staff opposition to the card was voiced at our meeting. Staff argued that the city would be issuing "breeder" documents that might be used to build false identities by terrorists. As with the concern at ICE over issuance of the Police General Order 06–2, this opposition would manifest itself in actions taken against the immigrant community later that summer. The city and its community-based allies kept going.

Opposition

Neither the launch of *Hablamos Español* nor the issuance of General Order 06–2 had resulted in much public debate. It would be clear later that the General Order had caught the attention of ICE, and that the prospective issuance of the Card had its detractors in the U.S. Attorney's Office, but the city's immigration initiatives were largely below the radar of real concern of public opinion in New Haven. That would change with the launch of the Card.

As is often the case with any initiative, the law of unintended consequences would now come into play. Vocal and aggressive opposition arose to New Haven's plan to issue the Card. So vocal and so loud was the opposition that it caught the attention of the broader New Haven community and caused it to think long and hard about immigration policy and also about what it meant to be a resident of New Haven. In time, the dispute and challenges to the Card and the city's undocumented community served to strengthen the social fabric of the city across neighborhoods and ethnic and racial groups in meaningful ways and to such an extent that few, least of all I, had imagined possible.

The first public forums on the Card were held at the city's legislative body, a 30-member Board of Alders (the Board), all of whom are elected by district. Since no taxpayer funding was involved, no city legislative action on issuance of the Card was required. Issuing the card was purely an administrative action by city hall. However, the Board would have to vote to accept the grant from FCFC to fund the cost of issuing the Card. While largely supportive of the Card, the Board held public hearings on acceptance of the FCFC grant which became the platform for proponents and opponents of the Card alike.

As opposition to the Card arose, the city and neighborhood advocates had decided to pursue and conduct a very public campaign supporting the adoption of the Card. Community organizing, lawn signs, t-shirts, and rallies were all part of the Card launch. What had started as an effort to provide an identification tool to the city's undocumented residents was very consciously turning into a campaign to define the meaning of membership in our community and a very visible test of social tolerance in New Haven. It was also becoming something bigger than just the city, as national opponents to immigration began to see New Haven as a platform to engage the larger national immigration reform debate. As more opposition presented itself, the proponents became more energized and the stakes even larger. And as the opposition to the Card became national, so too did its supporters.

This was especially true following adoption, in December, 2005, by the U.S. House of Representatives, of the so-called "Sensenbrenner Bill." Named for Wisconsin Representative Jim Sensenbrenner (R-WI), the bill would have, among its other provisions, imposed meaningful penalties on anyone assisting undocumented immigrants. The Sensenbrenner Bill offered a stark contrast to the direction New Haven was taking.

As public hearings commenced at the Board, an organized opposition emerged. Calling themselves the Southern Connecticut Citizens for Immigration Control (SCCIC), they wore yellow hard hats and were an all–white organization with links to John Birch-affiliated groups posted on their website. We were never able to identify any SCCIC members who were residents of New Haven, however.

Proponents of the Card came to see SCCIC as offering us an opportunity to characterize opponents as representing a set of values in stark contrast to those we believed represented those of New Haven residents in particular and the United States in general. The opponents were intolerant, not diverse, and they were definitely not staying true to the nation's traditional values. SCCIC was afraid of the future while Card advocates were rushing to embrace it. The opponents to the Card lived up to their hatreds. Their actions generated an unrivaled volume of hate communication, the likes of which I had not seen in my 20 years as Mayor of the city. Kica Matos received death threats. The opponents presented a clear, contrasting vision of citizenship–one that was narrow and driven by fear.

On June 4, 2007, and by a vote of 25 to 1, the Board voted to accept the FCFC grant to launch the Elm City Resident Card. A contest had been held to shape a more inclusive, more connected future for our city and that vision prevailed. And then the raids began.

The ICE Raids

About 36 hours after the Board voted, at 8:15 a.m. on June 6, 2007, a call came into the Mayor's Office. The citizen caller said that police raids were going on in the Fair Haven section of the city. Then a second call came in, also reporting raids in Fair Haven. Now this was unusual, for, as a rule I would receive notification from our police department of any police actions of any scale in New Haven. I called Chief Ortiz to ask about the raids. His response to me was: "what raids?" The Chief said he would call back. This was not good.

Subsequently what became known was that at 5:30 a,m. that morning raids were launched at seven locations in the city, mostly in Fair Haven. The raids were organized by ICE and virtually every federal law enforcement agency that had personnel authorized to carry a weapon participated in them. These were the first raids of this nature ever undertaken by ICE in the City of New Haven. ICE was not attempting to serve warrants. Rather, they were working off a dated list of 25 fugitives on whom orders to deport had been issued.

On that first day, some 29 residents were arrested. Three more were arrested in raids carried out the next day. 27 of the 32 arrested had no prior contact with ICE. These unlucky others just happened to be at the wrong place at the wrong time and had brown skin. They were categorized by ICE as collateral detentions.

Those arrested were detained throughout New England. Attempts to locate the 32 arrested before they were likely removed to facilities in Texas and Arizona were frustrated by ICE. CLC students were tasked with blind calling jails throughout New England on behalf of frantic family and community members.

Ultimately, 31 of the 32 arrested were located. The CLC represented those arrested and the community conducted a drive to raise the funds to cover the cost of their bail. 30 of our residents were released.

As it turned out, the NHPD was never notified or asked in advance to participate in the June 6th raids in its own city but carried out by ICE. Eventual notice of the ICE raids was passed along to New Haven Police but only after they had begun. That notice was made by ICE through a call to an operator on the city's 911 emergency hotline. No advance warning or notice to a command officer of the New Haven Police Department was ever given by ICE. Both failures were a violation of ICE's own internal policies. Self–evidently, ICE acted in a fashion to ensure that neither the city's police nor elected leadership would have the opportunity to object to the raids before they commenced.

The raids were clearly in retaliation by ICE for the city's immigrations policies, executed not on city leadership but its more vulnerable immigrant residents. ICE chose to take on those in New Haven whom they judged to be powerless.

On the day following the ICE raids of June 6th, some 1,500 people attended a rally at St. Rose of Lima Church protesting the unprecedented action by ICE. That same day some dozen businesses in New Haven did not open. Their workers were afraid to leave their homes.

On the same day as the raid, the CLC organized teams to go out to the street addresses where the raids had taken place to take depositions of witnesses to the raids. The following are select excerpts from the depositions of witnesses and family members of those seized in the raids:

In the words of a fourteen-year-old witness whose mother was taken away:

The officer told us that … my mother had to go with them for the paperwork. My mother asked if she could at least change her shoes before she left. The officer sent my father with another officer to look for the shoes. My father came down with shoes and socks … My father gave the shoes to my mother and she said they were the wrong shoes. One of the officers said, "It doesn't match her outfit" and then all the officers proceeded to laugh. My father went to get the other shoes, my mother put them on and got up and the officers handcuffed her. My mother said, 'Oh my god' and they took her away into a van … I was desperately crying after all the agents detained my mother.

A New Haven resident who witnessed the event states in her sworn testimony:

In the early morning hours of Wednesday, June 6, 2007, I was awakened by a loud knock on the door. It was about 6:30 a.m. My cousin opened the door. At that point eight men pushed their way through the door and entered the apartment. They were wearing uniforms; they were dark green, some of them read 'Police' in the back, others read 'POLICE-ICE.'

They did not identify themselves nor did they show any badges or paperwork of any kind (Staff Memo).

In 2007, ICE maintained that the raids were not retaliatory in nature, and they continue to do so to this day. It turns out that the raids had been scheduled for months and had been pushed back several times to the date on which they finally occurred. As ICE and related law enforcement e-mails obtained by the city subsequent to the raids demonstrate, the attitude of the officers participating in the raids is captured in the following exchange:

Hey Carmine … We have an op. scheduled for Wed, 05/02/07 in New Haven @ (redacted). I know you guys work nights, but if you're interested we'd love to have you! We have 18 addresses—so it should be a fun time!!

Let me know if you guys can play!! (redacted.)

In the days following the raids, the community and the city chose to escalate the matter, just as we had with SCCIC. The community did so with multiple rallies and marches. The city chose to engage its Congressional Delegation to challenge the Department of Homeland Security (DHS). Coming on the heels of adoption of the Card, the raids provided an opportunity to send a message in New Haven and to the nation just how arbitrary and uninformed federal immigration policy, and its execution, had become. The DHS and ICE denials that the raids were not retaliatory were, on their face, absurd. The denials framed the federal officials as dishonest and diminished their stature. Further, by breaking into homes without warrants or identification in some instances, they came across as bullies. In acting in this fashion, DHS and ICE provided immigration reform advocates a powerful message about the need for reform. And with the retaliatory raids we in New Haven became advocates for national reform.

In my telephone conversation with DHS Secretary Chertoff on the day following the raids, he denied that ICE's actions were in retaliation for the city's immigration policies and refused my request to suspend any future raids. After the intervention of New Haven Congresswoman DeLauro and Connecticut U.S. Senators Dodd and Lieberman, within days, DHS and ICE announced a temporary suspension of raids. And in fact they have not been back to New Haven since. In effect, and as a result of their own heavy handedness, ICE created the nation's first sanctuary city.

As it turned out, the opposition by SCCIC, national immigration reform opponents and ICE was key to promoting the broader success and meaning of New Haven's immigration initiatives. In New Haven, we started out on these efforts with relative acceptance by the New Haven public. The immigrant initiatives were home grown and had always focused on our community. However, the rhetoric and anger of some of the opposition caused many in Connecticut and in power to come to similar views concerning the importance of immigration to our economy and about immigration's immensely positive effect upon our vibrancy and values as a nation. The opposition did not change our message but it did change our audience.

The June 6th and 7th, 2007, ICE raids highlighted the incoherence as well as the arbitrary and abusive nature of the federal government on matters relating to immigration. In many ways, the overreaction of the federal government to the city's immigration policies did more to gain support for our immigration initiatives than any effort we undertook ourselves. For many, the federal government, by its own actions, lost whatever credibility it had on the matter.

The Elm City Resident Card

July 24, 2007, was the first day that the Card would be available to the public. Just as we had carefully considered in January 2007 what was the right room to hold our first mass meeting with New Haven's undocumented residents, so too was careful thought given to the public space that would be used to process Card applications. It was decided that the office would be located in a room just off the main entrance to New Haven's circa 1863 city hall, located on Church Street opposite the New Haven green.

As Mayor, I would generally arrive at city hall at 6:30 a.m. That day, right off the bat, it was clear as I drove down Elm Street to the city hall parking garage entrance that something was up that first morning. City hall did not open to the public until 9:00 a.m. but there were lines outside city hall at that early hour. By Friday of that first week, the line for Cards stretched from the front doors of city hall on Church Street, down to the corner at Elm Street and from there down to Orange Street. On the first day, hundreds waited for up to 12 hours to fill out applications for the Card. In time, the number of Cards issued in our City of 129,000 would total in excess of 10,000. That day, and in the years that followed, many of us who were issued a Card were, like me, born in New Haven.

Also on that first day, the SCCIC was organized in protest to the New Haven residents pouring into city hall. SCCIC filmed those in line that first day and one of their leaders, Ted Pechinski of North Branford, worked a bullhorn at the crowd with the words: "You are only 3% of the population. We are going to fight you all the way … .We are at war, we will fight you to the end to win that war." (Staff)

Places that are at war with themselves fail. Ted's war is not a fight to the finish but a fight to the bottom.

Immigration Reform from the Bottom Up

In 2013, the Connecticut General Assembly passed legislation that was signed into law granting the State Department of Motor Vehicles the authority to issue Connecticut Drivers Licenses to undocumented residents of the State of Connecticut. There was little opposition in the Assembly discussion about the proposed law. That fight had occurred years earlier in New Haven. In 2014, New York City authorized the issuance of identification cards for all its residents. New York, like the dozens of other communities that have acted out of a positive vision of immigration's impact on our nation's future, is practicing national policy from the bottom up.

Coming in an era when the national government is increasingly finding itself, like Ted Pechinski, at war with itself, an ever larger number of cities are acting in ever larger numbers of areas fashioning the future of the United States. These cities embrace a tremendously optimistic vision of our shared future. It is a future that is being shaped by a new coalition of civic architects. These builders are the residents, social and economic justice advocates, places of faith, businesses and local institutions, and officials who all share the aspiration of American exceptionalism, driven by core values of hard work, persistence, and social tolerance.

Citizenship is the work we do to ensure the continuity of and change for the better in the lives of ourselves, our families, and our neighbors. All these efforts are evolutionary, and the only constant is change. At the same time that change is the only constant, the only question is whether we will be ruled by change or whether we will rule the change we want. Engaging change guarantees our freedoms and ensures our future. Stand still and you die.

Community-based activities and participation are the work of citizenship. It is work toward a goal that, in a final sense, is never achieved but it is the work that moves us forward day by day, year by year. It is true that we in cities often measure our wealth in buildings and places that we display in colorful brochures. In reality, the wealth of cities is not to be found in buildings. Rather, it is measured by the strength of its citizenship, by a community's social capital. It is the shared reservoir of values, mutual support, mutual respect for, and personal knowledge of, one another that builds great cities.

Baseball stadiums do not do it. Neither do convention centers or shopping malls. Rather, it is the fabric of shared connections, interrelationships, and a community organized to a purpose that is an irresistible force. And, as has been said before, force creates friction. Without some tension and conflict, though, how else can genuine respect, alliances, and accomplishments emerge? And it is in the struggle that follows our organization and action to a purpose, that our freedom, our future and meaning, and purpose are to be found.

Issues such as immigration, school reform, police behavior, authentic economic competitiveness, and equality of opportunity are all areas where cities must put their social capital to work in the coming years. Deficits in social capital result in the kind of terrible outcome seen in Ferguson, Missouri, and with the assassination of two New York City police officers in 2014. That is Ted's war, a place at war with itself.

What started among JUNTA, ULA St. Rose of Lima Church, and the community in 2004 informed one city's immigration policies. But in a much bigger way it contributed to New Haven's reservoir of social capital. With many challenges before us and the future to be won, it is a capital that must increasingly be invested in the struggles that are shaping the future of America, a future ever more being shaped by our cities.

Works Cited

Alinsky, Saul. *Rules For Radicals*. New York: Vintage Books, 1971. Print.

Bailey, Melissa. "City Unveils A New ID." *New Haven Independent*, 2007. Web. 8 May 2007 <http://www.newhavenindependent.org/index.php/archives/entry/city_unveils_a_new_id/>.

Bass, Paul. "Immigrant's Wake." *New Haven Independent*. New Haven Independent, 2006. Web. 25 Oct. 2006 <http://www.newhavenindependent.org/index.php/archives/entry/immigrants_wake/>.

Belkin, Douglas. "Chinese Undergraduates Help Set New Record." *Wall Street Journal*, 2014. Web. 17 Nov. 2014 <http://www.wsj.com/articles/chinese-undergrads-help-set-new-record-1416200461>.

Cooper, Robertson & Partners. *Yale University A Framework for Campus Planning*. New Haven: Yale University, 2000. Print.

Dahl, Robert. *Who Governs*. New Haven: Yale University Press, 1961. Print.

Junta for Progressive Action, Inc. and Unidad Latina en Accion. *A City to Model*. New Haven, 2005. Print.

Lawler, Martin and Stock, Margaret. "Saying No Thanks to 87,500 High Skilled Workers." *Wall Street Journal*, 2014. Web. 7 May 2014 <http://www.wsj.com/articles/SB10001424052702303647204579544362260141496>.

Mills Brown, Elizabeth. *New Haven A Guide to Architecture and Urban Design*. New Haven: Yale University Press, 1976. Print.

New Haven Department of Police Service. "Disclosure of Status Information: Policies & Procedures." *General Order 06–2*. New Haven: New Haven Department of Police Service, 2006. Print.

Osterweis, Rollin. *Three Centuries of New Haven, 1638–1938*. New Haven: Yale University Press, 1953. Print.

Author redacted. "Op. 05/02/2007." Message to Camine Verno. 30 Apr. 2007. Email. Web.

Staff. "ID Card's A Hit." *New Haven Independent*, 2007. Web. 24 July 2007 <http://www.newhavenindependent.org/index.php/archives/entry/id_cards_a_hit/>.

Staff Memo. *Impact of Ice Raids of Families & Community in New Haven*. New Haven: Office of the Mayor 7 Apr. 2008. Print.

Chapter 9

Can Charter Schools Save Cities? Social Capital and Urban Public Education

John A. Lovett

Introduction

Over the last few years, encouraging high school graduation stories have begun to be commonplace in New Orleans. In May 2014, Cohen College Prep, one of the city's 77 charter schools, graduated its first high school class. Cohen's student body is 98 percent African American and more than 95 percent of its students qualify for free or reduced-price lunch (Cowen Institute Governance and School Guide). Yet 100 percent of Cohen's 54 graduating seniors were accepted to colleges and universities, and collectively the graduates earned more than $2 million in scholarship aid. Many Cohen students will be attending colleges and universities in Louisiana with 100 percent of their tuition covered by TOPS (Taylor Opportunity Program for Students) scholarships. Others will travel as far away as Beloit College in Illinois and Bates College in Maine (Abul-Alim 2014).

Across town, at the New Orleans Charter Science and Math Academy ("Sci Academy" or simply "Sci"), where 89 percent of the student body is African American and 91 percent of the students qualify for free or reduced lunch (Cowen Institute Governance and School Guide), 94 percent of the students in its first graduating class in 2012 were accepted by at least one four-year college, and 90 percent of Sci's first graduating class were first-generation college students. In addition, 80 percent of Sci's students passed Louisiana's graduate exit exam and 60 percent of its graduating senior class scored at or above 20 on the ACT, thus qualifying for a TOPS scholarship at a public university in Louisiana. At Sci, the acceptance letters poured in from Dillard University, Amherst College, Bard College, and Louisiana State University (Carr 282).

Back at Cohen College Prep, one of its first graduating students, Leonard Galman, a 17-year-old African-American young man whose father was murdered and whose mother gave birth to him at the age of 13, was offered a full scholarship to Yale. Although he had only been enrolled for one year at Cohen College Prep and split his senior year at an arts-based charter school in another part of the city, Galman credited his Cohen College Prep counselors for raising his college aspirations higher than he ever thought imaginable (Dreilinger May 23, 2014).

At the same time that these exhilarating stories emerge, others in New Orleans lament that neighborhood schools are now a thing of the past (Kamenetz). Before Hurricane Katrina, public school students in New Orleans travelled an average of 1.9 miles to school. But by 2011–2012, public school students traveled on average 3.4 miles from their home to school (Sims and Vaughan 17), and 25 percent travelled 5 miles or more to their school (Kamenetz). As a recent report demonstrates, the commuting paths that students take to charter schools in New Orleans now often span both banks of the Mississippi River and can extend from the suburban sub-divisions of New Orleans East to the city's oldest neighborhoods nestled along the east bank of the Mississippi River (Kamentz).

These stories and these new school commuting trajectories both suggest that a significant change in the social structure of New Orleans is taking place. According to many sociologists and urban studies experts, large ecological, demographic and economic forces exert the most important influences on the social structure of a city. Some of those forces spring from the city itself, and some have external sources. This chapter addresses one emerging source of urban change originating in an intellectual movement started by a handful of sociologists and education reformers working in the 1980s.

In 1988, James Coleman, a renowned sociologist based at the University of Chicago, published an article titled *Social Capital in the Creation of Human Capital* in the *American Journal of Sociology*. In his article, Coleman brought analytical rigor to an idea that had been circulating in American and French intellectual circles for many decades and applied the idea to the problem of how to improve educational outcomes of disadvantaged young people in American cities. Within several years of the publication of Coleman's article, the first law authorizing charter schools in the United States was enacted and the first charter schools began to operate in American cities. Today more than 2.5 million students in the United States attend more than 6,400 charter schools (National Alliance for Charter Schools 2014, 1).

The charter school movement is, of course, deeply controversial. Proponents view charter schools (along with voucher programs) as valuable institutions that give families and students meaningful educational choice and give American society the best chance to overcome the pernicious effects of racial and economic inequality and create bridges between the well-off and the disadvantaged (Ryan 286–91; Saiger). Opponents of charter schools argue that the schools represent an elitist attempt to privatize public education, subordinate racial minorities and divert attention away from the need for more systemic social, economic and political reform that could permanently change the lives of racial minorities and the poor (Black; DeJarnatt; James). This chapter cannot attempt to resolve that debate and cannot answer whether charter schools in particular and school choice more generally lead to improved educational outcomes for students.[1] It should be noted, however, that the most recent "gold standard" study of urban charter schools authored by Stanford University's Center for Research on Education Outcomes (CREDO) shows that "urban charter schools in the aggregate provide significantly higher levels of annual growth in both math and reading compared to their TPS [traditional public school] peers" (2). In effect, this recent study by CREDO demonstrated that students in urban charter schools at least are receiving "the equivalent of roughly 40 days of additional learning per year in math and 28 additional days of learning per year in reading" than students in traditional public schools (2).

This chapter instead seeks to explain how the charter school movement grew out of an intellectual milieu nourished by the ideas of social capital theorists like James Coleman. It suggests that the ideas of social capital theorists can help us understand the structure and

---

1   For a condensed summary of the national debate over whether charter schools and voucher programs enhance student academic performance, see Ryan (218–25). For a summary of the debate over whether charter schools have directly led to improved student performance in New Orleans, see Garda (3–5, 16–17). For a summary of recent research on public charter schools' academic effectiveness, see National Alliance for Public Charter Schools. For the most recent national study of the impact of urban charter schools, see CREDO.

purpose of charter schools and the controversial ambition of charter school advocates and organizers. To illustrate the transformative potential of the charter school movement, this chapter concludes by describing the reconfiguration of the public school system in New Orleans, Louisiana, in the wake of Hurricane Katrina, reporting on some initial studies of parent and community satisfaction with the new universal charter school system and recounting a powerful narrative of three charter schools in post-Katrina New Orleans.

As the rest of the chapters in this volume demonstrate, cities can accomplish a myriad of tasks that can improve the daily lives of their residents while simultaneously mitigating some of most deleterious effects of global climate change. They can lead the way in creating more effective transportation systems. They can build infrastructure that reduces energy consumption and fosters the development of renewable energy systems. They can create new forms of affordable housing that will ease some of the pressure of economic inequality. But even if cities do all of these things, they will still fail as human institutions unless their inhabitants are afforded the opportunity to develop their human capabilities—their human capital—to the fullest extent. Public schools are thus indispensable for human flourishing in cities. Without reliable and effective public schools, many urban residents—especially those at the bottom of the socio-economic ladder—will struggle to thrive in cities. Anyone interested in how cities can realize their potential for human development will want to understand the radical experiment in public school design that has taken place in New Orleans in the past decade and its roots in social capital theory.

Another reason that New Orleans' charter school experiment should matter to readers of this volume lies in the breadth of the structural reform adopted in the city. As Robert Putnam has demonstrated recently in his book, *Our Kids* (2015), a widening opportunity gap between children from families at the bottom and top of the US social structure has emerged in the last 50 years. As Putnam also notes, public K-12 schools in most American cities today tend to exacerbate the advantages and disadvantages created by family structure, parenting styles, and neighborhood effects that children from different ends of the socio-economic structure already bring with them to school (2015: 135–190). New Orleans's adoption of a universal choice model of charter schools in which every family has—at least in theory—an opportunity to choose a school that suits the needs of its children regardless of the family's geographic location in the city promises to break this cycle of public schools serving as an "echo chamber" of pre-existing social inequality (Putnam 2015: 182).

Finally, readers should be interested in this chapter because of the sheer rapidity in which this social innovation was achieved. After Katrina displaced the entire population of New Orleans, many urban planners viewed the city as a *tabula rasa* on which the latest theories of urban planning and sustainable development could be implemented in short order. However, land use patterns are very difficult to change, even after a disaster on the scale of Katrina. Indeed, most of the visionary urban planning ideas articulated in those early days failed to take hold and the city's land use patterns have only changed modestly in the 10 years since Katrina. As the second half of this chapter will show, however, a complete transformation in the city's public education structure was achieved in a decade—a remarkable fact in its own right.

James S. Coleman and the Foundations of Social Capital Theory

James Coleman was far from the first intellectual to use the term "social capital" to describe a phenomenon that enables certain individuals and communities to accomplish particular social ends. In 1916, Lyda Judson Hanifan, the superintendent of a rural school district in West Virginia, described the efforts of another school reformer (Lloyd T. Tustin) who organized adults in another rural school district to participate in meetings where school issues were discussed, which in turn led to meetings where other matters of local concern, such as road improvement and farming methods, were addressed (Fischel 113). Several years later, Hanifan observed that these activities built up the "social capital" of the rural communities, which she defined as "those tangible assets [that] count most in the daily lives of people, namely good will, fellowship, sympathy and social intercourse among the individuals and families who make up a social unit" (1920, 78). In subsequent decades, several other social observers, including notably Jane Jacobs, occasionally used the term, without defining it rigorously (Halpern 6).

In the 1980s and early 1990s, French education sociologist Pierre Bourdieu sought to expand the general notion of capital to include, not just economic capital, but also cultural and social capital. He defined the latter as "the sum of resources, actual or virtual, that accrue to an individual or a group by virtue of possessing a durable network of more or less institutionalized relationships of mutual acquaintance and recognition" (Bourdieu and Wacquant 119). At more or less the same time that Bourdieu was developing his theories, Coleman published his landmark paper.

Coleman sought to integrate the sociologist's tendency to see people as governed primarily by exogenous and impersonal norms, rules, and obligations with the economist's predilection for viewing individuals as autonomous, rational utility maximizers. He believed that both the sociologist's and economist's world views were incomplete and offered the concept of social capital as a way to unify them into a more general theory of social behavior.

Taking a functional approach, Coleman described social capital as an aspect of a social structure that helps actors within the structure achieve certain ends or achieve those ends in more efficient ways (96). Like physical capital and human capital, social capital "facilitates productive activity" (96). But unlike physical capital, which is created by changing raw materials into tools, machines, and productive equipment that is used to create things of value, and human capital, which is created by changes in persons that give them skills and capabilities to act in new ways and make new choices, social capital, per Coleman, "exists in the *relations* among persons" (100–101). Like an intangible lubricant, social capital helps the machinery of a social organization work more efficiently by reducing transaction costs and connecting the players within the structure through networks of trust and reciprocity. To illustrate social capital in practice, Coleman described how wholesale diamond merchants in New York City constantly exchanged their goods for inspection without formal contracts, observed the study circles of radical student activists in South Korea, and pointed to the close-knit merchants in the Kahn El Khalili market of Cairo, Egypt (99–100).

Drilling deeper into social capital, Coleman identified three constitutive elements or forms of social capital. First, there are obligations and expectations of mutual reciprocity, which Coleman illustrated by describing intangible "credit slips" that one person obtains by doing something for another person (102–104). The second element of social capital resides in information channels. Because people need information to be able to act, but information is often costly to obtain, persons with effective social capital can rely on their social relations to obtain the information they need in order to act more efficiently than people without social capital (104). The final piece of social capital, according to Coleman, resides in norms and effective sanctions. Norms can either set expectations for achievement that improve a group or community's ability to achieve some goal or they can sanction conduct that would undercut the chances of realizing the goal. Coleman observed that norms and sanctions can be internalized or supported by external rewards and punishments (104–105).

Another of Coleman's important theoretical insights concerns the nature of social networks that are most likely to contain a high level of social capital, especially in the form of effective norms. Coleman distinguished between two types of social networks: ones with "closure" and those without—what he called "open structures" (105). Coleman claimed that for effective norms to arise in a social structure, the structure must be one with closure. In a community without closure, persons with authority (for example, parents, school teachers or school administrators) will find it much harder to impose negative externalities to sanction those who require supervision or to impose positive externalities in the form of rewards. Coleman theorized that when parents of children have social ties with the parents of other children in the community, the parents will be able to establish more effective norms of behavior. In other words, when "the parents' friends are parents of the children's friends," the parents will achieve what Coleman calls "intergenerational closure" (106), a social fact which enables parents to monitor and guide behavior of all the children in a community more easily. Coleman also observed that it is easier to build trustworthiness and generate reciprocal obligations and expectations in a social structure with closure than in an open structure because in the latter defection cannot be penalized and reputation is harder to establish (107–08).

The remainder of Coleman's article probed data contained in a famous, long-term study of educational outcomes called *High School and Beyond*, to find support for his theory that enhanced forms of social capital could lead to the formation of more human capital. Among his key statistical findings were that: (1) children from families with two parents rather than one were less likely to drop out of high school; (2) children with only one sibling were less likely to drop out of high school than children with four siblings; (3) children whose mother expected them to attend college were less likely to drop out than children whose mother did not have this expectation; (4) the more frequently a child changed schools because the family moved, the more likely it was the child would drop out; and (5) children in religious-based high schools (primarily, but not exclusively Catholic high schools) had lower drop-out rates than children in public high schools or secular private high schools (111–14). Coleman asserted that the first three of these findings support the idea that a family's social capital could have a positive effect on the formation of children's human capital. He also claimed that the last two findings supported the notion that strong "intergeneration closure" in a community could improve the educational outcomes for a community's children (113–15).

Coleman's final theoretical insight also has important implications for the charter school movement. Social capital, he observed, is often a public, rather than private, good. Unlike the benefits of physical capital, which can be captured by the person who invests in the capital through property rights, and unlike the benefits of human capital, which can be captured by an individual seeking education through a higher paying job, more satisfying work, and greater understanding, the benefits of social capital, Coleman observed, often "do not benefit primarily the person or persons whose efforts would be necessary to bring them about, but benefit all those who are part of such a structure" (116).

Coleman illustrated this point by discussing a hypothetical school PTA, which provides just the sort of dense network of associations on which social capital depends. If one crucial member of the PTA withdraws, perhaps to take a full-time job or move to a new community, that decision, while entirely rational from the particular parent's perspective, can cause a significant loss of social capital for everyone else in that school community (116–17). Put differently, if someone lives in a neighborhood rich in social capital, that person can enjoy the added safety and security of the neighborhood even if he does not attend the barbecues, cocktail parties and community meetings that produce its social capital (Putnam 2007, 138). Because of its public goods quality, rational actors and communities can underinvest in social capital formation, and unless individuals or communities can be convinced to invest in social capital formation, this form of capital can easily dissipate over time (118). Based on these observations, Coleman worried that because traditional social structures that supply social capital—strong families and strong organic communities—were diminishing, we could expect to see "declining social capital embodied in each successive generation" (118). To counter this disheartening prospect, Coleman contended that society should substitute "some kind of formal organization for the voluntary and spontaneous social organization that has in the past been the major source of social capital available to the young" (118).

Components and Sub-Types of Social Capital

Building on Coleman's foundational work, many other social scientists have enriched our understanding of social capital and sought to test whether social capital, as formulated by Coleman and others, is in fact an important social phenomenon that is correlated with other desirable social outcomes. Most famously, in his sweeping book, *Bowling Alone*, Robert Putnam defined social capital as "connections among individuals—social networks and the norms of reciprocity and trustworthiness that arise from them" (19). Putnam argued that social capital, especially in the sense that it is aligned with "civic virtue" and "civic engagement," had, after rising steadily during the first six decades of the twentieth century, declined significantly in the last three decades of the century. *Bowling Alone* captured the imagination of many social observers and policy advocates, both nationally and internationally, but also prompted many social scientists to question the accuracy of Putnam's description of social capital, his empirical claims, and his explanation for the purported decline in social capital.

Despite the controversial nature of Putnam's work, consensus has emerged on the three constitutive components of social capital, regardless of whether the scope of analysis focuses at the micro-level of the individual or family, the meso-level of the community or organization, or the macro-level of a region or nation-state. These components closely mirror Coleman's elements. First, there is the social network or the set of relationships between members of a social structure. These networks can involve deep and strong bonds among family members and close friends or looser associations among neighbors, co-workers, and members of a voluntary or professional organization. The networks can be characterized by density (the proportion of people in the structure who know each other) or, as Coleman would say, by "*closure* (the preponderance of intra- versus inter-community links)" (Halpern 10).

The second component of social capital resides in the social norms that organize the networks of relationships. These social norms can be implicit or unwritten values or understandings. They can be habits or expectations. They can be formal codes, written rules, or even laws. The third widely recognized component of social capital consists of sanctions. Like norms, sanctions can be informal behaviors such as shunning, gossip or expressing disappointment, or they can take the form of more formalized punishments (Halpern 10–11).

The next crucial piece of social capital theory is the recognition by many social scientists that there are different "types" of social capital. Putnam himself drew this distinction in *Bowling Alone*, arguing that some forms of social capital are "inward looking and tend to reinforce exclusive identities and homogenous groups" (22). He called this "bonding social capital" and gave as examples the connections among members of a religious fraternal organization like the Knights of Columbus or exclusive country clubs. Other networks, which Putnam described as "bridging social capital," are "outward looking and encompass people across diverse social cleavages" (22). Examples would include the civil rights movement of the 1950s and 1960s, youth service groups, and ecumenical religious organizations. Putnam's own metaphorical description of the two forms of social capital has been repeated by many: "Bonding social capital provides a kind of sociological superglue whereas bridging social capital provides a sociological WD-40" (23).

As David Halpern remarks, the distinction between bonding and bridging social capital mirrors a distinction that Mark Granovetter drew earlier between "weak" and "strong" social ties. (Halpern 19). Weak ties—the kinds of connections that people have with business acquaintances—can help people acquire information and learn about new opportunities. Strong ties—like the bonds between close friends and family members—can provide "a more intense, multi-stranded form of support and as such might be expected to play a greater role in emotional well-being" (Halpern 20).

Halpern notes that while Granovetter's empirical work anticipated the bonding-bridging distinction in terms of networks, Hegel actually anticipated the distinction at the level of social norms. Hegel observed that "strong bonds of reciprocity and care"—what Halpern calls "normative bonding social capital"—are found among family members and inside small communities, but that more "self-interested norms" tend to predominate among relative strangers (Halpern 20). Because strangers might have less incentive to cooperate among themselves, Hegel suggested that societies would need to create norms of "impersonal altruism, notably through state action, through which relative strangers can cooperate successfully," a process that Halpern suggests we call "normative bridging social capital" (Halpern 20).

Although one might suspect that different groups or communities might be rich in one form of social capital but deficient in the other, Putnam's research shows that bonding and bridging social capital are not necessarily "inversely correlated in a zero-sum relationship" (Putnam 2007, 143–44), but often go together among individuals and groups (Halpern 21). This implies that researchers do not always have to worry about distinguishing between the two distinct forms of social capital: "If an individual or community is rich in one, then they will probably be rich in the other, too" (Halpern 21). On the other hand, the empirical work of Ronald Burt and others suggests that the bonding-bridging distinction is important because "bridging social capital has been found to decay at a much faster rate than bonding social capital" (Halpern 22).

Putnam and others have long been aware that social capital can also have an unpleasant, even pathological quality. To his credit, Putnam devotes an entire chapter of *Bowling Alone* to the "dark side of social capital" and confronts quite candidly "the classic liberal objection to community ties: community restricts freedom and encourages intolerance" (351). Putnam responds, in part, by pointing to substantial social science evidence establishing that the correlation between social participation and tolerance is, in fact, positive, especially when it comes to gender and racial tolerance. As Putnam puts it, "the more people are involved with community organizations, the more open they are to gender equality and racial integration" (355). At the same time, especially as he reflects on debates over the efficacy of public school integration efforts such as busing, Putnam acknowledges in *Bowling Alone* that to solve our most challenging social problems—particularly the inequalities created by the legacy of slavery and Jim Crow segregation—we will need to create not only the bonding social capital that forms a safety net for the weakest and most vulnerable members of society but also the bridging and opportunity-expanding social capital that is most difficult to build and sustain (363).

In response to this call, Halpern explains how a particular form of bridging social capital—what Woolcock, Szreter, and others call "linking social capital"—can help people escape the shadows and constraints of too much bonding social capital. If we examine a particular individual's social networks and identify the specific resources available to the individual, we see that sometimes an individual's social networks are much richer and resource-filled than another person in the same general community (Halpern 25). When a particular individual's social networks are generally weak, that person will have a much better chance of overcoming the asymmetrical power relationships of her society if that person can take advantage of networks that link people of unequal power or resources. As Halpern explains it:

> Given the inherent asymmetry of such bonds, linking social capital may necessarily involve norms of mutual respect or moral equality that counterbalance the narrow, rational self-interest of the resource rich ("normative linking" social capital). Hence linking social capital may be provisionally viewed as a special form of bridging social capital that specifically concerns power—it is a vertical bridge across asymmetrical power and resources (Halpern 25).

Halpern uses this concept of linking social capital as a lens to characterize different kinds of societies:

> [H]igh levels of bridging and linking indicate a society that is highly interconnected, thereby sharing power and resources through a never-ending and evenly spun web of connections. On the other hand, if the society is characterized by low levels of bridging and linking then . . . the society is fragmented into relatively disconnected personal networks or strata. In such a society, we will see power and resources heavily clustered into segregated (and protected) club goods, cliques, and "protected enclaves" (Halpern 25).

This seems like an apt description of much of contemporary America, especially the current organization of public education. Americans often express the desire to live in a society well endowed with bridging and linking social capital, yet they are often reluctant to invest significant resources to build that kind of social capital, especially if this would require any personal sacrifice or risk of their own social capital.

The public school system in many American cities—including New Orleans prior to Katrina—can also be characterized in these terms as a social structure comprised of clusters of rich bonding and bridging social capital in suburban school districts, where middle- and upper-class children are exposed to strong social norms and expectations of high academic achievement, and inner city or urban districts where bonding social capital may be strong but linking and bridging social capital is much less prevalent and educational expectations are much lower (Ryan 164–70). As we shall see, however, the universal charter school system that has emerged in post-Katrina New Orleans could be characterized as a social structure containing intentional networks that (1) aim to link persons between different strata, (2) are based on high academic expectations and little or no toleration for deviant behavior, and (3) rely on formal sanctions to enforce these rigid norms of behavior with strong forms of punishment and exclusion.

Education and Social Capital

Many social scientists have attempted to evaluate the intersection between social capital and education. At every level of analysis, social capital appears to have a powerful effect on educational outcomes and learning in a multitude of settings. While other forms of capital, particularly the human and financial capital of parents, have a significant effect on educational achievement, numerous studies have shown that when parents' economic status and education levels are controlled for social capital explains much of the remaining variance in educational outcomes (Halpern 142–43).

While the volume of research in this area is too vast to explore in detail here, a few conclusions are important for understanding why education reformers have focused so intensively on charter schools and other vehicles of school choice as a mechanism to increase the amount of social capital available to disadvantaged children and their families. First, many studies have confirmed that, at the micro-level, the size and quality of a child's immediate social network—principally the child's family—has a large effect on educational attainment. When parents are present and attentive, children score better on tests and are more likely to complete high school and attend college (Halpern 143). Second, numerous studies show that not only do parents' education levels predict their children's levels of educational success, but that when parents set high expectations for their children, the children also tend to achieve more (Halpern 144). While numerous studies have attempted to explore whether more parental involvement always equals more educational success for children or whether certain kinds or qualities of parental involvement matter more than the quantity of involvement per say, the general consensus is that: (1) two parent families yield more structural or network social capital, which in turn benefits children;[2] (2) having more siblings or being born later is associated with lower levels of educational achievement (Halpern 145–146); and (3) households where both parents work, especially where the mother works part time outside the home, tend to produce higher education attainment than families in which only one parent works outside the home, apparently because of the higher educational aspirations the working mothers impart to their children and because working parents seem to put greater emphasis on "quality" time (Halpern 146–47). Finally, studies have also shown that family social capital plays a consistent role across various ethnic and racial settings. In other words, differences in parent-child and parent-school social capital, not membership in any particular ethnic or racial group per say, appear to explain differences in education achievement levels across ethnic and racial lines (Halpern 148–49).

At the meso-level of analysis—that of the school or neighborhood—social capital also appears to play a very significant role in educational outcome. In fact, studies have confirmed Coleman's social capital findings that when "[p]arents, teachers and students were bound together in a network of shared values and with a high degree of 'closure'"—as in a typical Catholic school—educational outcomes improve, especially as measured by

---

2   Interestingly, the reason that single parents struggle to produce social capital is not attributable to lack of effort but rather to the fact that they move locations more often, thus disrupting the family's social capital and causing children to change schools more often (Halpern 146).

reduced drop-out rates (Halpern 151–53). Not surprisingly, smaller schools also typically produce stronger academic achievement and fewer behavior problems than large schools (Halpern 151–52). Research by Halpern and others shows, however, that the high levels of educational aspiration communicated to students in a particular school setting may be more important than parent-to-parent closure as a factor in predicting educational achievement. (Halpern 153) Indeed, Halpern's own studies of 15–17 year-old students in England revealed that parents' education levels predicted student educational achievement levels at both the individual and neighborhood level and that parent-to-parent socializing only reinforced, positively or negatively, the educational attainment of the children (Halpern 154). Transnational comparative analyses of students in Columbia, South America and India also show that bonding social capital, without bridging social capital, can lead children to poor educational outcomes. In these environments, when there is closure among parents, but parents come from a relatively low socio-economic strata and have low educational aspirations, tight bonding social capital without bridging social capital "can trap many young people in a cycle of low aspirations and low lifetime achievement" (Halpern 155).

Moving beyond the social structure of the school itself to the surrounding neighborhood, many carefully constructed sociological studies have confirmed that the average socio-economic status of a child's class or school, or, put differently, the neighborhood characteristics of the surrounding community, have a powerful predictive effect on educational outcomes and that the effects of this "neighborhood social capital" are just as powerful as that of a child's family or school itself (Halpern 155). The multi-level statistical analyses of Robert Sampson and Yongmin Sun, in particular, show the community or ecological effects on educational outcomes that are produced by merely living in an area with certain socio-economic conditions. By including "control," "structural" and "process" variables, Sun was able to identify the "community level effects" of living in a community with lower family income and higher minority enrollment in schools. Sun found that "poor families living in poor areas suffered from a double disadvantage, first from being poor themselves, and second from the poverty of their neighborhoods" (Halpern 156). According to Halpern, Sun's work confirms Coleman's hypothesis that "community characteristics affect educational performance" and reinforces his insight that "closure matters" (Halpern 157).

Sampson, Morenoff, and Earl's study of Chicago neighborhoods also confirmed Coleman's "closure" hypothesis by showing that "the best predictors of closure (parent-parent and parent-child connectivity) and neighborly exchange were concentrated affluence, low population density and residential stability" (Halpern 157; Sampson et al.). Concentrated disadvantage, on the other hand, did not inhibit parent-to-parent exchange but it did lower expectations for the informal control of children. Although strong community social capital cannot be reduced entirely to income levels because some lower-income communities can still create effective social networks and establish higher educational aspirations, higher income levels certainly make it easier to achieve beneficial forms of social capital (Halpern 158). Quoting Willms (2000), Halpern concludes that community-level social capital can have a "Double-Jeopardy" effect on educational outcomes for "low social capital, disadvantaged families who are in turn disproportionately concentrated in poor, low social capital communities" (Halpern 158).

On a more encouraging note, a number of studies have shown that strong social capital within a school's teaching staff can have a significant positive effect on educational outcomes for students. The cross-national comparative analyses of Hargreaves have shown that in countries like Japan, where teachers participate in frequent professional development activities, where teachers mentor one another, collaborate on lesson plans, and share successful techniques with one another, the teachers' collective social capital pays significant educational dividends. Conversely, cross-national studies have shown that unsuccessful schools are often characterized by high staff turnover and rigid work rules that restrict out-of-hours contacts between teachers, both of which tend to reduce teacher-to-teacher social capital, trust, and collaboration (Halpern 158–59).

Social Capital and Charter Schools

A handful of studies have attempted to measure specifically the degree to which charter schools build social capital and, if so, whether this additional social capital leads to improved educational outcomes. In one study, Cara Stillings makes a strong argument that charter schools do create social capital for certain families and students and these associational ties could be leading to improved educational outcomes for students.

At the outset, Stillings notes that the first major linkage between social capital and educational outcomes was found in the *Coleman Report* (1966), authored by none other than James Coleman. That report upended the long-standing assumption that differences in school resources and other characteristics were the dominant causes of different educational outcomes. Instead, analyzing national surveys and test score data, the *Coleman Report* found that (1) racial segregation in schools was still a major problem in the United States, (2) African-American and Hispanic children, who generally had lower test scores than white children, performed better when they attended schools that were majority white, and (3) family and community background had a much greater impact on educational outcomes than previously thought (Stillings 60).

The *Coleman Report* inspired court-ordered busing to combat de facto desegregation, increased funding for impoverished schools, and much additional research into the effects of family background, race and class on student achievement. As Stillings notes, one branch of that research, derived largely from ethnographic studies, found that certain habits—parents reading to children, helping with homework, communicating with teachers and school officials and setting high expectations for school achievement—can have positive impacts on educational attainment and that these habits occur more often in middle class homes than elsewhere. This research thus reinforced the belief that socio-economic status was highly correlated with the amount of social capital within families (60–61). The cumulative weight of the post-*Coleman Report* research, according to Stillings, shows that social capital is related to educational achievement and that socio-economic and cultural background can affect how parents behave towards schools and how schools behave towards parents and children. It follows that connections and associational ties between families and schools should be improved so that children from disadvantaged backgrounds can achieve more and society can interrupt cycles of poverty that trap children from these backgrounds (61–62).

Stillings sees charter schools as an intentional strategy to cultivate this missing social capital. According to Stillings, the charter school movement makes three primary claims about social capital. First, charter schools cultivate social capital by giving parents and community members a choice of where to send their children to school, and the exercise of this choice serves to forge ties between schools and communities while empowering parents to become more involved in schooling. Second, charter schools can produce better outcomes for students by making education a community endeavor. Because charter schools are, in theory at least, initiated by members of civil society and the local community must commit to the charter schools by giving them space and sending their children, charter schools should build more community ties from the start.[3] In other words, the crucial role of voluntary associations in forming charter schools should encourage social capital formation. Third, charter schools can generate more parental involvement by requiring parents to sign binding volunteer and participation contracts (62–63).

Stillings' first quantitative claim is that charter schools are activating demand; that is, they are cultivating ties to families and communities by bringing them into the school community. She finds evidence for this in two facts: (1) more and more students who are racial minorities are enrolling in charter schools and (2) the number of charter schools has grown exponentially. Unfortunately, Stillings does not consider whether the growth in charter school enrollment and the number of charter schools merely reflects a conscious decision on the part of school district officials to close traditional schools and offer charter schools in their place or the circumstances of a portfolio district like New Orleans where the only choices available are charter schools. To her credit, though, Stilling, just like James E. Ryan, the current Dean of Harvard's Graduate School of Education, acknowledges the possibility that self-selection may undermine the claim that charter schools increase the total amount of social capital in a community because parents who choose charter schools may already have higher amounts of social capital than other members of the community (Stillings 65–66; Ryan 219–23).

Stillings' second quantitative claim concerns parental participation and involvement in schools. "Participation" refers to simple actions like attending school events and supporting schools. Involvement, for Stillings, means activities like donating time and expertise to a school environment, including making contributions to curriculum, instruction, and school governance. Pointing to a 2003 study, Stillings notes that parental participation in charter schools is markedly higher than parental participation in traditional public schools when it comes to activities like open houses, back-to-school nights, and parent-teacher conferences. Similarly, charter schools also benefitted from higher levels of parental involvement than traditional schools as measured by indicators such as written parent-school contracts, involvement in school governance, and volunteering in schools. Stillings acknowledges that some of the parental involvement in charter schools may be coerced or "forced" to the extent that such schools require written parent-school contracts and thus some of this latter data could be biased. Nevertheless, whether this involvement is voluntary or partly forced, Stillings claims that "charter schools are forging reciprocal ties with the families and communities they serve" (67).

The third branch of Stillings' quantitative claims concerns constituent satisfaction with charter schools. Reporting on data from Vanourek (2005), Stillings notes that parents are generally "very satisfied" with charter schools (Stillings 68). Parents report that charter schools are open and accessible and are not shutting students out. Students generally rate charter schools more favorably than the previous schools they attended, especially when the former school was a public school. Teachers also report to be either very satisfied or somewhat satisfied regarding educational philosophy, teaching decision-making, community relations, parental involvement, and administrators (Stillings 68–69).

---

3   As shown later in this chapter, not all charter schools originate in a bottom-up, community-led process that Stillings envisions. In post-Katrina New Orleans, for instance, the vast majority of the city's charter schools were established by the state's Board of Secondary and Elementary Education (BESE) in a top-down fashion.

Although her analysis is generally supportive of charter schools as institutions that are capable of generating positive social capital, Stillings cautions that charter school researchers need to take into account the effect of selection bias and recommends that charter school organizers should work to overcome the possibility of selection bias by actively recruiting parents who have been alienated from traditional public schools and by streamlining charter school application processes to make them easier to understand and use by parents with less social and human capital. She also encourages charter schools to reconsider the use of parent-school contracts because they could well be shutting out parents with low amounts of social capital whose children could benefit the most from what charter schools offer (Stillings 61–62).

One of the most sophisticated statistical studies of the link between social capital formation and charter schools was recently performed by Tedin and Weiher (2011). They start from the premise that parental involvement in schooling improves student achievement, especially for students from low-income and minority households, and reduces the frequency of truancy, dropping out of schools and behavioral problems. Although many researchers equate parental involvement with schools as social capital, Tedin and Weiher distinguish between two types of social capital—general social capital and education-related social capital. Their research attempts to measure the degree to which the parental choice to send children to a charter school actually increases education-related social capital, which they define as involvement by parents in school activities.

Like so many others, Tedin and Weiher start with James Coleman and note his claim that school choice, if properly crafted, could be used to overcome the decline in the family and family-based institutions that put so many children at risk because school choice would catalyze parental involvement in schooling. While Tedin and Weiher acknowledge some evidence to support the suggestion that school choice increases parental involvement, they acknowledge the existence of the alternative hypothesis—that the observed school choice effect is mostly due to self-selection—and thus attempt to measure whether self-selection explains increased parental involvement in charter school settings.

In addition, Tedin and Weiher seek to determine what factors, if any, contribute to an increase in education-related social capital. Is it, as advocates of "family sovereignty theory" postulate, the very act of making a choice itself—particularly in an "option demand system," in which parents must first "choose to choose" because there exists a default placement for every child based on geographic proximity to a public school—that increases education-related social capital? This theory is based on the rational actor assumption: because school choice involves an investment in time and energy (and maybe even money), parents will be more likely to engage in behaviors that increase the value of this investment. Thus parents who choose charters will rationally feel more responsible for and have a sense of ownership in their children's schooling. Or, is any increase in education-related social capital mainly due to the second stage of the choice structure—the institutional structure and incentives inherent in the charter school itself? As Tedin and Weiher note, many charter schools are intentionally designed to create "functioning communities" (611). The schools intentionally force more links between parents and schools through more face-to-face contact and other forms of interaction. If this explanation is true, then the more years a family is connected with a charter school, the greater would be the level of education-related social capital.

Tedin and Weiher acknowledge that the available evidence suggests that families who choose schools have higher levels of general social capital (voluntary associations and activities that are not specifically education-related) than those who are not choosers. Not surprisingly, Tedin and Weiher admit that some of the difference in general social capital is due to selection bias, especially in an option-demand system. In such a system, it is often interpersonal contacts and general social capital that enable some parents to "choose to choose" (611). Tedin and Weiher also admit that there is some evidence that choosing parents sometimes have higher levels of education-related social capital as well. Thus, again in option-demand systems, some critics allege that choice schools are "creaming off" from traditional public schools precisely those parents with higher levels of education-related social capital and thus depleting this commodity from traditional public schools (611).

Tedin and Weiher discuss the studies of Schneider and his colleagues (1997), which showed that controlling for selection bias, choosing parents were more likely to be PTO members, more likely to volunteer at a school, talk to other parents about school issues, and more likely to trust teachers (Tedin and Weiher 611–12). Thus, Schneider found that in an option-demand system, school choice increased education-related social capital. Significantly, Schneider's finding was supported by similar results obtained by Kevin Smith (2005) for universal choice districts (where school choice is mandatory), thus ruling out selection bias as an alternative hypothesis.

Spurred on by this research, Tedin and Weiher designed a study to test whether school choice makes an independent contribution to the formation of education-related social capital that cannot be attributed to selection bias, nonrandom attrition, or the level of general social capital that choosing parents possess initially. Their data was derived from a Texas survey of thousands of parents of children who attended demographically similar charter and traditional public schools in choice-rich environments (for example, traditional and charter schools were located close enough to each other that parents had a choice to send their children to one or the other) (612–13). The parents in the survey were asked whether they: (1) attended a PTO meeting, (2) did voluntary work for the school, (3) attended a school board meeting, (4) helped with school curriculum decisions, (5) helped with school fund raising, or (6) attended parent teacher conferences (613–14).

Tedin and Weiher's analysis of the Texas survey data produced several significant findings. First, and perhaps most surprisingly, the mean level of education-related social capital for traditional public school parents was higher than for parents of students who had attended charter school for just one year (614). This was particularly surprising given that the parents of the charter school students had significantly higher levels of general social capital than traditional public school parents. This finding has two important implications: (1) the act of choosing schools does not, by itself, create education-related social capital; and (2) establishing a system of choice schools does not result in "creaming off" parents most involved in schools, at least not as an initial matter (614).

The second crucial finding of Tedin and Weiher has implications in a different direction. Essentially, Tedin and Weiher found that parents of children who have attended charter schools for two or more years acquired more education-related social capital the longer their children remained in a charter school. In particular, parents with children in charter schools for two years had more education-related social capital than not only parents with children in traditional public schools for the same amount of time but also parents with children in charter schools for just one year. Further, Tedin and Weiher found that parents of children who attended a charter school for three years or more have the most education-related social capital. This finding suggests that it is experience and time with a charter school, rather than the act of choosing itself, that leads to increased education-related social capital (614–15).

Interestingly, Tedin and Weiher's findings with respect to general social capital are consistent with what other social scientists have found and predicted about charter schools.

Based on eight indicators of general social capital, Tedin and Weiher found that parents of charter school children have higher levels of general social capital than parents of children in traditional public schools, but the level of general social capital does not increase appreciably the longer the children remain in a charter school (615–16).

Another implication of this research is that perhaps experience with charter schools raises the level of education-related social capital (but not general social capital) because the general social capital lays a foundation for later increases in social capital of all kinds. Noting that general social capital and education-related social capital appear to be correlated, Tedin and Weiher wanted to test whether the relationship is spurious. In other words, are parents with high levels of general social capital simply the kind of people who are more likely to take advantage of school choice and thus bring with them the potential to be especially active in the sort of voluntary school activities that constitute education-related social capital (616)?

Controlling for general social capital through a regression analysis, however, Tedin and Weihir were able to conclude "that charter schools perform a socialization function, where parents are socialized over time to become more involved in school activities—that is, the institution is able to inculcate gains in education-related social capital that is independent of the general social capital that parents bring to the charter schools" (618). This finding is significant for it shows that a government policy—establishment of charter schools—can modestly raise levels of education-related social capital.[4]

After reporting on an even more complicated multivariate regression analysis employed to test the impact of a variety of control variables, Tedin and Weiher confirmed that "experience with a charter school increases the level of education-related social capital among parents" (623). Nevertheless, Tedin and Weiher concluded that the many independent variables mentioned—in particular parental home resources and parental education—have a larger positive impact on education-related social capital than two or more years of experience with a charter school (624).

New Orleans Public Schools: Before and After Act 35

With this background on social capital in mind, we must now consider what has widely been recognized as the most sweeping charter school experiment in the United States—the almost universal adoption of charter schools in New Orleans over the course of the last 10 years since Hurricane Katrina forced the closing of all public schools in New Orleans in the fall of 2005. Before Katrina hit New Orleans, the public schools operated by the Orleans Parish School Board (OPSB) were populated almost entirely by African-American children. In 2004, 94 percent of the system's students were black and only 3 percent were white. Moreover, at 114 of the city's 129 public schools, more than 90 percent of the students were black and at 65 of the schools there was no racial diversity at all (Carr 63).

Although some OPSB schools had begun to show modest improvement before Katrina, a majority of the OPSB's schools were still considered "failing" under the state's school accountability system (Carr 64). Moreover, drop-out rates were high and the number of

---

4   Notably, Tedin and Weiher also found that the increases in education-related social capital among charter school parents over time were not the result of selective attrition—that children of parents with weaker levels of general social capital simply drop out of charter schools and return to traditional public schools (618–19).

students who achieved even satisfactory scores on state mandated performance tests was shockingly low (Dreilinger, Aug. 31, 2014). Media stories focused on the most tragic examples of academic under-performance like the valedictorian at Fortier High School in uptown New Orleans who was unable to pass the state's graduate exit exam for high school seniors (Carr 19).

As a governing entity, the OPSB itself was plagued with corruption and dysfunction. Between 1998 and 2007, the system was led by no less than eight different superintendents—two regular and six interim. Unable to provide an accurate financial statement, subjected to an FBI investigation that led to convictions of more than 20 school district employees and vendors, and effectively bankrupt, the OPSB had been forced by the State of Louisiana to turn over its financial operations to a private accounting firm. Although a majority of the district's schools were characterized as failing under Louisiana's school accountability system, only five schools had been taken over by the Recovery School District (RSD), the entity designated by state law to turn around failing schools, and handed over to charter school operators (Carr 64).

The physical and financial damage produced by Hurricane Katrina's flood waters deepened the system's misery. The storm displaced approximately 60,000 public school students from New Orleans and caused $800 million in damage to public school buildings (Garda 9). With no students and no money to operate schools, the OPSB soon placed thousands of its teachers and staff on unpaid leave (Oliver v. OPSB 2). It was obvious that drastic state action of some kind would be needed to resuscitate public education in the city.

The crucial actor in this dire situation was the Louisiana Legislature. After lobbying by Cecil Picard, the acting Superintendent of Secondary Education, and Leslie Jacobs, an insurance executive who had become active in education reform and won a seat on the state's Board of Secondary and Elementary Education (BESE), the legislature passed a landmark law that came to be known as Act 35 (Carr 65). Effective on November 30, 2005, Act 35 was probably the most important state law in New Orleans' post-desegregation public school history and certainly one of the most significant laws in the history of the charter school movement.

With a single stroke, Act 35 amended Louisiana's pre-Katrina school accountability legislation to designate 102 of the OPSB's 117 public schools as "failing schools" and made them immediately subject to transfer to and takeover by the RSD (Carr 65; Garda 9).[5] Of the 24 remaining public schools in New Orleans, seven were closed as uninhabitable, 12 became charter schools, and five were reopened and directly managed by the OPSB (Oliver v. OPSB 4).

In January 2006, the RSD began to open schools in New Orleans. Although it had no choice but to directly operate many of its schools at first, the RSD moved as swiftly as

---

5   Under Louisiana's earlier accountability legislation, a "failed school" was defined as one that is "academically unacceptable under a statewide program of school accountability established pursuant to rules established by [BESE.]" La. Rev. Stat. 17:10.5. Legislation enacted in 2004 allowed the state to impose restrictions on a local school board once it was determined to be "academically in crisis," that is, if more than 50 schools in the district were designated as "academically unacceptable" or more than 50 percent of the district's students attended schools that were academically unacceptable. La. Rev. Stat. 17:10.6. Act 35 added criteria by which a school could be designated as "failing" and immediately transferred to the RSD by specifying that a school is "failing" if it has a baseline school performance below the state average (though not actually failing itself) and is located in a district that both has been declared to be "academically in crisis" and has at least one school eligible to be transferred to the RSD. La. Rev. Stat. 17:10.7(A)(1).

it could to close the failing schools and replace them with charter schools. Although the complete transformation took the RSD longer than it had initially planned, by August 2014 the transformation was essentially complete. Today, the RSD no longer directly operates any traditional public schools. After it closed its last five direct-run schools in the summer of 2014, the RSD became the first "all charter" urban public school district in the United States. All 57 of its schools are charter schools. Importantly, though, all but 13 of these charter schools are affiliated with a charter management organization (CMO), an entity that will typically operate anywhere from two to seven schools. (Sims and Vaughan 8; Cowen Institute Governance and School Guide).

Although the OPSB still operates six direct-run public schools, it too has embraced charter schools. In fact, the OPSB now operates 14 charter schools, two of which are affiliated with a CMO. For its part, BESE also operates four charter schools in New Orleans. Finally, the Louisiana Legislature directly supervises another public school, New Orleans Center for Creative Arts (NOCCA), open to students from parishes across the state and that functions essentially like a charter school (Sims and Vaughan 3; Cowen Institute Governance and School Guide). In summary, 77 out of the 83 public schools currently operating in New Orleans today are charter schools, the highest percentage of any major city in the United States (National Alliance).

In 2013–2014, 91 percent of the 44,791 public school students in New Orleans attended a charter school (Sims and Vaughan 8). This is the highest percentage of charter school enrollment of any urban school district in the country, 36 percentage points higher, in fact, than the city with the next highest percentage, Detroit (Sims 1). With the closing of the last five RSD direct-run schools, the percentage of public school students attending a charter school will now be even higher.

Although some charter school critics still claim that the charter school transformation in New Orleans has not yet produced evidence of significant improvement in educational outcomes (Garda 3–5, 16–17), there are undeniable signs of academic progress on many fronts. In 2004–2005, the Orleans Parish public school district ranked second to last out of the 68 public school districts in the State of Louisiana. Taken as a whole, today the public school system in Orleans Parish ranks 38 out of 70 districts in the state. Since 2004–2005, per pupil spending in New Orleans has actually increased from $6,509 to $8,859 even while the percentage of students qualifying for free or reduced-price lunch has increased from 77 percent to 84 percent.

In 2004–2005, only 35 percent of public school students were scoring at basic or above on state education exams, far below the state average of 58 percent. Today, 63 percent of public school students in New Orleans are scoring at basic or above on state exams, just slightly behind the state average of 67 percent. While only 32 percent of black students scored at basic or above on state exams in 2004–2005, which was below the state average of 40 percent, today 58 percent of black students are scoring at basic or above, surpassing the state average of 53 percent.

In 2004–2005, only 25 percent of graduating seniors in New Orleans public schools qualified for the Taylor Opportunity Program for Students (TOPS), the program that provides full tuition at a Louisiana public university to any Louisiana student who scores at or above the state average on the ACT and who maintains a 2.5 grade point average in high school. In 2013–2014, 38 percent of graduating seniors at public schools in New Orleans qualified for TOPS.

In 2004–2005, 64.3 percent of the public schools in New Orleans were characterized as "failing" schools at the start of the term. In 2014–2015, only 10 percent of public schools in New Orleans were considered failing at the beginning of the term, including four schools that have now been closed and transformed into charter schools (Dreilinger Aug. 31, 2014; Sims and Vaughan 20–24). Finally, the recent CREDO study of the performance of urban charter schools in 41 urban centers across the United States found that students enrolled in charter schools in New Orleans between 2006 and 2012 showed more academic progress than peers enrolled in traditional public schools in both math and reading and, in fact, showed some of the strongest gains of any of the 41 urban centers studied (CREDO 8–15).

Key Features of New Orleans's Portfolio School Choice Model

Although New Orleans is not the only urban school district with large numbers of charter schools, it is safe to say that no other school district has gone as far as New Orleans in creating a universal choice system for public education. Besides the sheer number of charter schools in New Orleans, however, several other characteristics make the city's public education landscape distinctive.

First, while some characterize the governance structure of public schools in New Orleans as a "fractured system" (Garda 12), others see a growing coherence. It is true that at the beginning of the 2014–2015 school year, there were 44 different governing bodies running 82 public schools in New Orleans. Yet acting as an intermediary between the RSD and OPSB and most of the charter schools in the city are 12 non-profit CMOs that operate multiple schools with a shared mission or vision (Cowen Institute Governance and School Guide). Indeed, 60 percent of all charter schools in New Orleans are now part of CMOs and 60 percent of all public school students are enrolled in a CMO-directed charter school. Some of New Orleans' CMOs actually enroll more students than many parish-wide school districts in Louisiana (Sims 2015, 1–2).

These CMOs are crucial players in the new landscape of public education in New Orleans. Although one CMO, Knowledge is Power Program (KIPP), is affiliated with a national organization, the other CMOs in New Orleans, like the Algiers Charter School Association or First Line Schools, are locally based and were founded by educators who were working in New Orleans public schools well before Hurricane Katrina. In 2013–2014, the three largest CMOs each operated six different schools serving a total population of more than 3,000 students per CMO. The next two largest CMOs operated five and four schools each, serving 2,800 and 2,000 students respectively. Five more CMOs operated either two or three schools and served between 1,900 and 1,000 students (Sims and Vaughan 8). In other words, each of these CMOs functions much like a small suburban school district, capable of achieving economies of scale and specialization, while also providing close oversight of each individual school in its network.

Those who see coherence in the public school system in New Orleans describe it as a "portfolio district" (Sims and Vaughan 2). This term refers to the fact that the two over-arching governing entities, the RSD and the OPSB, oversee a system of independent schools and CMOs that operate under performance contracts. Much like administrators of a private school, individual school and CMO administrators can hire teachers and staff of their own choosing, can determine their own budgets, and can contract for their own services, without having to work through a central school district administration. And yet the charter schools receive public funding based on a per-pupil allocation model just like other public schools.

A crucial feature of the charter school bargain nationally is that if a charter school fails to meet the system's performance-based accountability standards, the governing authority can revoke the charter and close the school or turn its administration over to another charter school operator. The evidence to date indicates that this part of the charter school bargain is being satisfied in New Orleans, especially between 2012 and 2014, when the RSD closed four charter schools and reopened three more under new charter operators (Sims and Vaughan 2, 9).[6]

Although charter schools in New Orleans still enjoy a great deal of administrative autonomy, there has been some effort to centralize and standardize a few key procedures across the entire system. First, a Cooperative Endeavor Agreement (CEA), signed in March 2014, defines the relationship between the RSD and the OPSB and seeks to promote cooperation on a number of fronts, including addressing the needs of special education students, facility improvement, and shared financial resources (Sims and Vaughan 4–5). As a result of the new CEA, changes in the state's Minimum Foundation Program (MFP), cost sharing of funds provided under the Individuals with Disabilities Education Act (IDEA), and a new special needs funding program at the RSD, schools serving the district's neediest students will begin receiving additional funding. All of these changes should help alleviate inequities among charter schools in addressing the neediest students in the system (Sims and Vaughan 15). A recent federal court consent decree resolving a lawsuit brought by the Southern Poverty Law Center against the OPSB and the State also promises to bring greater oversight on how charter schools in this portfolio district meet the needs of special education students.[7] On a different front, all public schools in New Orleans have also now adopted common expulsion policies and procedures, including a centralized Student Hearing Office (Sims and Vaughan 4).

In a portfolio district like New Orleans, parents theoretically enjoy a wide range of options to choose from in selecting the most appropriate school for their child. In practice, the range of options is more restricted because some schools are more popular choices than others and there is not room in every school for every student who seeks to attend. Before 2012, parents could fill out a common application for many of the city's public schools, but they still had to apply directly to each public school to which they sought admission. Enrollment decisions were made at the school level, mostly on a "first-come, first served" basis. This decentralized system frustrated many parents because the most popular schools filled up quickly and admission practices lacked transparency. School administrators were also unhappy with the system because parents could apply to more than one school, and thus administrators did not know what their enrollments would be until after the school year started.

To remedy this confusing enrollment situation, a new common application process called *OneApp* was launched in 2012. Using *OneApp*, parents and students rank up to eight schools in order of their preferences or choose to stay in their current school. School assignments are then made using an algorithm, similar to that used to match graduates of US medical schools with residency programs, that assigns students based on a variety of priorities. Students are then offered one seat in a school that best matches their preferences and has an available seat.

---

6 In December 2014, the RSD announced that it would close one more troubled school, Miller-McCoy Academy, at the end of the 2014–15 school year. Miller-McCoy had received a failing grade earlier in fall 2014 under BESE's accountability system (Dreilinger, Dec. 17, 2014).

7 For more information, see Dreilinger's report in "Landmark New Orleans Special Education Case is Settled, Parties Say," which provides a link to the settlement consent decree. For an even more recent assessment of New Orleans's charter schools' dramatic improvement in addressing the needs of special education students, see Dreilinger, "Has Special Education changed in New Orleans? Success so far."

In its first year (2012–2013), 25,000 students submitted a *OneApp* application, and 84.2 percent of entering kindergarten and rising ninth grade applicants were offered a seat in one of their top three school choices, with 75 percent of applicants receiving a placement in their top choice. Eighty percent of students elected to stay in their current school. For the most recent school year (2014–2015), 80 percent of applicants were offered a seat in one of their top three schools, and 90 percent of kindergarteners and rising ninth graders were placed in one of their top three choices. About 75 percent of families did not submit a *OneApp*, indicating their intent to keep their children in their current school (Sims and Vaughn 17–18). While the *OneApp* system cannot guarantee perfect universal choice and while some detractors continue to lament the limited number of A- and B-rated schools available for parents to choose from (Kamenetz 6–8), *OneApp* has brought greater trust and transparency to the charter school application process and has enhanced parent satisfaction with the system as a whole.

Schools officials have also sought to make the *OneApp* process more comprehensive by including as many schools as possible, even some schools that have selective admissions criteria, such as language immersion schools or schools with audition-based music and performing arts programs. By the fall of 2014, 72 of the city's 83 public schools participated in *OneApp*, with the exceptions being nine OPSB charter schools and NOCCA (the *conservatory style* arts school overseen by the legislature). The nine charter schools that do not yet participate in *OneApp* will be required to join when they renew their charters in coming years (Sims and Vaughan 4).

In October 2014, the reach of the *OneApp* process expanded even further by allowing New Orleans families to apply for spots in not only the 72 participating public elementary and high schools, but also 50 preschools and 16 private schools that accept vouchers under a state-subsidized voucher program (Sims and Vaughan 17–18).

A recent survey conducted by the Center on Reinventing Public Education (CRPE) of 4,000 parents and guardians (collectively "parents") in eight different cities in the United States with "high choice" public education models (Baltimore, Cleveland, Denver, Detroit, Indianapolis, New Orleans, Philadelphia, and Washington, D.C.) provides a very detailed picture of the extent to which the choice system employed in New Orleans is building trust and confidence in parents (Jochim). The most important findings regarding New Orleans were the following:

- Of the eight "high choice" cities, New Orleans had by far the highest percentage of parents (87 percent) who chose a non-neighborhood-based school (9–10).
- New Orleans had the highest percentage of parents choosing a non-neighborhood school regardless of the parents' education levels. Indeed, parents with a bachelor's degree or more and parents with a high school degree or less were both choosing non-neighborhood schools at levels above 80 percent (10–11).
- 92 percent of surveyed parents in New Orleans reported being satisfied with the quality of education their child receives and 89 percent agreed their child's school was responsive to their concerns (11–12).
- 44 percent of parents in New Orleans stated that available schools were not a good fit for their child (12).
- Parents of children with special needs in New Orleans reported that available schools were not a good fit at almost the same rate as parents of children without special needs (13).

- 52 percent of parents in New Orleans reported having at least one good public school option available to them other than the school their child currently attends (13–14).
- 56 percent of New Orleans parents stated that public schools in the city were getting better. This was the third highest positive outlook rate of the eight cities surveyed. Only Washington, D.C. (65 percent) and Denver (59 percent) had higher positive outlook rates (15).
- Only 42 percent of parents in New Orleans reported a great deal or fair amount of trust or confidence in the school system to provide all neighborhoods with great public schools. The misalignment between New Orleans' parents' generally positive outlook and their relative lack of trust and confidence in the school system is difficult to interpret, however, because the question posed by CPRE did not discriminate between the RSD and the OPSB (15).
- In New Orleans, 68 percent of white parents were optimistic that public schools were getting better, compared to 53 percent of black parents and 39 percent of Hispanic parents (16).
- In New Orleans, 23 percent of parents stated that they struggled to get the information they needed to choose the best school for their child, a percentage that was slightly less than the average rate for the high choice cities surveyed (18).
- New Orleans parents were the most likely to report that their child travels to school on a school bus and least likely (only 19 percent) to identify transportation as a problem. These findings are not surprising given that New Orleans is the only one of the sample cities where charter schools are required to provide students with transportation. Nevertheless, the most common means of transportation reported by New Orleans parents was driving their child to school (19–20).
- 35 percent of New Orleans parents reported difficulty understanding which schools their child was eligible to attend. This eligibility confusion rate tied with Detroit as the highest among the sample cities. Eligibility confusion rates were also higher for parents with less formal education and for parents of children with special needs (21–22).
- Although a significant percentage of parents in New Orleans reported having difficulty with application deadlines (27 percent), CPRE acknowledged that surveyed parents may be reporting their experience with the decentralized application system before *OneApp* was implemented (23, 31).

Another, even more recent survey of New Orleans residents' perceptions of public education conducted by Tulane University's Cowen Institute for Public Education Initiatives produced similar results. The Cowan Institute survey found that a strong majority (72 percent) of the 600 respondents polled in April 2015 support New Orleans' open enrollment policy, compared to 23 percent who favored assigning students to schools based on geography alone. A plurality (37 percent) of the respondents also stated that public schools in the city were better after Hurricane Katrina, compared to 21 percent who believed they were worse. Of all parents surveyed, 50 percent stated they were now more likely to send their child to a public school than before Katrina, while only 30 percent said they were less likely. Finally, the vast majority of public school parents (76 percent) stated they felt they had sufficient information about schools and the school selection process to make a choice for their child (Rossmeier and Sims 2–5).

Voter Support for the New System

Is the new charter school dominated public education system in New Orleans garnering support among the voting public? According to the results of a recent city-wide tax referendum designed to provide long-term funding for facilities construction and maintenance in New Orleans public schools, the answer is, yes; public support for the system exists and is growing.

Several years after Hurricane Katrina, FEMA determined that New Orleans's public school system was owed $1.8 billion to repair and reconstruct flood and wind damaged school buildings. After a long planning process that required cooperation of the RSD and the OPSB, a School Facilities Master Plan was finalized in October 2011 to govern the reconstruction and renovation of most of the public school buildings in the city. Unfortunately, by 2014 it became apparent that the FEMA funds would not be sufficient to pay for all of the planned projects and that a $330 million shortfall existed. Further, no funds were dedicated for maintenance of the buildings and facilities that have been or will be reconstructed or repaired.

In response, the Louisiana Legislature enacted legislation giving Orleans Parish the authority to dedicate funds from a pre-existing ad valorem property tax millage to long-term maintenance and preservation of the newly constructed and renovated school facilities. To become effective, the property tax millage rededication required approval by Orleans Parish voters (Sims and Vaughan 16). On December 5, 2014, a strong majority of Orleans Parish voters (59 percent) approved the millage rededication, guaranteeing up to $15.5 million per year by 2021 for capital replacements, repairs and improvements (Dreilinger, Nov. 28, 2014, Dec. 6, 2014).

The positive outcome on this referendum is a powerful statement of public support for the new charter school dominated public school system in New Orleans. A precinct-by-precinct analysis shows that support for the millage rededication was strong across the entire city, in both predominantly white and black districts, with particularly strong levels of support in many of the precincts in "the sliver by the river," the older, more affluent areas near the Mississippi River that did not flood after Katrina and that repopulated most quickly after the storm.[8]

Charters Sticking with the RSD

One additional indicator of support of the current system is found in the decisions of the individual boards governing the various charter schools currently under the supervision of the RSD. Current state law allows charter schools that achieve acceptable academic ratings and demonstrate academic gains for two consecutive years to return to "local control." State law, however, gives the governing boards of RSD charters the authority to decide whether to remain under RSD control or transfer to the OPSB (Sims and Vaughan 4). Although a number of RSD critics assumed that charter schools would return to OPSB control as soon as the schools were no longer deemed to be academically at risk, so far only two charter schools operated by local boards have elected to return to the OPSB. In fact, in late fall 2014 and early 2015, boards governing 34 of the 36 charter schools eligible to

---

8   For the precinct-by-precinct analysis, see http://thelensnola.org/2014/12/05/voters-to-decide-whether-to-extend-existing-tax-in-order-to-maintain-school/.

return to local control opted to remain with the RSD, citing concerns over the uncertainty of superintendent leadership at the OPSB as well as affinity with the mission of other RSD charter schools (Dreilinger Dec. 22, 2014, Jewson).

The decisions of these governing boards mirror the views of the vast majority of New Orleans voters. Indeed, when polled in May 2014, only 18 percent of likely voters in New Orleans indicated they wanted "takeover schools" to return to the OPSB in the next two years and only 11 percent wanted to see these schools return to the OPSB in the next three to five years. Fifty-seven percent were content with the status quo (Cowen Institute, Voters Perceptions).

A Qualitative Account of Social Capital in New Orleans Charter Schools

Although no academic studies have yet attempted to measure the extent to which the new charter schools in New Orleans are producing new social capital that in turn could lead to improved educational outcomes for the city's youth, education journalist Sarah Carr has produced a remarkably detailed and insightful narrative recounting the experiences of students, teachers and administrators at three charter schools in post-Katrina New Orleans. Carr's account in *Hope Against Hope: Three Schools, One City, the Struggle to Educate American's Children* shows that at their best charter schools in New Orleans have succeeded in creating extensive new bonding social capital within the school community—especially among students and teachers—and significant bridging and linking social capital for the students who attend those schools and aspire to greater academic achievement.

In Carr's account, New Orleans Charter Science and Math Academy ("Sci Academy" or simply "Sci") demonstrates the promise of charter schools. She tells its story primarily through the eyes of an idealistic teacher, Aidan Kelly, who joined the teaching staff at Sci Academy in 2010, after two years of teaching at another charter school in New Orleans. A graduate of Harvard College, Kelly was recruited to New Orleans by Teach for America. Kelly represents the youthful, idealistic, well-educated, and often maligned teachers who came to New Orleans after Katrina to staff its new charter schools. Unlike some Teach for America recruits who spent only two years in New Orleans and then left the city or teaching altogether, Kelly committed himself to a career in public education and remains in New Orleans. The crucial factor in Kelly's success as a teacher is the leadership of Sci Academy.

Like several other charter schools in New Orleans, Sci Academy was incubated by New Schools for New Orleans, a non-profit formed after Katrina to support the growth of charter schools. Sci was led by an intensely driven Yale graduate, Ben Marcovitz, who started the school at the age of 28, after several years of teaching in Boston and New Orleans, a Master's degree in Education from Harvard, and a brief stint as an assistant director at another charter school in New Orleans.

Carr vividly depicts the teaching and learning culture that Marcovitz created at Sci and shows how, through persistent coaching, close attention to data, rigorous focus on teaching techniques, and incredible commitments of time and energy, Sci Academy molded Kelly into an effective Social Studies and Language Arts teacher of African-American students from poor families in New Orleans. Although Kelly struggled at times to motivate and relate to his students, at the end of the day Kelly succeeded as a teacher and helped his students achieve impressive academic gains. In the spring of 2012, although only two of his AP English students scored a 3 on the English AP exam, his students' AP scores approved across the board. More importantly, as noted in the beginning of the chapter, an astounding 94 percent of Sci's first graduating class in 2012 were accepted by at least one four year college and 60 percent of its graduating class qualified for a TOPS scholarship at a public university in Louisiana (Carr 282). In another symbol of the burgeoning social capital created at Sci, two spin-off schools opened in the summer of 2012 led by former Sci teachers, while Kelly himself became the Director of Curriculum and Instruction at Sci, and Marcovitz became the CEO of a CMO called Collegiate Academies (Carr 283).

What enabled Kelly and Marcovitz to achieve their hard-won success? First, the leadership team at Sci intentionally created a social structure within the school that was rich in all three components of social capital. From the outset, Marcovitz established strong networks among all of the individuals in the school community. Carr recounts the numerous meetings among teachers, the frequent visits by Marcovitz and other teachers to Kelly's classroom and the intense collaboration among all staff at the school, constantly focused on how to achieve better results (88–91, 131). Not coincidentally, it is precisely these types of practices that education experts have observed helping to produce strong education outcomes in other countries (Halpern 159).

Next, Carr's account of Kelly's formation as a teacher at Sci demonstrates the powerful effects of strong norms of behavior and high expectations for students. Literally from the moment they first set foot on the school campus, Sci students receive the message that attending college is the expectation for every student, that every action at school should be directed toward achieving this end (73–76, 176–77). Charter schools like Sci are often called "no excuses" schools. Although "no-excuses" schools certainly have many detractors who complain that they stifle creativity and innovation (Kamenetz), Carr's account allows you to see why so many parents—especially overwhelmed, working-class families—love the rigor, strict discipline, and accountability of these schools.

Finally, Carr's reporting on Sci shows the powerful effect of sanctions (both positive and negative) in the social structure of the school. She recounts the swiftness of teachers' sanctions of inappropriate student behavior—for instance, confiscation of student cell phones—and the swiftness of Marcovitz and other administrator's frequent assessments of Kelly's teaching effectiveness (131–133). Carr details not only the rigorous, time-consuming process Marcovitz uses to hire his teachers, but also the clear reward system Marcovitz established for his teachers. Rather than give collective bonuses for overall gains in student improvement, Marcovitz gives bonuses to teachers who complete specific tasks that teachers might perceive as drudge work but that nevertheless advance its "core mission" (178).

In a particularly telling passage, Carr describes Marcovitz's fundamental goal at Sci. Rather than reflect the values, social code, and aspirations of his students' community, Sci intentionally aims to "supplant them" (81). As one of Marcovitz' teachers explains, the first week at Sci is "about creating this feeling that everything is different" (81). Sci intentionally aims to undo the debilitating neighborhood effects created by pervasive violence, poverty, and social and economic disadvantage by creating an alternative universe where more time is available to help students overcome prior education deficiencies and where there is a feeling of safety, the sense that adults are in charge (Carr 81).

Carr's account also illustrates the effect of social capital "closure" as James Coleman would describe it. She observes that "[c]harter schools like Sci expect a certain degree of compliance—or mission alignment—from their teachers as well as their families" (86). During student orientation, teachers are required to dress like their students to show solidarity. Like their students, they walk on the right side of the hallway. Teachers greet Marcovitz every morning just as the students do with an affirmation of their purpose for being at Sci. Charter school leaders like Marcovitz constantly encourage and motivate their teachers, reminding them that they are performing "the most important work in the world" (87). Indeed, Carr notes how easy it is for observers of charter schools like Sci to compare them to "a traditional Catholic school minus the religion, with the teachers playing the role of the nuns minus celibacy and seclusion" (87). These similarities to Catholic schools, which have been found to generate social capital and collective efficacy in some urban neighborhoods, no doubt play a role in Sci's success (Brining and Garnett 890).

But Carr also shows that not all charter schools are equal. Despite lofty ambition, sometimes a charter school's rigid rules and adversarial relationship with the culture of its students can lead to a downward spiral of distrust and frustration. Carr's account of KIPP Renaissance High School reveals all that can go wrong with this kind of well-intentioned, but culturally insensitive charter school initiative. Carr tells the story of KIPP Renaissance through the eyes of Geraldlynn Stewart, a student who first attended KIPP Believe, a middle school started by KIPP soon after Katrina. Geraldlynn started at KIPP Renaissance in the fall of 2010, when it took in its first ninth grade class. Adding more texture to the portrait of KIPP Renaissance are frequent reports on the lives and perceptions of Geraldlynn's parents and the struggles of its principal, Brian Dassler.

KIPP Renaissance began its inaugural year with many of the same attributes as Sci Academy—a "no excuses" approach to behavior and homework that appealed to its no-nonsense parents who knew the stakes of failing to live up to rigid rules in the workplace, expectations that teachers would be available to students and parents at all hours, and expectations that all graduates would attend college—and some that are special to all KIPP schools such as long school hours (Carr 9–10, 16). Further, even before Geraldlynn had started at KIPP Renaissance, KIPP had established strong social capital with her family through her successful three years at KIPP Believe (Carr 24–28). This was a potentially important asset because, as Carr recounts, Geraldlynn's family, like many low-income African-American households in New Orleans, had seen its informal social safety net diminished by the Katrina diaspora (19). After accompanying Geraldlynn on one of KIPP Believe's many class trips to other cities, Geraldlynn's mother, Racquel, picked up on the school's attempt to create linking social capital when she remarked, "I've just never seen anything like it. . . . They take them from world to world, you know" (25).

Given all of this admirable social capital, why did KIPP Renaissance struggle so much more than Sci Academy? Based on Carr's account, it seems that KIPP Renaissance's young, inexperienced leadership team and teachers were not able to maintain the norms and sanctions they aimed to establish and, critically, were not able to establish effective bonds of trust between staff and students. In the end, the school's strict behavioral rules caused resentment and resistance among the students and led, finally, to the collapse of order and the eventual departure of the principal, Dassler (287–88). As early as end of their first quarter of teaching, KIPP Renaissance's teachers realized that the "emotional bank accounts" they maintained with their students (strangely reminiscent of Coleman's "credit-slips") were badly overdrawn because of the innumerable demerits they had given out (Carr 145). Likewise, the principal tended "to sweat the small stuff" too much, while failing to address more chronic problems such as fights and failure to do homework (112). Although Dassler hired a black, New Orleans native to serve as the school disciplinarian, an "us versus them mentality" had already developed between teachers and students and even between the teachers and the principal (145). In Carr's words, KIPP Renaissance "had an abundance of rules . . . but little discipline" (157).

Another problem at KIPP Renaissance actually stemmed from one of its admirable characteristics—its willingness to take all-comers, including students with clear histories of behavior problems like Geraldlynn's cousin Maurice, who was frequently "benched" for speaking out of turn (97), and one particular student named "Brice," who admitted to Carr that he was "being pulled into a scene of drugs, fast money, and guns" (99), but who also exerted a powerful influence over other students in his classes and sometimes over the entire school (155–57). Further, rather than accept only students who had attended a KIPP middle school and had become accustomed to KIPP's rules and rewards, KIPP Renaissance recruited students from many different schools across the city, thus making it much harder for the school's staff and leaders to create the intense social norms that its strong social capital structure requires (152–53).

While Carr bluntly acknowledges the failure of KIPP Renaissance to overcome the cultural gap between its predominately white, well-educated teachers from out of town and its virtually all black, largely poor student body (148–50), she also shows compassion for Dassler and his beleaguered staff. She admits that they were thrown into an almost impossible situation, with little support from KIPP's national leaders, with little advance time to prepare curricula and train themselves, and were dealing with students whose socio-economic, mental health, and family problems often overwhelmed the teachers' one-dimensional focus on discipline and college preparation (160–61).

And yet despite all of the admitted shortcomings of KIPP Renaissance, Geraldlynn's experience at KIPP Renaissance is not entirely negative either. Geraldlynn visited colleges (Dillard, Florida State and Florida A&M). She and her family learned how to plan for the financial demands of college. Finally, Geraldlynn even began to visualize herself as a college student, especially after she visited Florida A&M University and was inspired by its diversity and embrace of African-American culture (Carr 214–18, 275–79). Despite a temporary drop in her grades in the beginning of her sophomore year, Geraldlynn's grades rebounded and she finished her sophomore year with a mixture of A's, B's and only one C, after she received individual tutoring sessions on Saturdays from one of her favorite teachers. Just as importantly, her ACT score shot up at the end of her sophomore year to 17, putting her on track to earn a 20 (the all-important TOPS qualifying score) by her senior year (Carr 289).

The third charter school featured by Carr, O. Perry Walker High School, presents the most ambiguous story of all. Rich in bonding social capital like some traditional neighborhood public schools, part of a home grown network of charter schools on New Orleans's "West Bank," and led by an inspiring New Orleans bred and educated principal, Mary Laurie, Walker provides a strong web of social support for its students and their families—a warm and comforting safety net that nurtures its students through countless struggles and tragedies. Yet even the principal Laurie comes to realize that O. Perry Walker has not achieved as much bridging social capital as perhaps it could have. Although O. Perry Walker sent 73 percent of its graduating seniors to college in 2011—a not inconsiderable achievement (237)—Laurie and some of her colleagues openly worried that they had "not done enough to grow their students wings" (238).

Conclusion

In the past, most parents of public school students in New Orleans simply had to accept the fact that their children were assigned to a failing school. That is no longer true. As Michael

Stone, the co-CEO of New Schools for New Orleans, a non-profit that incubates and funds charter schools, recently remarked, "I think we are starting to see that ZIP Code doesn't equal destiny" (Kamenetz 9).

The movement to establish a universal charter school system in New Orleans will continue to be controversial. It will take years before we know whether the charter school experiment will lead to sustainable educational gains for its students.

But we do know this: the charter schools that have emerged in New Orleans are designed to supply their disadvantaged and minority students with many of the habits, high educational aspirations, and social capital relations that are typically found in solidly middle class families and communities—what one KIPP Renaissance teacher calls "the middle-class code" (Carr 160). At the same time, unless these schools and their leaders become more sensitive to their environment, they will continue to be criticized for implying that poor and minority communities (especially African American ones) are impoverished when it comes to social capital.

Charter schools may not be able to solve all of the social and educational inequities that poor, minority communities face in so many of our major U.S. cities today. Nevertheless, they have the potential to be a significant part of the solution, especially if education leaders and public officials learn from the dramatic experiment currently underway in post-Katrina New Orleans. Moreover, everyone interested in the fate of cities will benefit by appreciating the ways in which charter schools can build or destroy all forms of social capital.

Works Cited

Abul-Alim, Jamaal. "Educators focusing on preparing students for college." *Advocate* 20 Aug. 2014: n. pag. Web. 18 Aug. 2014 <http://theadvocate.com>.

Black, Derek W. "Charter Schools, Vouchers, and the Public Good." *Wake Forest Law Review* 48 (2013): 445–88. Print.

Bourdieu, P., and L.J.D. Wacquant. *Invitation to Reflexive Sociology.* Chicago: U. of Chicago, 1992. Print.

Brining, Margaret F., and Nicole Stelle Garnett. "Catholic Schools, Urban Neighborhoods, and Education Reform." *Notre Dame Law Review* 85.3 (2010): 887–954. Print.

Carr, Sarah. *Hope Against Hope: Three Schools, One City, and the Struggle to Educate America's Children.* New York: Bloomsbury, 2013. Print.

Coleman, James. "Social Capital in the Creation of Human Capital." *American Journal of Sociology* 94 (1988): S95–S120. Print.

"Cowen Institute Governance and School Guide." Tulane University Cowen Institute for Public Education Initiatives (2014). Web. 30 Dec. 2014 <http://www.speno2014.com/governance-school-guide/>.

DeJarnatt, Susan L. "Community Losses: The Costs of Education Reform." *University of Toledo Law Review* 45 (2014): 579–600. Print.

Dreilinger, Danielle. "A 13-year-old mother, a murdered father and a scholarship to Yale." *Times Picayune* 23 May 2014: n. pag. Web. 28 Aug. 2014 <http://blog.nola.com/education_impact>.

———. "Has special education changed in New Orleans? Success so far." *Times Picayune,* 26 May 2015: n. pag. Web. 28 May 2015 <http://www.nola.com/education/index.ssf/2015/05/new_orleans_special_education_2.html>.

———. "Landmark New Orleans special education case is settled, parties say." *Times-Picayune* 19 Dec. 2014: n. pag. Web. 28 Dec. 2014 <http://blog.nola.com/education_impact>.

———. "Miller-McCoy Academy to close in the summer." *Times-Picayune* 17 Dec. 2014: n. pag. Web. 30 Dec. 2014 <http://blog.nola.com/education_impact>.

———. "New Orleans parents need more help choosing a public school, report says." *Times-Picayune* 24 Dec. 2014: n. pag. Web. 28 Dec. 2014 <http://blog.nola.com/education_impact>.

———. "New Orleans school maintenance tax easily approved by voters." *Times-Picayune* 6 Dec. 2014: n. pag. Web. 30 Dec. 2014 <http://blog.nola.com/education_impact>.

———. "New Orleans school property tax would raise $15.5 million for building maintenance." *Times-Picayune* 21 Nov. 2014: n. pag. Web. 30 Dec. 2014 <http://blog.nola.com/education_impact>.

———. "Number of New Orleans schools now stands at 83, down from 124." *Times-Picayune* 31 Aug. 2014: B-3. Print.

———. "OPSB fails to attract schools." *Times-Picayune* 22 Dec. 2014: B-1+. Print.

Fischel, William A. "Why voters veto vouchers: public schools and community-specific social capital." *Economics of Governance* 7 (2006): 109–32. Print.

Garda, Robert. "Searching for Equity Amid a System of Schools: The View from New Orleans." *Fordham Urban Law Journal* 42 (forthcoming Spring 2015). Print.

Garnett, Nicole Stelle. "Are Charters Enough Choice? School Choice and the Future of Catholic Schools." *Notre Dame Law Review* 87.5 (2012): 1891–1916. Print.

Halpern, David. *Social Capital*. Cambridge: Polity, 2005. Print.

Hanifan, Lyda J., "The Rural School Community Center." *Annals of the American Academy of Political and Social Science 67* (1916): 130–38. Print.

Hanifan, Lyda J. *The Community Center*. Boston: Silver, Burdette, 1920. Print.

"The Health of the Public Charter School Movement: A State-By-State Analysis." National Alliance for Public Charter Schools (2014). Web. 30 Dec. 2014 <http://publiccharters.org/publications>.

James, Osamudia R. "Opt-Out Education: School Choice as Racial Subordination." *Iowa Law Review* 99 (2014): 1083–1135. Print.

Jewson, Marta. "Second New Orleans Charter School Opts for Return to Local Control." *New Orleans Advocate*: 5 Jan. 2015: n. pag. Web. <http://www.theneworleansadvocate.com/news/11237604–123/second-new-orleans-charter-school>.

Jochim, Ashely, et al. "How Parents Experience Public School Choice." Center on Reinventing Public Education (2014). Web. 30 Dec. 2014 <cpre.org>.

Kamenetz, Anya. "The End of Neighborhood Schools." National Public Radio (2014). Web. 30 Dec. 2014 <http://apps.npr.org/the-end-of-neighborhood-schools>.

"Nola by the Numbers: Leap & iLeap Test Results, 2014." Tulane University Cowen Institute for Public Education Initiatives (2014). Web. 30 Dec. 2014 <www.coweninstitute.org>.

Oliver v. Orleans Parish School Board, No. 2014-C-0329 c/w 2014-C-0330 (La. Oct. 31, 2014).

"Public Charter School Success: A Summary of the Current Research on Public Charters' Effectiveness at Improving Student Achievement." National Alliance for Public Charter Schools (2013). Web. 30 Dec. 2014 <http://publiccharters.org/publications>.

Putnam, Robert D. *Bowling Alone*. New York: Simon & Schuster, 2000. Print.

———. "*E Pluribus Unum*: Diversity and Community in the Twenty-first Century." *Scandinavian Political Studies* 30.2 (2007): 137–74. Print.

————. *Our Kids: The American Dream in Crisis.* New York: Simon & Schuster, 2015. Print.

Rossmeier, Vincent, and Sims, Patrick. "K-12 Public Education through the Public's Eye: Parents' and Adults Perception of Public Education in New Orleans." Tulane University Cowen Institute for Public Education Initiatives (2015). Web. 14 May 2015 <http://www.coweninstitute.com/2015Publicpoll>.

Ryan, James E. *Five Miles Away, A World Apart; One City, Two Schools, and the Story of Educational Opportunity in Modern America.* Oxford: Oxford U, 2010. Print.

Saiger, Aaron. "What We Disagree About when We Disagree About School Choice." *Iowa Law Review Bulletin* 99 (2014): 49–60. Print.

Sampson, Robert J., Jeffrey D. Morenoff, and Felton Earls, "Beyond Social Capital: Spatial Dynamics of Collective Efficacy for Children." *American Sociology Review* 64 (1999): 633–60. Print.

Schneider, Mark, et al. "Institutional Arrangements and the Creation of Social Capital: The Effects of Public School Choice." *American Political Science Review* (1997) 81: 82–93. Print.

Smith, Kevin. "Data Don't Matter? Academic Research and School Choice." *Perspectives on Politics* 3 (2005): 285–99.

Sims, Patrick, "Charter Management Organizations in New Orleans." Tulane University Cowen Institute for Public Education Initiatives (2015). Web. 13 May 2015 <http://www.coweninstitute.com/wp-content/uploads/2015/04/CI_Policy_Brief_No1.pdf>.

Sims, Patrick, and Debra Vaughn. "The State of Public Education in New Orleans: 2014 Report." Tulane University Cowen Institute for Public Education Initiatives (2014). Web. 30 Dec. 2014 <www.coweninstitute.org>.

Stillings, Cara. "Freedom of Education and Charter Schools." *International Journal for Education Law and Policy* 1.2 (2005): 59–72. Print.

Tedin, Kent L. and Gregory R. Weiher. "General Social Capital, Education-Related Social Capital, and Choosing Charter Schools." *Policy Studies Journal* 39.4 (2011): 609–29. Print.

"Urban Charter School Study Report on 41 Regions 2015." Stanford University Center for Research on Education Outcomes (2015). Web. 12 May 2015 <urbancharters.stanford.edu/summary.php>.

Vanourek. G. "State of the Charter School Movement 2005: Trends, Issues and Indicators." Charter School Leadership Council (2005). Print.

"Voters' Perceptions: Public Education in New Orleans, Spring 2014." Tulane University Cowen Institute for Public Education Initiatives (2014). Web. 30 Dec. 2014 <www.coweninstitute.org>.

Willms, J.D. "Three hypotheses about community effects relevant to the contribution of human and social capital to sustaining economic growth and well-being." Paper prepared for the OECD International Symposium on "The Contribution of Human and Social Capital to Sustained Economic Growth and Well-Being" (2000). Print.

Chapter 10
Cities Seeking Justice:
Local Government Litigation in the Public Interest

Kathleen S. Morris

Introduction[1]

What do you think of when you hear the term "public interest lawyer"? Do you think of lawyers who work at non-profit organizations like the American Civil Liberties Union, the National Association for the Advancement of Colored People, or the Natural Resources Defense Council? Or maybe you think of lawyers who file class action cases to vindicate environmental, consumer, or employment rights? If it occurs to you that some government lawyers are "public interest lawyers," do you think of the United States Department of Justice or the nation's 50 state attorneys general?

Chances are the term "public interest lawyer" does not make you think of city law offices. There are good reasons for that. City law offices—by which I mean both in-house and outside civil counsel for cities—are not generally thought of at all. They are largely invisible, not only to the public at large, but even to lawyers and law schools. Moreover, most city law offices historically have not pursued "public interest cases," that is, cases designed to vindicate the general public interest rather than the interests of the city-as-public-corporation. City law offices typically serve two distinct functions: providing advice to city officials on legal issues and defending the city when it is sued.

However, over the past few decades, an ever-growing number of city law offices have begun to develop public interest case dockets that vindicate not only their own corporate interests, but also broader public interests in economic, environmental, public health, and civil rights (Brescia I 167; Morris I 1–45).

Progressives in the U.S. should welcome this shift towards local enforcement of civil laws as part of a broader effort to rebalance the current imbalance between the commercial activities of private corporations and public and private oversight of those activities. This chapter accepts as true that civil laws in the U.S., particularly those that most directly impact the corporate bottom line—such as consumer protection, environmental health, wage-and-hour, and industrial safety regulations—are dangerously under-enforced (Surowiecki, Fine 815, Duhigg A1). Indeed, as one commentator has written, "it's hard to think of a recent disaster in the business world that wasn't abetted by inept regulation." (Surowiecki). This chapter also accepts as true that permitting civil laws to go under-enforced disrupts the nation's economic health by loosening the natural tension between the public and private sectors (Surowiecki). The primary responsibility of the private sector is to find new ways to maximize profits. So, unsurprisingly, corporations push the envelope not only creatively,

---

1 For helpful comments and discussions, I thank Vikram Amar, Owen Clements, Heather Gerken, Darien Shanske, and Christine Van Aken. This chapter is dedicated to Dennis J. Herrera (San Francisco City Attorney, 2001–present).

but also legally. As a counter-balance, federal and state officials, backstopped by private plaintiff's lawyers, are charged with watchdogging corporations closely enough to dissuade illegal activities. In a perfect world, the balance between private and public would result in robust corporate creativity and profits, a stable economy, thriving communities, and a well-tended natural environment.

But high-profile events over the past decade—like the 2008 financial collapse, the 2010 British Petroleum oil spill in the Gulf, and the clean water crisis—strongly underscore that private activities and regulatory activities are dangerously out of balance (Surowiecki, Duhigg A1). Consider the particular example of the 2008 financial collapse. It took many years for banking and public trading practices to slowly build up to what became the most devastating financial crisis in almost a hundred years. Yet the rising tsunami escaped the attention of every institutional watchdog in the U.S. until after it swamped the economy. U.S. federal agencies, state agencies, private law and policy organizations, and the media all failed to adequately sound the alarm until disaster was upon us (Chan, Krugman, Starkman). Meanwhile, relatively little about Wall Street practices has changed since 2008 (Cohan 23–5). Indeed, banks and other corporations continue to cry foul and lobby to unravel laws aimed at putting safeguards in place to protect the public (Cohan 23–5). As a nation, we must build a level of public and private oversight that gives us confidence that public actors are reliably identifying and challenging illegal corporate activities.

This chapter argues that our nation should bolster civil law enforcement by inviting and encouraging city law offices to enter the fray. This is not to suggest that city law offices are a panacea. However, they are a geographically and institutionally vast, largely untapped resource in the struggle to establish adequate oversight over corporate activities in the U.S.

An increasing number of cities have already taken up this challenge. Taken together, the cities of Baltimore; Buffalo; Cleveland; Gary, IN; Los Angeles; New York; San Diego; and San Francisco are actively litigating a wide range of economic, environmental, civil rights, and public health matters. Their complaints typically demand not only court orders halting unlawful corporate practices, but attorney's fees and penalties to back-fund the litigation and fund future public interest cases. The most ambitious cities—Los Angeles, New York, and San Francisco—have dedicated units within their offices for the investigation and litigation of public interest cases (Morris II 51–66).

A few law schools have jumped in to help cities seeking justice by permitting law students to work on public interest cases alongside city lawyers in exchange for law school credit. The most ambitious and established city-law school partnership is the San Francisco Affirmative Litigation Project ("SFALP") at Yale Law School, co-founded in 2006 by Professor Heather Gerken and this author (then a San Francisco Deputy City Attorney) (Yale Law School I 1, Morris III 1920). Through SFALP and similar projects, cities could get much-needed legal support, while students learn about and experience every stage of the litigation process (Yale Law School II 1). If we encourage every law school to "plug in" to at least one city pursing public interest cases, we could ultimately see dozens or even hundreds of formal city-law school partnerships.

Achieving this vision would not be easy. For starters, it is likely that not all cities have a majority of constituents who would want their city law offices to pursue public interest cases. It is also possible that many cities, though open to the idea of public interest cases, would rather make different use of their resources. Finally, even assuming some cities are open to developing public interest dockets, they will likely face one or more of three barriers: cultural, legal, and financial.

The primary cultural barrier is that most city lawyers do not see themselves as champions of the public interest (Morris II 53–6). The primary legal barrier is that cities that seek to sue on behalf of the public may not meet mandatory legal requirements known as "standing" and "causation." The primary financial barrier is that litigation costs money. Cities may feel they cannot afford to create public interest units or even pursue the occasional public interest case. These barriers are formidable. Any discussion of cities seeking justice must address them.

This chapter proceeds as follows: Part I describes city law offices and their typical functions. Part II describes how a handful of cities have recently engaged in public interest litigation. Part III examines the cultural, legal, and financial barriers to "cities seeking justice," and suggests changes that would enable willing cities to develop or expand their capacities to bring public interest cases.

I. What are City Law Offices and What Functions Do They Traditionally Serve?

The law refers to cities as "public corporations," which is to say, not-for-profit, publicly funded corporate entities. Like for-profit corporations and private non-profits, cities essentially come into being by applying to state governments for permission to exist. Once created, cities have constituents, a tax base, and often-growing responsibilities.

Today, larger cities manage budgets ranging from millions to billions of dollars and run a multitude of public systems that may include power and water systems, mass transit systems, ports, airports, road and bridge systems, housing structures, libraries, museums, and parks. This is an astoundingly wide-ranging management portfolio, and cities need sophisticated legal advice to help manage it all. To meet that need, larger cities create in-house legal departments while smaller cities hire outside counsel. As mentioned previously, this chapter includes in-house and outside legal counsel in its definition of "city law offices."

Because cities identify legally as public corporations, the traditional focus of city law offices has been to serve the needs of the city as a corporation. In this role, a city law office has two functions: (1) to provide legal advice to city officials regarding what the law requires, so the city can minimize the risk of litigation; and (2) to defend the city in litigation when it is sued. These functions are roughly comparable to those of a corporate law firm that serves for-profit companies. But since a public corporation does the public's business, the subject matter a city law office tackles is somewhat different from that of a private corporate firm. While private firms advise their clients on topics like business, finance, and intellectual property, city law offices advise their clients on topics like civil liberties, energy, ethics, land use, and municipal operations. Both types of firms advise on subjects of mutual interest like tax, contracts, and employment, and defend the corporation when it is sued in tort, contract, property, and the like.

Most of the time, then, city law offices (like private corporate law firms) are playing defense. Their primary goal on the advice side is to avoid litigation. Their primary goal on the litigation side is to defend the city in litigation. In recent decades, however, several city law offices have added a third function. In addition to providing advice and defending against litigation, these offices have been pursuing litigation to vindicate the broader public interest (Brescia I 167). The underlying mental shift this move requires is for a city's lead counsel to embrace the city-as-public-corporation and the city-as-unit-of-representative-democracy (Morris II 52–3). With respect to the former role, a city attorney necessarily focuses on corporate well-being. But with respect to the latter, he or she may also consider constituent well-being.

This mental shift is small but powerful. It is small because it seems intuitively right, perhaps even obvious, that a city should look out for its constituents. Yet it is powerful because, over the long haul, this mental shift has the potential to open the door to dozens, or even hundreds of city law offices across the U.S. developing robust public interest case dockets. The next part of this chapter provides a small sample of cities that have made this mental shift and are pursuing public interest cases.

II. A Sample of City Public Interest Cases

In recent years, cities have pursued a range of public interest cases, sometimes successfully and sometimes unsuccessfully. This part of the chapter briefly summarizes a range of selected cases selected.[2]

A. San Francisco's Attack on Collusive Credit Card Arbitrations

Only a public entity could bring this case. As of 2008, as a condition of receiving a Bank of America ("bank") credit card, the bank required customers to sign a binding arbitration agreement. The agreement stated that a bank-chosen arbitrator must resolve any contract disputes that arose between the bank and the consumer. Both parties waived all rights to challenge the bank in any court of law. To handle any such disputes, the bank hired a private arbitration company, the National Arbitration Forum ("NAF").

As it turned out, NAF had two faces. To the general public, NAF advertised itself as a provider of bona fide, neutral arbitration services characterized by the highest ethical standards. For example, the city alleged NAF's website stated its arbitration services would ensure "all parties a fair, unbiased dispute resolution process" that is "not beholden to any company or individual that utilizes [its] services" (*People v. National Arbitration Forum* 18–19). But to corporations that might employ it to provide comprehensive arbitration services, NAF advertised itself as a reliable friend to corporations that had overwhelmingly ruled against consumers and in favor of corporate clients (*People v. National Arbitration Forum* 18–19). NAF's message to corporations was essentially: "Choose us. You'll win."

This message was alarmingly accurate. A San Francisco City Attorney investigation revealed that on a consistent basis, NAF knowingly engaged in illegal and unfair practices that gave the bank an advantage against consumers. Those practices included failing to provide notice of arbitration, granting the bank monetary awards against consumers in excess of the amount permitted by law, failing to forward on communications from

---

2 A brief word about process: I researched city public interest cases by scouring the city websites for the 50 most populous cities. My rationale was that the largest cities were most likely to be pursuing public interest cases, and if they were, would likely broadcast them via their websites. I searched these city websites for any indication that they engaged in public interest cases, consumer cases, plaintiff's cases, affirmative litigation, or the like. I also searched legal scholarship pertaining to cities and/or consumer protection; searched the national print and web media for mentions of city cases; and reached out via email to the nation's city attorneys via the International Municipal Lawyer's Association, the only non-profit organization that entirely consists of the nation's local public lawyers (imla.org). This chapter does not seek to be a comprehensive summary of all city public interest cases, but rather a discussion of some of the more interesting, recent cases.

consumers to arbitrators (including messages that the disputed charges were incurred by identity thieves), labeling and submitting to arbitrators as "uncontested" disputes that were in fact contested, providing arbitrators with financial incentives to decide matters quickly and with little review (which advantaged the bank), granting more business to arbitrators who most often ruled in favor of the bank, firing arbitrators who ruled in favor of consumers, and applying its own procedural rules unevenly to favor the bank (*People v. National Arbitration Forum* 2–20).

In the end, San Francisco's evidence showed that NAF arbitrators ruled in favor of the bank and against consumers in approximately 96 percent of credit card disputes (*People v. National Arbitration Forum* 2, Berner). The city's complaint alleged that the bank had hired NAF to operate as "an arbitration mill, churning out arbitration awards in favor of debt collectors and against California consumers, often without regard to whether consumers actually owe the money sought by the debt collectors" (*People v. National Arbitration Forum* 2). As for NAF, the city alleged that company "abdicated its role as the sponsoring organization of neutral arbitrations. Instead, NAF's business model consisted of assisting debt collectors in recovering money against consumers–such that it was no longer acting as an arbitration sponsoring organization at all, and its arbitration panelists could no longer be said to be engaging in arbitrations" (*People v. National Arbitration Forum* 16).

Why could only a public entity have brought this case? Why could the affected consumers not hire counsel to sue NAF and the bank? Because, recall, the bank's credit card customers had signed mandatory arbitration agreements. Under those agreements, any disputes related to the credit card agreement–including a dispute over whether NAF arbitrators were "fixing" arbitrations–could only be brought before a NAF arbitrator.

In 2008, San Francisco sued NAF and the bank. The suit alleged violations of California's unfair business practices statute, California Business and Professions Code section 17200 ("section 17200"). Section 17200 broadly prohibits persons and corporations from engaging in "unlawful, unfair, or fraudulent business act or practices." That statute grants a handful of large California cities (including San Francisco) standing to enforce its provisions on behalf of "The People of the State of California." (The "People" standing-to-sue privilege is typically reserved for state attorneys general and district attorneys.)

One year after San Francisco filed its lawsuit against the bank and NAF, the State of Minnesota also sued NAF. In a settlement with the Minnesota Attorney General, NAF agreed to completely shut down its consumer arbitration business. Subsequently, after three long years of active litigation against San Francisco, the bank agreed to pay the city and consumers a total of $5 million in penalties and suspend all credit card arbitrations for two years.

B. Cleveland's Lawsuit Against Mortgage Lenders to Remediate Blight After Widespread Defaults

The financial crisis of 2008 devastated multiple U.S. cities, neighborhoods, and residents. As former Federal Reserve Chairman Ben Bernanke has explained:

> [F]oreclosures can inflict economic damage beyond the personal suffering and dislocation that accompany them. Foreclosed properties that sit vacant for months (or years) often deteriorate from neglect, adversely affecting not only the value of the individual property but the values of nearby homes as well. Concentrations of foreclosures have been shown to do serious damage to neighborhoods and communities, reducing tax bases and leading to

increased vandalism and crime. Thus, the overall effect of the foreclosure wave, especially
when concentrated in lower-income and minority areas, is broader than its effects on
individual homeowners (Bernanke Speech).

A bit of background: in the decades prior to the relatively recent emergence of
"subprime" loans, major banks made only "prime" loans, that is, uniformly-priced loans
to borrowers with solid credit. The emergence of a major subprime mortgage market made
credit more broadly available to those with shakier credit, though at much higher interest
rates. This development benefitted poor and minority families and neighborhoods when
mortgage lenders acted responsibly.

Unfortunately, however, many mortgage lenders acted irresponsibly. Specifically, they
aggressively marketed low "teaser" loan rates (and thus monthly payments) that would
later balloon into unaffordable monthly payments. Before the borrowers could default, the
original lenders sold the mortgages to other lending institutions so they would not be on
the hook by the time borrowers defaulted (California Reinvestment Coalition 1–51, Apgar,
Duda, and Gorey 1–57). Once the loans came due, millions of borrowers defaulted and
entire city neighborhoods collapsed financially. Research by the California Reinvestment
Coalition and others has described this practice (California Reinvestment Coalition 1–51,
Apgar, Duda, and Gorey 1–57).

Of all U.S. cities, the financial crisis hit Cleveland the hardest. Of the top 20 zip codes for
foreclosure filings, in the crisis' wake, four of them (that is, one-fifth of all neighborhoods in
the 20 worst cities) were in Cleveland (Cutts 1409). Cleveland sued 22 financial institutions
in an attempt to hold them responsible for their respective roles in the local financial crisis.
Cleveland alleged that the named defendants in the case had sold literally thousands of
so-called "subprime" loans to millions of homeowners who later defaulted on those loans,
devastating its neighborhoods and tax base.

Cleveland brought its case on a legal theory known as "public nuisance." A "public
nuisance" is generally defined as an unreasonable interference with a right common to
the public at large (*City of Cleveland I* 502–507). Cleveland alleged that the subprime
loans triggered a devastating local economic slide and that beyond the individual rights
of homeowners, the general public has a common right not to have banks turn a profit by
engaging in business practices that financially upended entire neighborhoods.

Cleveland's lawsuit against the lenders proved a noble failure. Cleveland's case
failed not because its allegations were factually incorrect, but because the law of public
nuisance could not provide a remedy. The trial judge dismissed Cleveland's complaint in
part on the ground that the city could not prove the loans directly rather than indirectly
caused must of Cleveland's housing market to collapse (*City of Cleveland I* 513–536).
The court of appeal affirmed (*City of Cleveland II* 496–507). The City of Buffalo has
filed a similar case. Buffalo's case is still pending (Buffalo City Attorney High Profile
Matters).

C. Los Angeles' Battle Against Racially Discriminatory Lending Practices

The economic crisis also triggered city claims against lenders for race discrimination in
violation of the federal Fair Housing Act (sometimes alongside public nuisance claims)
(Pindell 169). Los Angeles filed a Fair Housing Act action against several major banks,
alleging that institutionalized racism was one of the forces that drove the banking industry
to make millions of dollars' worth of toxic-from-the-start loans in minority neighborhoods.

Los Angeles claimed the lenders' practices, as described in the Cleveland case (subpart B), also illegally discriminated against racial minorities ("City of Los Angeles I" 1–51; "City of Los Angeles II" 1–53; "City of Los Angeles III" 1–69).

A bit of background: the story of racially discriminatory lending practices in Los Angeles stretches back for decades. By the 1990's, major banks had an established practice of "redlining" in U.S. cities, that is, refusing to lend in majority-minority neighborhoods (Entin and Yazback 757). The banks' redlining systems collected reliable data on the most economically vulnerable households in Los Angeles. Later, when market practices shifted, that data became useful for an entirely new purpose. Beginning in the late 1990's, the banks began to flood minority neighborhoods with high-cost subprime loan offers (Bocian, Li, and Reid 8). This new business practice reflected a nationwide trend. Between 1996 and 2006, the national subprime loan market grew from $97 billion to $640 billion (Bocian, Li, and Reid 8, Ernst, Bocian, and Li 6).

Los Angeles's complaint alleges that subprime lenders offered minorities loans on worse terms than those offered to Caucasians with the same credit and income profiles. As described in subpart B, the lenders' plan was to issue mortgages then immediately sell the potentially bad debt so they would not be on the hook when borrowers defaulted. Their business model allowed them to maximize profits by making loans without considering whether borrowers had the ability to pay. Because minority borrowers had worse loan terms than offered to Caucasian borrowers, when the Los Angeles housing market collapsed, minority borrowers were much more likely than Caucasians to lose their homes. Then, adding insult to injury, after the economic meltdown the same lenders disproportionately refused minority borrowers' requests to refinance the original loans so they could stay in their homes.

This "renewed redlining" practice meant that majority-minority neighborhoods saw more foreclosures than majority-Caucasian neighborhoods. The racial statistics are stark. In Los Angeles a loan in a majority-minority neighborhood is more than twice as likely to result in foreclosure as a loan in a majority Caucasian neighborhood. And since home equity represents a disproportionately high percentage of overall wealth, these actions will impact generations (Schwemm and Taren 382).

In addition to harming their constituents, Los Angeles alleges that race discrimination in lending damaged the city-as-corporation. The city has had to devote additional personnel time to inspecting vacant properties, issuing orders for unsafe conditions, removing excess vegetation, hauling away debris, boarding vacant properties, and condemning and demolishing dangerous abandoned structures. The city alleges it lost hundreds of millions of dollars as a result of the defendants' lending practices (*City of Los Angeles v. JPMorgan Chase & Co.* 42–7).

The lenders filed a motion asking the court to dismiss Los Angeles' case for lack of standing to sue, arguing only a minority borrower could bring a Fair Housing Act case. The city responded that, because the city-as-corporation had been indirectly injured by the discriminatory lending practices, the city had standing to sue as an "aggrieved person" under section 3602(i) of the Fair Housing Act.

The trial court accepted Los Angeles' argument, but notably, not all courts would have. The City of Baltimore filed a nearly identical Fair Housing Act case (*Baltimore v. Wells Fargo Bank* 847, Morgenson). A federal court dismissed Baltimore's case on the grounds that the city lacked standing and could not prove causation (*Baltimore v. Wells Fargo Bank* 849–51). As in the Cleveland public nuisance case, the Baltimore court held that city failed to prove the banks directly rather than indirectly caused Baltimore's injuries, and that the

Fair Housing Act requires proof of direct causation. This position finds support among some scholars (Marll 255).

Los Angeles has dropped its Fair Housing Act case against JP Morgan, but cases against the other banks are still pending in federal court. Los Angeles has also filed complaints under state anti-discrimination law in state court. The city's complaints generally seek: (1) orders prohibiting the lenders from engaging in redlining and reverse-redlining; (2) orders that the lenders must disgorge (that is, forfeit) all profits made through acts of race discrimination; and (3) orders that the lenders must pay restitution to the city for the $1.2 billion in damages it sustained as a result of racially discriminatory lending practices. The city of Miami is also pursuing race discrimination cases against major lenders. If either or both of these cities prevail, they will provide a legal roadmap for any other city seeking to challenge racially discriminatory lending practices that harm not only individual families but the city as a whole.

D. Los Angeles' Campaign against Hazardous Waste Dumping

Under-enforcement of environmental laws hits poor communities hardest. Empirical studies comparing disparities of exposure to industrial pollution in the U.S. show: (1) it is possible to objectively measure inequality in the distribution of environmental hazards; and (2) low-income and high-minority communities are at a disproportionate risk of being exposed to industrial pollutants (Boyce). Leading researchers have pressed federal and state regulators to identify and rectify the health and environmental impacts of pollution in poor and minority communities, and have encouraged local communities to directly engage with, and where necessary confront, private sector actors that are creating the pollution (Boyce).

One sub-category of industrial pollution is "hazardous waste." This term captures a broad array of pollutants that include airborne acids and gases that may leak from manufacturing sites, liquid waste at risk of being poured into common drains, chemicals at risk of explosion from being stored improperly, and hazardous and corrosive materials at risk of being mishandled in transit. Examples of hazardous waste include medical supplies, ignitable liquids, household chemicals, aerosols, electronic parts, and electronic devices.

For decades, federal and state governments have regulated how corporations track, handle, transport, and dispose of hazardous waste. Enforcing hazardous waste regulations is a relentless and difficult regulatory challenge. In California, regulators recently openly admitted they have not kept up with polluting corporations (Garrison). Recent news articles document ongoing specific examples of enforcement failures, like the battle over arsenic emissions from a Vernon, California, battery recycling plant and the fight over improper and potentially deadly storage of cyanide by a Richmond, California, company (Garrison, Christensen, and Poston 1–2). In 2014, the Los Angeles Times reported that approximately 25 percent of California's major hazardous waste facilities are operating on expired permits (Garrison, Christensen, and Poston 2).

The Director of the California Department of Toxic Substances Control recently admitted in a public hearing that, for decades, her department has allowed private corporations handling hazardous waste to skate by on "smoke and mirrors" (Garrison I). In 2014, residents from poor communities across California provided legislative testimony about relatives who died from cancer potentially caused by exposure to toxic chemicals from nearby factories (Garrison I). The ineffectuality of government enforcement in California has shifted the battleground over hazardous waste in that state almost entirely to non-governmental forces including neighborhood activists, environmental researchers, and dogged journalists (Garrison, Christensen, and Poston 1–10).

But in Los Angeles, city lawyers have entered the fray. The Los Angeles City Attorney's Office has created a groundbreaking internal division known as the "Environmental Justice and Protection Unit." That unit recently aggressively investigated, and quickly resolved a toxic waste dumping case against the supermarket chain Albertsons LLC ("Albertsons"). The complaint Los Angeles filed, along with five other California localities, alleges as follows: for years, up and down California, Albertsons has engaged in a routine practice of illegally dumping hazardous waste in trash bins or transferring it to illegal third-party waste processors (*People v. Albertsons* 7–14). Albertsons' dumped waste includes over-the-counter medications, pharmaceuticals, aerosol products, ignitable liquids, batteries, electronic devices, and pool chemicals. The complaint accused Albertsons of violating several hazardous waste statutes and California's broadly worded section 17200 (*People v. Albertsons* 14–18).

The case was over almost as soon as it began. Although it officially denied fault, faced with the city's evidence, Albertsons quickly agreed to expend more than $3.3 million to settle the city's lawsuit. The funds will go towards (1) civil penalties, (2) recouping the taxpayers for the investigatory and litigation costs, and (3) developing and implementing in all California Albertsons stores a formal environmental program that includes employee training, dumpster audits, and progress reports (Koerner).

E. San Diego, Oakland, and San Francisco's Lead Paint Case

From the 1890's to the 1970's, U.S. paint companies used lead as an additive in household paint. Lead is a heavy, soft, ductile, and anti-corrosive metal. Adding lead to paint makes it "bright, durable, flexible, fast-drying, and cheap" (Beam). Unfortunately, lead is also a poison. All paint deteriorates, leaving behind chips, flakes, and dust. When residential lead paint deteriorates, it leaves lead-laden residue on floors, windowsills, and in front and back gardens where children learn to crawl, walk, and play (Centers for Disease Control and Prevention 1).

In the US, lead paint is the leading cause of lead poisoning in children. Because lead is a poison, even very low levels of exposure can have disproportionately severe health consequences. Low blood lead levels in children are associated with adverse effects on development, delayed onset of puberty, decreased growth, decreased hearing ability, lower intelligence, and delinquency (Proposed Statement of Decision 16).

A California court has found that lead paint industry has known about the problem of lead poisoning, especially in poor children, since at least the 1930's. Gatherings of paint industry association leaders in 1930, 1937, and 1958 included open discussions of what to do about the lead poisoning problem (Proposed Statement of Decision 22). Yet, notwithstanding their awareness of the dangers of lead paint, leading paint corporations continued not only to make, but to actively promote lead paint for household use, publicly denied and concealed the dangers of lead paint, and aggressively campaigned against government regulation of the paint industry.

Today, millions of acres of privately owned residential property in California are still coated in decades-old lead paint. The key law and policy question is: who must pay to remove it? Private property owners? The State? Cities and Counties? Or paint companies? Officials for several California cities and counties including San Diego, Oakland, and San Francisco strongly felt the corporations should bear the burden of removing lead paint from private property. Their lawyers have spent 13 years litigating the issue against five of the U.S.'s largest manufacturers of lead paint (Proposed Statement of Decision 1).

The cities' case, like Cleveland's case (discussed in subpart A), rests on the legal theory of "public nuisance." Under California law, a "nuisance" is anything that "is injurious to health . . . or is indecent or offensive to the senses, or an obstruction to the free use of property, so as to interfere with the comfortable enjoyment of life or property" (Nuisance, Cal. Civil Code). It defines a "public nuisance" as a "nuisance" that "affects at the same time an entire community or neighborhood, or any considerable number of persons, although the extent of the annoyance or damage inflicted upon individuals may be unequal" (Public nuisance).

After years of litigation battles between the cities and the companies, the case went to trial. At trial, the defendants made the following two legal arguments (among others): (1) the cities and counties had waited too long to sue, so they could not recover; and (2) the companies responsible for creating, promoting, and selling lead paint had essentially vanished through decades of corporate mergers and acquisitions. The descendant companies, they argued, were innocent and should not be required to remedy the problem.

The trial court rejected the defendant's arguments. It ruled as a factual matter that, for decades, all five defendants (or their predecessor corporations) knew or reasonably should have known they were selling a product that was detrimental and possibly fatal (Proposed Statement of Decision 92–3). It ruled as a legal matter that the successor companies had to take responsibility (Proposed Statement of Decision 93–6). The court ordered the defendant paint companies to fully fund a publicly supervised lead abatement plan across California to the tune of $700 million (Proposed Statement of Decision 96–110). The case is still pending on appeal.

F. San Francisco's Campaign Against High-Caffeine Beverages

Federal law requires companies that make food additives to affirmatively prove those additives are "generally recognized as safe" (or "GRAS") before putting them in food or beverages offered for sale (Definitions I). An additive can only satisfy the GRAS standard if there is a "reasonable certainty in the minds of competent scientists that the substance is not harmful under the intended conditions of use (Definitions I). The manufacturer bears the burden of proving the additive is safe for its intended use based on published scientific literature (Definitions I).

In 2013, a group of scientists, professors, clinicians and other health professionals ("the scientists") sent the U.S. Food and Drug Administration ("FDA") a letter of concern (Scientists' Letter to FDA 1–6). The letter contends that a company called the "Monster Beverage Corporation" ("the company") is selling highly caffeinated beverages that do not meet the GRAS standard. Specifically, the FDA GRAS standard approves added caffeine in beverages in concentrations up to 200 parts per million, but high-energy drinks exceed that level.

Moreover, the scientists warned, the beverages at issue, which are being sold under the brand names Monster Energy, Rockstar, and Red Bull, are being aggressively marketed not only to adults, but also to children. They explain that medical researchers and professionals had identified a number of health risks to children tied to highly caffeinated beverages. Those risks include seizures, cardiac arrhythmia, altered heart rates, elevated blood pressure, sleeplessness, anxiety, and obesity. Moreover, they write, the National Poison Data System indicates that these products are associated more than any other in accidental ingestion by very young children. About half of all emergency calls to the National Poison Data System for energy-drink-related caffeine toxicity concern children five years old or younger.

The scientists' overall message to the FDA is that no scientific consensus supports the safety of high levels of added caffeine in energy drinks even for adults, let alone children, and that the drinks placed not only young children, but also teenagers and young adults at an unacceptably high risk of severe cardiac events, seizures, and related neurological problems (Scientists' Letter to FDA 1–6).

Several San Francisco officials share the scientists' concerns. The San Francisco City Attorney, who, with the help of city health officials, had been investigating energy drinks since 2012, sent the FDA a similar letter of concern (Herrera Letter to FDA 1–2). (On October 31, 2012, the City Attorney had sent a letter directly to the company, asking it to substantiate its public claims that its high-caffeine beverages were "completely safe" for adolescent and adult consumption [Herrera Letter to Monster Energy Corp. 1–4]. The company did not respond to the October 31 letter). The City Attorney's letter to the FDA urged that agency to address the scientists' concerns. It also urged, at a minimum, that the FDA require the manufacturers of high-caffeine beverages to publish caffeine content on product labels (Herrera Letter to FDA 1–2).

The company responded to the growing pressure by preemptively suing San Francisco in federal court, seeking an injunction ordering the city to shut down its investigation. The company argued the city (1) had unfairly singled out its products in particular, (2) was trying to "dictate who may and may not consume" its products, and (3) was threatening to restrict what it could say on product labels and marketing materials (*Monster Beverage Corp. v. Herrera* 1–28). The company argued that complying with San Francisco's demands would force it to breach or disrupt various private contracts yet would not solve any of the problems the City Attorney had identified.

A few days after the company filed its complaint against San Francisco, the city filed suit against the company in state court (*People v. Monster Beverage Corp.* 1–23).[3] San Francisco's complaint alleges that the company intentionally targets children and young adults with marketing that, for example, encourages consumers to "pound down" and "chug down" its products and with assurances that Monster Energy consumers "can never get too much of a good thing." San Francisco's complaint points out that the company's marketing includes its "Monster Army" website, which uses children as young as six years old to promote the Monster brand. The complaint alleges that the Monster Energy Corporation aims it marketing directly towards children despite scientific findings that such products may cause "significant morbidity in adolescents" from elevated blood pressure, brain seizures, and severe cardiac events (*People v. Monster Beverage Corp.* 8).

San Francisco alleges that, taken together, the company's practices violate California's consumer protection law (section 17200, discussed in part B) and the Sherman Food Drug and Cosmetics Law. The city seeks a court order forbidding the company to cease all illegal conduct and pay civil penalties and restitution (*People v. Monster Beverage Corp.* 22–23). The federal court hearing the company's complaint against the City Attorney dismissed that case on December 13, 2013 (Minute Order Granting Motion to Dismiss 1–21). San Francisco's case against the company is still pending.

3 The same day San Francisco sued the company, the FDA announced it had decided to investigate the entire high-caffeine beverage industry (FDA Investigation Announcement 1).

G. Gary's Case Against the Firearms Industry

Gun violence in the U.S. is a seemingly intractable problem that kills about 30,000 people per year. The personal and social costs of gun violence are profound and immeasurable. In addition, some cities have argued that illegal gun sales cause significant harms not only to the victims of gun violence, but to the city itself and the general public (*City of Gary v. Smith & Wesson Corp.* 1243, *City of New York v. Berretta U.S.A. Corp. I* 262).

The more aggressive cities have filed suit against gun manufacturers, wholesalers, and/or dealers. These suits allege (among other things) that the defendant firearms corporations knowingly sell or aid in selling firearms to persons who are forbidden by law from owning them. (Former New York City Mayor Michael Bloomberg launched perhaps the most high-profile effort in this regard (Stout). Cities that filed cases against firearms industry defendants sought court orders requiring them to stop intentionally engaging in, or contributing to, illegal gun sales.

Gary, Indiana, filed one such case. In the months leading up to its lawsuit against the gun industry, Gary conducted a series of local "sting" operations against gun dealers (*City of Gary v. Smith & Wesson Corp.* 1228). In a typical operation, an undercover officer posing as a purchaser would tell a Gary gun store clerk that he could not lawfully purchase a gun, but that he had a friend who could buy it for him. Then, a second undercover officer would (with the sales clerk's full knowledge) buy a gun for the first "customer." Gary's investigation also revealed that gun stores were routinely violating the law by failing to conduct required background checks and selling multiple firearms to a single purchaser (*City of Gary v. Smith & Wesson Corp.* 1228).

Gary has sued not only gun dealers, but also wholesalers and manufacturers. Gary alleges that wholesalers and manufacturers know which dealers in Gary are engaging in illegal gun sales and have the ability to change the distribution system to prevent them from selling guns illegally, but have intentionally failed to do so. Gary alleges that over a three-year period the defendant gun dealers illegally sold more than one-third of all handguns used to commit crimes within city limits. In terms of harm, Gary alleges that these illegal business practices by the gun industry drive up gun-related injury and deaths as well as the cost of taxpayer-funded investigative and emergency services in Gary (*City of Gary v. Smith & Wesson Corp.* 1228, 1243).

From the start, the most potent threat to Gary's case was the federal Protection of Lawful Commerce in Arms Act ("PLCAA"), 15 U.S.C § 7901–7903. The PLCAA bars most claims against the firearms industry. A narrow exception in the PLCAA's statutory scheme permits lawsuits to go forward when the plaintiff alleges that the defendant knowingly violated state or federal law in gun sales or marketing practices (Definitions II).

After Congress passed the PLCAA, plaintiffs litigating against the gun industry tried to save their cases by arguing that because their complaints alleged violations of state law—public nuisance law—the PLCAA did not bar their suits. Most courts rejected this argument, ruling that the PLCAA's statutory exception only applies when the defendant is alleged to have violated a state law that explicitly, by its stated terms, regulates firearms sales or marketing. Public nuisance law, these courts held, is too general to fall within the PLCAA's exception (*Ileto v. Glock, Inc.* 1126–1163, *City of New York v. Berretta U.S.A. Corp. II* 384–408).

In the Gary case, the Indiana Courts of Appeals disagreed with the federal courts' reading of the PLCAA. The Indiana court ruled that the PLCAA's "violation of state law" exception was broad enough to protect a lawsuit alleging a violation of state public nuisance claim (*Smith & Wesson Corp v. City of Gary* 422–435). Gary's lawsuit against the gun industry is still pending.

III. What Would Help More Cities Seek Justice?

Not all city leaders will want to pursue public interest cases. They may philosophically disagree with such cases or prefer to make alternative uses of city time and resources (Gottlieb 13–17). But even if they choose to pursue such cases, city officials interested in pursuing public interest cases will likely face one or more of three formidable barriers: cultural, legal, and financial. Overcoming these barriers will require three major policy changes. We must (1) develop a more expansive view of the role of city law offices in American life; (2) grant cities standing to enforce the nation's consumer protection statutes; and (3) provide resources to cities that wish to engage in public interest litigation.

A. Expanding Our View of City Law Offices

When I say that "we" must expand how we think of city law offices, I have at least six audiences in mind: (1) elected officials who direct the activities of city counsel (whether mayors, city council members, or elected city attorneys); (2) deputy city attorneys and other city lawyers; (3) private law and policy specialists who might advise and even work with city lawyers; (4) federal and state officials who might welcome and open the door to local enforcement of state and federal consumer protection laws; (5) scholars who write about cities and the public policy issues they face; and (6) law students who are interested in working on public interest cases.

Although U.S. cities have great potential as agents of civil law enforcement, this potential is almost entirely untapped. Of the top 50 most populous cities, only three—San Francisco, New York, and Los Angeles—publicly hold themselves out as platforms for public interest cases. Meanwhile, law schools that are training and mentoring the next generation of public interest lawyers almost entirely overlook city law offices in favor of more traditional public interest jobs in non-profit law firms; for profit, plaintiff-side firms; or the U.S. Department of Justice.

An even deeper psychological barrier is that law students and lawyers who view themselves as "progressive" will sometimes recoil at the idea of working for cities. Indeed, 100 years of U.S. history has taught us that city governments often oppose progressive values. And, of course, because cities "run the Constitution" (Ewald 3)—that is, routinely interpret and apply the Bill of Rights to citizens by (for example) engaging in police practices, placing restrictions on public protests, enforcing jail policies, and the like—cities will be routinely and forever defendants in civil rights cases. For these reasons, progressives often see cities as intractable defenders of awful power structures rather than champions of the public interest.

We must remove these blinders; they are costing us far too much. Cities are public corporations and sometime defenders of the status quo. But, they are also units of representative democracy. If we embrace the idea that city law offices should act as agents for the public good, it would open up great possibilities for civil law enforcement and public interest law jobs. Over time, city law offices could develop a formidable army of publicly-funded public interest law units. This would result in greater civil law enforcement, while creating hundreds of public interest law jobs. But to accomplish this goal, we must let go of our current view of city law offices as, at best, corporate lawyers only and, at worst, the enemy. City officials are charged with actively protecting and defending the public good. Their lawyers should be front and center in these efforts.

B. Granting Cities Standing to Enforce Consumer Protection Statutes

1. Common legal barriers to city public interest cases

As Cleveland's public nuisance case and Baltimore's Fair Housing Act case illustrate, cities that bring public interest cases often face formidable legal barriers. The best "general" legal theory available to all cities is public nuisance law, which prohibits corporations from engaging in any acts that interfere with rights common to the general public. However, to succeed on a public nuisance theory, a city must often prove that: (1) the corporate acts have resulted in an injury ("the injury requirement"); (2) the injury is being felt not only by constituents but by the city itself ("the standing requirement"); and (3) the nuisance is the direct rather than indirect cause of the injury ("the causation requirement"). More specific statutory vehicles, like the Fair Housing Act, have similar legal requirements.

These requirements significantly limit cities' ability to challenge illegal corporate activities, and have been criticized for that reason (Brescia II 8–9, 39–52). The injury requirement prevents challenges to acts that likely will, but have not yet, caused a palpable harm. When the plaintiff is a public actor and the illegal acts threaten the public good, this limitation seems unwise. The standing requirement means that, unlike state attorneys general, city lawyers are not permitted to bring cases if the only harms are to their constituents. This places an unnecessary limit on our elected agents' ability to protect and defend the public good (Brescia 39–52).

Finally, the causation requirement significantly limits cities' ability to halt and remedy corporate malfeasance. The Cleveland case (part II (B)) illustrates this point. Recall that Cleveland sued a number of financial institutions that made thousands of bad loans in poor and minority neighborhoods, knowing the borrowers would likely default. When the financial crisis hit, four Cleveland neighborhoods were financially devastated, leaving multiple communities and the city on the hook. The trial court dismissed that case, not because the banks did not engage in illegal lending practices, but because Cleveland could not prove that those practices directly (as opposed to indirectly) caused the foreclosures. At most, the court held, Cleveland could prove the banks "set up" clients for likely foreclosure. Proving this would not satisfy the causation requirement.

The causation requirement may ultimately pose a threat to Gary's case against the gun industry (part II (G)) because, applying the Cleveland court's analysis, Gary may have trouble proving that illegal gun sales directly caused higher crime and gun deaths. Had San Francisco pursued its arbitration case (part II (A)) or its caffeinated beverages case (part II (F)) on a public nuisance theory, both of these important public health cases would have run aground as lacking sufficient legal basis. Fortunately, San Francisco has a rare arrow in its quiver: the ability to enforce California's consumer protection statute.

2. The untapped power of consumer protection statutes

Federal and state legislatures in the U.S. have enacted far-reaching consumer protection statutes (Morris III 1903–05). The federal law, known as the "FTC Act," contains typically far-reaching language. The Act prohibits "unfair methods of competition" and "unfair or deceptive acts or practices" (15 U.S.C. § 45(a)(1)). The Federal Trade

Commission ("FTC") has explained what factors should be considered to decide whether a practice is "unfair":

> (1) whether the practice, without necessarily having been previously considered unlawful, offends public policy as it has been established by statutes, the common law, or otherwise—whether, in other words, it is within at least the penumbra of some common-law, statutory, or other established concept of unfairness; (2) whether it is immoral, unethical, oppressive, or unscrupulous; (3) whether it causes substantial injury to consumers (or competitors or other businessmen) (16 C.F.R. pt. 408).

The 50 state consumer protection statutes are similarly broad (Morris III appendix). California's section 17200, for example, broadly prohibits corporations from engaging in any "unlawful, unfair, or fraudulent business act or practices." Section 17200 grants certain public entities standing to file suit to halt an unlawful, fraudulent, or unfair business practice regardless of whether that practice is directly causing any harm. For section 17200, it is enough that the practice is unlawful, fraudulent, or unfair; the statute seems to presume that such practices have caused, or will cause, harm to competitors or the general public. Section 17200 has been a boon to San Francisco, which has relied on it to challenge unfair bank arbitration practices (part II (A)) and unlawful use of high levels of caffeine in beverages marketed to children (part II (F)).

When state legislatures grant cities statutory standing to enforce consumer protection laws, cities are suddenly able to halt illegal and unfair corporate practices within their jurisdictions at the earliest stages. Cities do not face the same standing, injury, and causation hurdles under consumer protection statutes. This has implications not only for what types of cases can be pursued, but also for how the litigation will unfold. When enforcing public nuisance law, cities must devote enormous resources to developing a record that establishes standing, harm, and causation. By contrast, in a case brought to enforce a consumer protection statute, often the only major factual question—and thus almost the entire focus of discovery—is whether the corporation is engaging in the act or practices alleged. If it is, and the court finds the act or practices to be illegal or unfair, the corporation must disgorge profits earned through malfeasance and pay penalties to the city.

In this way, consumer protection cases are more prosecutorial than public nuisance or other types of cases (such as Fair Housing Act cases). The focus is on the defendant's behavior alone. Thus the defendant cannot as easily bury its opponent in aggressive counter-discovery. There is no need to certify a class and prove actual damages, because the monetary remedy cities seek takes the form of disgorgement of profits and penalties. Further, corporations cannot hide behind mandatory arbitration agreements because public entities are not signatories to those agreements. Instead, when challenged by a city, the challenged corporation must present itself to a jury.

For all of these reasons, this author has called for Congress and the state legislatures to grant all cities with populations over 50,000 (that is, an additional 2,000 cities) standing to enforce our nation's 51 consumer protection statutes (Morris III 1903–1928). This would be a radical change. At present, the FTC Act is only enforceable by the FTC itself (Morris III 1910). As for state law, only seven of the 50 states allow some or all city law offices to enforce their states' consumer protection statutes (Morris III 1910–1911). This should change. The nation needs its corporations more effectively watchdogged, and cities are uniquely well positioned to help in that effort (Morris III 1904–1905).

C. Providing Resources to Cities Seeking Justice.

The federal and state governments could most directly incentivize cities to engage in public interest litigation by making grants to cities for that purpose. Of course, this may not be politically feasible. But there are other, powerful ways to provide resources to cities to engage in public interest litigation. One is to permit cities to use outside contingency counsel to help share the burdens and risks of such cases. A second is to encourage law schools to partner with cities in supporting public interest cases.

1. Permitting cities to use contingency counsel
Not all cities can realistically afford to establish public interest law units or even litigate public interest cases without help from contingency counsel. Contingency counsel are lawyers hired to work on cases in exchange for a percentage of the ultimate recovery instead of being paid an hourly rate. The issue of whether public entities should be permitted to use contingency counsel in public interest cases is hotly debated (Griffis 22–4; Stern 2–5). Suffice it to say that cities would be in an infinitely better position to add public interest case dockets if they could engage contingency counsel.

Consider the example of the lead paint case, discussed in part II (E). This case was a major victory for public health. Yet it was so vast, both legally and factually, and the downside of losing was so great, that no public entity would likely have pursued it without outside contingency counsel. The paint company defendants understood this, so they spent literally a decade litigating whether the public entity plaintiffs could legally use contingency counsel in the case. In 2010, the California Supreme Court issued a sweeping opinion on that issue (*County of Santa Clara v. Superior Court* 35). The court held that public entities in California are permitted to use contingency counsel in public interest cases as long as city lawyers oversee the litigation and retain complete control over the course and conduct of the case (*County of Santa Clara v. Superior Court* 63–64). This decision made the lead paint victory possible. The use of contingency counsel is critical to the ability of cities to pursue public interest litigation.

2. Encouraging law schools to partner with cities seeking justice
The nation's law schools are well positioned to help cities seek justice, and a few already do. The most ambitious city-law school collaboration of this kind is the San Francisco Affirmative Litigation Project at Yale Law School ("SFALP") (Yale Law School II 1). SFALP is a clinical program through which law students work alongside deputy city attorneys to conceive, develop, and litigate public interest cases that have local dimensions but often nationwide impact (Yale Law School I 1).

In the eight years since its founding, SFALP students have worked on matters involving a wide range of substantive issues including fairness in arbitration, wage theft, sexual orientation discrimination, reproductive rights, internet privacy, housing conditions in poor neighborhoods, environmental protection, childhood health and nutrition, illegal lending practices, and immigrant rights. SFALP students even helped San Francisco take its Proposition 8 challenge through the trial and appellate courts and on to the U.S. Supreme Court.

Through law school projects like SFALP, law students across the nation can help cities brainstorm new cases, draft pleadings and motions, help with discovery and witness preparation, research thorny legal issues, and even draft briefs. For their part, city lawyers have the opportunity to become teachers and mentors by taking their knowledge and experience into the classroom. It would be a win-win-win: cities obtain pro bono legal help, law students obtain practical legal training, and law schools obtain bragging rights.

Conclusion

At the start, this chapter highlighted a worrying imbalance between private corporate activities and public oversight. The idea that cities might join in civil law enforcement efforts through public interest litigation is not a panacea. But city law offices are a sleeping army: a vast, untapped resource for civil law enforcement at a moment where we need all hands on deck. We as a nation should encourage and support cities seeking justice.

Works Cited

Apgar, W., M. Duda, and R. Gorey. "The Municipal Costs of Foreclosure: A Chicago Case Study." *NeighborWorks America*. 2005. 1–57. Web. 27 Feb. 2005 <http://www.nw.org/ network/neighborworksProgs/foreclosuresolutions/documents/2005 Apgar-DudaStudy -FullVersion.pdf>.

Baltimore v. Wells Fargo Bank. 677 F. Supp. 2d 847. 847–852. United States District Court, District of Maryland. 2010. Print.

Beam, Christopher. "Why Would A Toy Manufacturer Ever Use Lead Paint?" *Slate* 15 Aug. 2007: n. pag. Web. 27 Jan. 2015 <http://www.slate.com/articles/news_and_politics/ explainer/2007/08/why_do_they_put_lead_paint_in_toys.html>.

Bernanke, Ben S. Operation HOPE Global Financial Dignity Summit. Atlanta. 15 Nov. 2012. Address. <http://www.federalreserve.gov/newsevents/speech/bernanke20121115a.htm>.

Berner, Robert and Brian Grow. "Banks vs. Consumers (Guess Who Wins)." *Bloomberg Business Week* 5 June 2008: n. pag. Web. 27 Jan. 2015 <http://www.bloomberg.com/ bw/stories/2008–06–04/banks-vs-dot-consumers-guess-who-wins>.

Bocian, Debbie Gruenstein, Wei Li, and Carolina Reid. "Lost Ground, 2011: Disparities in Mortgage Lending and Foreclosures." *Center for Responsible Lending*, 2011. Web. Nov. 2011 < http://www.responsiblelending.org/mortgage-lending/research-analysis/ Lost-Ground-2011.pdf>.

Boyce, James K. *Economics, the Environment, and Our Common Wealth*. Northampton: Edward Elgar 2013. Print.

Brescia, Raymond H. "Subprime Communities: Reverse Redlining, the Fair Housing Act and Emerging Issues in Litigation Regarding the Subprime Mortgage Crisis" ("Brescia I"). *Albany Government Law Review* 2 (2009): 164–216. Print.

Brescia, Raymond H. "On Public Plaintiffs and Private Harms: The Standing of Municipalities in Climate Change, Firearms, and Financial Crisis Litigation" ("Brescia II"). *Notre Dame Journal of Law, Ethics & Public Policy* 24 (2010): 7–52. Print.

Buffalo City Attorney High Profile Matters. 1. Web. 27 Jan. 2015. <https://www.ci.buffalo. ny.us/Home/City_Departments/LawDepartment/HighProfileMatters>.

California Reinvestment Coalition. "From Foreclosure to Re-Redlining." *California Reinvestment Coalition*, 2010. Web. Feb. 2010 <http://www.community-wealth.org/ sites/clone.community-wealth.org/files/downloads/report-stein-gwynn.pdf>.

Centers for Disease Control and Prevention. "Preventing Lead Poisoning in Young Children." *U.S. Department of Health and Human Services Centers for Disease*

Control and Prevention, Aug. 2005. Web. 27 Jan. 2015 <http://www.cdc.gov/nceh/lead/publications/prevleadpoisoning.pdf>.

Chan, Sewell. "Financial Crisis Was Avoidable, Inquiry Finds." *New York Times* 25 Jan. 2011: Web. 27 Jan. 2015 <http://www.nytimes.com/2011/01/26/business/economy/26inquiry.html>.

City of Cleveland v. Ameriquest Mortgages Securities, Inc. ("City of Cleveland I"). 621 F. Supp. 2d 513. 513–536. United States District Court, Northern District of Ohio. 2009. Print.

City of Cleveland v. Ameriquest Mortgage Securities, Inc. ("City of Cleveland II"). 615 F. 3d 496. 496–507. United States Court of Appeals, Sixth Circuit. 2010. Print.

City of Gary v. Smith & Wesson Corp. 801 N.E. 2d 1222. 1222–1249. Supreme Court of Indiana. 2003. Print.

City of Los Angeles v. Citigroup Inc. ("City of Los Angeles I"). United States District Court, Central District of California. 1–51. Web. 2014 <http://atty.lacity.org/stellent/groups/electedofficials/@atty_contributor/documents/contributor_web_content/lacityp_027426.pdf>.

City of Los Angeles v. JPMorgan Chase & Co. ("City of Los Angeles II"). United States District Court, Central District of California. 1–53. Web. 30 May 2014 <http://atty.lacity.org/stellent/groups/electedofficials/@atty_contributor/documents/contributor_web_content/lacityp_028287.pdf>.

City of Los Angeles v. Wells Fargo & Co. United States District Court ("City of Los Angeles III"), Central District of California. 1–69. Web. 2014 <http://atty.lacity.org/stellent/groups/electedofficials/@atty_contributor/documents/contributor_web_content/lacityp_027426.pdf>.

City of New York v. Beretta U.S.A. Corp. ("City of New York v. Beretta U.S.A Corp. I"). 315 F. Supp. 2d 256. 256–286. United States District Court, Eastern District of New York. 2004. Print.

City of New York v. Beretta U.S.A. Corp. ("City of New York v. Beretta U.S.A Corp. II"). 524 F. 3d 384. 384–408. United States Court of Appeals, Second Circuit. 2008. Print.

Cohan, William D. "Wall Street Rises Again." *The Atlantic Magazine* Jan. Feb. 2015: 23–25. Print.

County of Santa Clara v. Superior Court. 50 Cal. 4th 35. 35–42. Supreme Court of California. 2010. Print.

Cutts, Kyle. "City on the Brink: The City of Cleveland Sues Wall Street for Public Nuisance." *Case Western Reserve Law Review* 58 (2008): 1399–1421. Print.

Definitions ("Definitions I"), 21 C.F.R. 170.3 (2015). Web.

Definitions ("Definitions II"), 15 U.S.C.A. 7903(5)(A)(iii) (2015). Web.

Duhigg, Charles. "Clean Water Laws are Neglected, at a Cost." *New York Times* 13 Sept. 2009: A1. Print.

Entin, Jonathan L, and Shadya Y. Yazback. "City Governments and Predatory Lending." *Fordham Urban Law Journal* 34 (2007): 757–783. Print.

Ernst, Keith, Debbie Bocian, and Wei Li. "Steered Wrong: Brokers, Borrowers, and Subprime Loans." *Center for Responsible Lending*, 2008. Web. 8 Apr. 2008 <http://www.responsiblelending.org/mortgage-lending/research-analysis/steered-wrong-brokers-borrowers-and-subprime-loans.pdf>.

Ewald, Alec C. *The Way We Vote: The Local Dimension of American Suffrage*. Nashville: Vanderbilt U Press, 2009. Print.

Fine, Janice. "Solving the Problem from Hell: Tripartism as a Strategy for Addressing Labour Standards Non-Compliance in the United States." *Osgoode Hall Law Journal* 50.4 (2013): 813–844. Web. <http://digitalcommons.osgoode.yorku.ca/ohlj/vol50/iss4/3>.

FDA Investigation Announcement. Web. 3 May 2014 <http://www.fda.gov/ForConsumers/ConsumerUpdates/ucm350570.htm>.

Garrison, Jessica. "State Toxics Department Seeks $4.5 Million Increase in Funding." *Los Angeles Times* 15 Jan. 2014: n. pag. Web. 27 Jan. 2015 <http://articles.latimes.com/2014/jan/15/local/la-me-toxics-20140116>.

——— "State Agency Slow to Protect Public From Toxic Waste." *Los Angeles Times* 15 Jan. 2014: n. pag. Web. 27 Jan. 2015 <http://articles.latimes.com/2014/jan/15/local/la-me-toxics-20140116>.

Gottlieb, Richard and Andrew J. McGuinness. "When Bad Things Happen to Good Cities, Are Lenders to Blame?" *Business Law Today* 17 Aug. 2008: 13–17. Print.

Griffis, Carson R. "Should States Ban Contingency Fee Agreements Between Attorneys General and Private Attorneys?" *Professional Lawyer* 20–3 (2010): 22–24. Print.

Herrera Letter to FDA. 1–2. Web. 19 Mar. 2013 <http://sfcityattorney.org/modules/showdocument.aspx?documentid=1241>.

Herrera Letter to Monster Energy Corp. 1–4. Web. 31 Oct. 2012. <http://sfcityattorney.org/modules/showdocument.aspx?documentid=1070>.

Ileto v. Glock, Inc. 565 F. 3d 1126. 1126–1163. United States Court of Appeals, Ninth Circuit. 2009. Print.

Koerner, Claudia. "Albertsons Settles Hazardous Waste Lawsuit for $3.3 Million." *Orange County Register* 26 June 2014: n. pag. Web. 27 Jan. 2015 <http://www.ocregister.com/articles/albertsons-627020-hazardous-waste.html>.

Krugman, Paul. "How Did Economists Get It So Wrong?" *New York Times* 2 Sept. 2009: n. pag. Web. 27 Jan. 2015 <http://www.nytimes.com/2009/09/06/magazine/06Economic-t.html?pagewanted=all>.

Marll, Samuel. "Do Municipalities Have Article III Standing to Sue Mortgage Lenders Under the Fair Housing Act?" *University of Pennsylvania Journal of Business Law* 15 (2012): 253–304. Print.

Minute Order Granting Motion to Dismiss. 1–21. Web. 16 Dec. 2013. <http://sfcityattorney.org/modules/showdocument.aspx?documentid=1451>.

Monster Beverage Corp. v. Herrera. U.S. District Court, Central District of California. 1–28. Web. 29 Apr. 2013 <http://sfcityattorney.org/modules/showdocument.aspx?documentid=1250>.

Morgenson, Gretchen. "Baltimore is Suing Bank Over Foreclosure Crisis." *New York Times* 8 Jan. 2008: A12. Print.

Morris, Kathleen S. ("Morris I"). "The Case for Local Constitutional Enforcement." *Harvard Civil Rights-Civil Liberties Law Review* 47 (Winter 2012): 1–45. Print.

——— ("Morris II"). "San Francisco and the Rising Culture of Engagement in Local Public Law Offices." *Why the Local Matters: Federalism, Localism, and Public Interest Advocacy.* Eds Kathleen Claussen, Adam Grogg, Ethan Frechette, Rachel Deutsch. New York: Columbia, New Haven: Yale, 2010. 51–66. Print.

——— ("Morris III"). "Expanding Local Enforcement of State and Federal Consumer Protection Laws." *Fordham Urban Law Journal* 40 (2013): 1903–1928. Print.

Nuisance; what constitutes, Cal. Civil Code § 3479 (2015). Web.

People v. Albertsons, LLC. Superior Court of the State of California, County of Orange. 1–21. On file with author. May 2014.

People v. Monster Beverage Corp. Superior Court of the State of California, County of San Francisco. 1–23. Web. 2014 <http://www.sfcityattorney.org/modules/showdocument. aspx?documentid=1296>.

People v. National Arbitration Forum. Superior Court of the State of California, County of San Francisco. 1–26. Web. 22 Aug. 2008 <http://www.sfcityattorney.org/Modules/ ShowDocument.aspx?documentid=257>.

Pindell, Ngai. "The Fair Housing Act at Forty: Predatory Lending and the City as Plaintiff." *Journal of Affordable Housing and Community Development Law* 18-Winter (2009): 169–178. Print.

Proposed Statement of Decision. 1–110. Web. 16 Dec. 2013. <http://www.sfcityattorney. org/modules/showdocument.aspx?documentid=1454>.

Public nuisance, Cal. Civil Code § 3480 (2015). Web.

Schwemm, Robert and Jeffrey Taren. "Discretionary Pricing, Mortgage Discrimination, and the Fair Housing Act." *Harvard Civil Rights-Civil Liberties Law Review* 45 (Summer 2010): 375–433. Print.

Scientists' Letter to FDA. 1–6. Web. 19 Mar. 2013 <http://sfcityattorney.org/modules/ showdocument.aspx?documentid=1241>.

Smith & Wesson Corp. v. City of Gary. 875 N.E. 2d 422. 422–435. Court of Appeals of Indiana. 2009. Print.

Starkman, Dean. *The Watchdog That Didn't Bark: The Financial Crisis and the Disappearance of Investigative Journalism.* New York: Columbia U Press, 2014. Print.

Stern, William L. and Nicholas A. Roethlisberger. "The 'Con' Side: Outsourcing Justice: The California Supreme Court's Decision in County of Santa Clara v. Superior Court (Atlantic Richfield)." *Journal of the Antitrust and Unfair Competition Law Section of the State Bar of California* 19–2 (Fall 2010): 2–5. Print.

Stout, David. "Justices Decline New York Gun Suit." *New York Times* 10 Mar. 2009: n. pag. Web. 27 Jan. 2015 <http://www.nytimes.com/2009/03/10/washington/10guns.html?_ r=0>.

Surowiecki, James. "The Regulation Crisis." *The New Yorker* 14 June 2010: n. pag. Web. 13 May 2015 <www.newyorker.com/magazine/2010/06/14/the-regulation-crisis>.

Yale Law School (cited as "Yale Law School I"). "Frequently Asked Questions." Web. 25 Jan. 2015 http://www.law.yale.edu/academics/SFALPfaq.htm>.

Yale Law School (cited as "Yale Law School II"). "San Francisco Affirmative Litigation Project." Web. 25 Jan. 2015 http://www.law.yale.edu/ academics/SFALP.htm>.

Chapter 11
Strategic Code Compliance Enforcement: A Prescription for Resilient Communities

Kermit J. Lind, J.D

Introduction

Communities seeking neighborhood stability need a new and better approach to improving housing and neighborhood environments, whether those communities face dystopian conditions or are endeavoring to replace unusable, abandoned residential property with new and rehabilitated homes. This chapter presents the case for recognizing strategic code compliance and enforcement as the essential approach to sustaining neighborhoods against current destabilizing forces in the housing market and for a future of more resilient residential habitats.[1]

This chapter's thesis is that improved housing and neighborhood code compliance and enforcement are essential for urban neighborhoods' recovery and resilience. The improvement required is code compliance and enforcement that functions as a system of connected and coordinated parts focused on common goals. This new enforcement system rejects the old, but unfortunately still extant, code enforcement regime characterized by a conglomeration of unconnected local government institutions, agencies, departments, and programs each pursuing its own objective. Instead, the new system is integrated and generally requires agencies, departments, and institutions dealing with real property conditions and titles to align in pursuit of common objectives attainable only by mutual effort. Informed by a comprehensive data system hosting all available parcel-level real property data within local jurisdictions, this kind of strategic code policing facilitates deployment of public and private resources for maximum community benefit. The result is a code compliance and enforcement operation that is nimble and flexible, allowing for strategic engagement with chronic problems instead of a series of reactions to individual complaints.

Part 1 describes weaknesses and inadequacies of traditional policing policies and programs for maintaining residential dwellings and neighborhoods in safe, healthy, and secure conditions. Part 2 describes how the devastating impact of the last 20 years of abuse to the housing and mortgage markets culminated in a crisis laying waste to whole neighborhoods and communities. Part 3 identifies how local reformers armed with comprehensive data on real property transactions and conditions are responding to the housing crisis. Part 4 offers a vision of strategic code compliance as an alternative to the failed and outdated traditional policies and programs. Part 5 offers examples of data driven strategic code compliance enforcement as practiced in real, on-the-ground situations.

---

1   The case for the priority of code compliance and much of the material in this paper is derived from my 30 years as an advocate, lawyer, and clinical professor. There are earlier code compliance recommendations explored in previous publications including Lind 2008; Lind 2011; Lind 2012a; Lind 2012b; Lind 2015; and Keating and Lind 2012.

Part 1. Assessing Challenges to Conventional Residential Property Maintenance

To understand the old code enforcement regime employed by local governments nationwide is to appreciate the incredible diversity of American local governments. Some preliminary clarification of terminology may be helpful.

The standards and enforcement procedures for the construction, use, and maintenance of structures is a matter of state and local law. Use of common terms is almost impossible due to wide variation among jurisdictions in both the legally adopted standards and in the terms describing them. Thus, for our purposes, the term "housing maintenance code" refers to the standards for lawful use and condition of residential structures whether single or multi-family buildings. "Neighborhood code" or "neighborhood environmental code" refers to the required standards for the exterior of buildings, either single-family dwellings or collections of dwellings that may be a block of houses or a neighborhood, including common and public areas. References to "building codes" in both legislation and in descriptive literature requires special clarification. In some circumstances, "building codes" means codes to regulate the workmanship and materials for constructing, rehabilitating, or repairing structures, while in other circumstances "building codes" includes the standards for the use and maintenance of built structures.

It is also important to appreciate that codified building or housing code standards are usually accompanied by codified procedures for enforcement of the standards. Here, we will use "code enforcement" or "code policing" to mean the procedures, both administrative and judicial, for enforcing legislative or administrative orders to comply with adopted regulations.

Finally, it should be noted that enactment and enforcement of standards for the condition of structures and neighborhood environments is based upon the exercise of the state's police power, which is the fundamental power of government to protect the public health, safety, and welfare. Defining in each situation what is the legal exercise of police power to make, enforce, and interpret police powers is a political matter; but political and judicial determinations are rooted in the inherent power of government to protect the public health, safety, and welfare from actions or omissions that will do harm to people and the social order (Listokin and Hattis 4, 12).

Times Have Changed for Regulating Residency Maintenance

Historic housing and neighborhood environment codes are not effectively meeting the current challenges in typical city neighborhoods. Standards and policing methods for real property maintenance were developed decades ago in economic and market conditions far different from those in the contemporary housing markets. For 50 years, students of housing maintenance have proposed improvements with little success. (Listokin and Hattis 5; Marco and Mancino *passim*; Harvard Law Review *passim*). Historically, local government enforcement of codes applicable to single–family dwellings presumed that resident-homeowners completed regular maintenance and necessary repairs. Housing consumers generally saw their dwellings as homes, not as investments or commodities for wealth accumulation or for speculation. Their self–interest generally inclined them to voluntarily comply with locally adopted codes. Social and cultural norms in homogeneous neighborhoods of single–family dwellings produced peer pressure for keeping residential property in good repair. Absentee institutional investors were not in the picture.

These old realities about occupancy, ownership, property values, and neighborhood stability are being displaced. Most cities now have neighborhoods plagued with high–risk borrowing, predatory lending, underwater loans, loan defaults, high foreclosure rates, abandonment, chronic vacancy, distressed sales, speculative investing, get–rich–quick house flipping, and massive population shifts leading to the decommissioning of neighborhoods (Beckman 3). Most of this new wave of lending and ownership activity is controlled remotely by persons having no personal stake in the community, while the number of resident-owners and their influence over the condition of their neighborhood declines. Further, in an era when the income of families who depend on steady employment to sustain homeownership is declining while job insecurity is rising, the capacity to pay for the normal home maintenance and repair continues to diminish (Alexander 13; Schwartz and Wilson 1–12; Peck 60).

Neighborhood landscapes in legacy cities underscore the inadequate results of historic modes of code compliance and enforcement. These urban landscapes are replete with examples of how the historic standards and procedures relied upon to keep ordinary homeowners secure from harm and loss are failing to do that job. The decline in municipal capacity to police housing conditions and neighborhood environments is greatest in communities with households of average or below-average incomes, older housing stock, and with declining populations. As Judge Raymond, Cleveland's Housing Court judge observed, "[t]here has never been a time in the memory of most people now living in this nation when the stability and fabric of neighborhoods and communities have been so threatened in so many places all at once" (Pianka 44).

Fragmentation of Functions and Responsibilities

The obsolescence of historic approaches to housing and residential environment code enforcement is further evident in the insularity of the various departments of government that play a role in policing compliance. Consider that most city codes have traditionally placed building standards for construction and standards for housing maintenance in different code sections. Further, inspection for compliance with *building* codes is initiated by issuing permits for construction through a building official, while inspection for compliance with *maintenance* codes is usually initiated by a distinct code enforcement office in response to complaints. Especially in large cities, this division of the permitting and compliance function results in separate local government departments with different procedures and managers. Review of submitted plans and inspection of municipally authorized construction and repairs are usually given more status and priority than inspections done in response to complaints about high grass, litter, broken sidewalks, junk cars, sagging porch roofs, or crumbling foundations.

Frequently, maintenance inspection responsibilities are divided into subject areas—health, fire, structural safety, environmental, interiors, rental, and single family—with virtually no coordination or contact between the departments whose officers carry out the inspections. In turn, these maintenance inspections may or may not be followed by post–inspection investigations of ownership history, whether the occupant is the title holder of record, legal condition of the title, property tax records, unpaid assessments, pending violations notices, pending legal actions affecting the title holder, or other parties holding legal claims to the property. In short, the initial inspection and citation of violations and the subsequent investigation of noncompliance with notice of violations are functionally different, requiring different skills, aptitudes, and training, and thus did not historically occur in a unitary code standards, inspection, and enforcement system.

Prosecution of code violations has presented another occasion for fragmentation of local government code inspection and enforcement efforts. It is rare to find inspectors and prosecutors working closely together, even to prosecute repeat violators.

Further, it is rare that those in adjudicative positions as hearing officers and judges are experts in housing and neighborhood environmental matters. Municipal courts may assign housing cases to a single judge or magistrate, but separately elected or designated judges presiding exclusively over a housing docket are very rare. That means adjudicators do not have the expertise that comes with concentrated experience trying housing and residential issues. Municipal judges usually reach the bench because of their reputation for hearing cases involving drug dealers, thieves, and felons rather than first-hand housing knowledge. Most municipal courts lack the necessary jurisdictional authority found in special purpose housing and environmental courts that are specifically empowered to fully adjudicate issues in challenging cases of defective real property conditions (Pianka 44–49).

Code compliance cannot be effective when those who are enforcing it operate in separate bureaucratic silos with separate objectives. The various program elements and agencies involved in code enforcement do not generally operate as a system in which each does its part to achieve a common goal. Rather, each department, division, or institution defines the objectives of its work in terms of its limited function rather than a mutual goal of protecting the health, safety, and security of people and their property. Instead of focusing on the goal of improving quality of life for the community of neighborhood residents, these fragmented code compliance agencies often focus narrowly on bare statistics concerning processing and disposing of individual housing, health, sanitary, or zoning code violations.

Part 2. Mortgage Crisis Challenges to Residential Health, Safety, and Stability

The Disconnection of Mortgage Servicing from the Care of the Collateral

Mortgage lending and servicing drastically changed with the introduction of financing that turned mortgage loans into commodities to be packaged and sold in bulk to investors as securities. Securitization changed the relationship between lenders and the collateral securing the debt (Eggert 753–61; Engel and McCoy 43–68; United States. Cong. Senate 116–43). It is designed to create new investment products using the stated value of mortgage payments to sell to investors. This new mortgage lending business aims to aggregate large numbers of mortgages for assembly in batches. The batches are then repackaged for resale to investors as shares of a trust. In this new arrangement, the securitization of loans unhinges the lenders' direct connection with the collateral (Levitin and Twomey 45–57). Further, once the loan is made, the mortgage and loan are sold separately from the debt servicing rights and added to a pool of mortgage loans held in trusts. The work of servicing mortgages is a profitable byproduct separate from the value of the mortgage collateral (Collins). Borrowers then may find themselves in a business relationship with one or more computerized avatars, digitally engineered voices, with no prospect of a face-to-face encounter with anyone responsible for receiving their payments or capable of solving problems or making a business decision relating to these mortgage loans. To compound the problem, defaulting borrowers who are being hounded to make mortgage payments do not have funds to maintain their houses. Indeed, the misfeasance, malfeasance, and nonfeasance in the mortgage servicing industry has made headlines for years and continues to be the subject of both regulatory and criminal investigations (Federal Reserve White Paper 2; Porter 43). Traditional code compliance programs and methods were not designed to work with the new business model employed by mortgage debt holders and their servicers.

In this new arrangement, borrowers and code compliance officials almost never encounter anyone representing the mortgagee agents of houses owned or controlled by banks. In other words, mortgagees rarely have staff with actual knowledge or concern about the condition or value of the collateral. Loan originators and brokers, as well as investors holding securities representing an interest in loan pools, are looking only for payments in the amount of the balance owed on the debt regardless of the condition—and sometimes even the existence—of a structure on the premises. They are insulated from responsibility for liabilities derived from the collateral's physical condition. As those who seek to clear titles on vacant lots know well, houses can disappear in calamities or by demolition but still remain as assets with "book value" for lien holders in the amount owed at the time of default plus penalties, fees, and interest. Mortgage insurance and other guarantees, including taxpayer bailouts, matter more than collateral value for protecting investments in mortgage–backed securities.

Likewise, no holder of a mortgage packaged into a mortgage trust need be concerned about legal responsibility for maintaining the collateral's condition in compliance with local law. Responsibility for lenders' interest in the collateral is assigned by contract to servicers whose identities and functions are not part of the public title record available to title record research. Records searches find no documentation on the publicly recorded mortgage itself identifying the current mortgage servicers. Although servicers complain about failure to get prompt notice of adverse municipal actions against noncomplying properties, the servicers—who were delegated authority and responsibility for properties—do not record the assignment of their legal authority in the public records. In fact, all who benefit from mortgage payments or the asset value to back securities sold to investors are shielded from any actual knowledge, burden, or responsibility arising in connection with the condition of the collateral until they are tracked down by diligent investigators. Municipal officials charged with policing neighborhood and residential housing conditions go to great expense to engage mortgagees and their servicers in communication about the condition of their collateral (Schilling 2009, 124–27). Mortgage servicers cannot be counted on to respond to notices, warnings, or other communication indicating that low-value collateral in their portfolio is being wasted, has become a public nuisance, is threatening the value of other collateralized housing in the neighborhood, or is being condemned for demolition (Lind 2011, 102; Ford 54).

Securitization of mortgage loans reduces the incentive for mortgage lenders and servicers to mind the condition of homes that constitute the collateral. (B. White 1107–19, 124–29; Engel and McCoy 43–65) After all, there is no individual liability for ignoring the collateral and no profit or benefit to be realized from protecting it when its present market value is less than the amount owed on the mortgage loan. The business interests and economics of mortgage securitization and servicing itself tend to justify abandoning responsibility for wasted collateral when its loss of value is recognized (Theologides 77). In fact, code enforcers and community advocates contending with noncompliant mortgagee entities and servicing companies are told that the servicers' business plan does not work if mortgagees or banks holding legal title to houses purchased at sheriff sales are required to make the repairs or abate nuisance conditions required by law. In essence, mortgage holders and banks consider their institutions exempt from municipal housing maintenance laws.

Meanwhile, neighborhood residents are not only subject to enforcement of the same health and safety code provisions, but they must suffer the increased risks to health and safety, diminished property values, increased insurance rates, and unsightly appearance associated with the spoiling bank collateral.

The Nature and Spread of Abandonment and Vacancy

Abandoned vacant dwellings are the great symbol of the current mortgage crisis (Ford). Natural disasters like Hurricane Katrina or the earthquake in Haiti suddenly turned whole neighborhoods into solid waste; but earthquakes, floods, and other natural disasters are confined to one fairly contiguous geographic area. The mortgage disaster is man made and has spread out of control in cities and regions all over the country. While it proceeds at a slower pace than a hurricane or flood, it nevertheless turns whole neighborhoods into wastelands (Williams A15).

Abandonment and blight are concentrated in neighborhoods that are home to low–wage earning persons of color whose savings were in their home equity (Kneebone and Homes). Urban "legacy cities" in the Mid-west where the scourge of abandoned vacant housing has been an issue for more than 15 years are especially vulnerable. Looking at city maps showing abandoned houses by location reveals what national- and regional-level data obscures; namely, that within older suburbs and cities like Chicago, Detroit, Cleveland, Buffalo, Toledo, Youngstown, Cincinnati, and St. Louis abandoned houses are clustered in poor and predominantly minority neighborhoods. The hardest hit neighborhoods are not yet recovering from recession. Many cities are spending all the federal and bank settlement funding they can get and as much local tax money as they can spare to demolish by the tens of thousands inventories of abandoned housing. The abandoned and vacant property epidemic is so severe that these cities have difficulty knowing how many homes should be demolished or even when the local governments will be able to suspend their demolition programs. Despite this uncertainty and despite mass demolitions' impact on inner-city neighborhoods, the importance of demolition for neighborhood recovery is not merely to remove harmful health and safety nuisance conditions; it also serves to preserve values in neighboring properties threatened by proximity to unmarketable abandoned houses (Garvin 412–26; Griswold 57, 71). Cleaning up housing blight shows neighboring owners that they and their neighborhood have not been totally abandoned by their city.

Poor maintenance of homes or their complete abandonment not only destroys the physical integrity of structures; it also inflicts long–term damage on housing market values (Duke 3–6, 19; Whitaker and Fitzpatrick 31). The downward neighborhood spiral triggered by dilapidation and abandonment is vicious. When homes are abandoned and become vacant, it reduces the value of the subject house and surrounding houses to such an extent that otherwise habitable homes cannot be repaired and sold for a price that will pay for the repairs that bring them into compliance with housing maintenance codes (Government Accountability Office 2011, 3–79). To make matters worse, abandoned homes often fall into the hands of unscrupulous home rehabilitation firms. These firms frequently do poor work using only cheap materials that do not meet code standards. They then sell the homes to inexperienced consumers or large absentee corporate investors in unstable neighborhoods. Houses that fall prey to this type of abuse can go through multiple cycles of foreclosure, abandonment, repossession, sale, improper repair, resale, unsustainable financing, and the cycle repeats (Kotlowitz).

Mortgage modification programs to keep resident owners in their homes have not yet stemmed the tide of defaults by borrowers whose home mortgage loan balance exceeds the current value of their house. The term describing mortgages in that situation is "underwater." Studies show that most borrowers are still making their payments on their underwater mortgages, however, this does not mean that those homeowners will continue making their payments (B. White 7–15, 24–30). The ongoing reports by RealtyTrac, a national company providing real estate data on the number and concentrations of underwater mortgages, suggest the foreclosure and abandonment due to loan defaults will not subside quickly in the hardest-hit communities (Pohlman).

The Mortgage Crisis Requires New Approaches to Code Compliance

It is ironic that the primary protectors of the collateral for the mortgages on which the world economy depends for economic stability are housing inspectors—those Joe Schilling calls the "first responders" to the early signs of abandonment (2009, 149). It is these code compliance inspectors and enforcers, using local laws, police power, and public resources who end up caring for the economic interests that mortgagees and investors have in their mortgage collateral. The business structures and practices described above have made that task vastly more difficult and expensive for municipal code officers and enforcers. Municipalities find those costs unsustainable, especially as they see their tax revenues declining and traditional state support diminishing (Weinstein 7–18). These new challenges to property maintenance code compliance in the mortgage crisis are forcing communities to look for new approaches to the preservation and protection of housing property rights and neighborhood health, safety, and security. These new approaches will require new capacities—capacities for changing to a new way of policing housing and neighborhood conditions more strategically (Mallach 13–15).

Part 3. Building Capacity For Strategic Policing

Having described the shortcomings and inability of local governments' historic codes and code enforcement policing to protect residential neighborhoods from the dystopian realities in the housing market, I turn now to responses to from those at the local level. In hindsight, it is clear that the mortgage and financial crises in the first decade of this century were foreseeable, but were not recognized or validated by many before 2007. Wall Street's growing appetite for mortgages and the products and marketing practices developed to feed it was a change that produced consequences at the neighborhood level that could not be ignored. The housing mortgage debt crisis had to be recognized and understood as a new kind of challenge in order for a meaningful response to be made.

Recognizing a Crisis and Responding

In July 2007, a delegation of officials and residents from Cleveland appeared before the Congressional Joint Economic Committee to report on the mortgage crisis in the city and its nearest suburbs (U.S. Cong. 12–21, 44–54). County Treasurer Jim Rokakis focused attention on a Cleveland neighborhood called Slavic Village, where real estate speculation and predatory mortgage practices that began there in the 1990s were threatening a decade later to destroy the entire neighborhood. He also cited other cities around the country with neighborhoods confronting the same calamities. City Councilman Anthony Brancatelli, a long–time community development director in Slavic Village before becoming a city councilman, explained how people in his ward were fighting back against blight, crime, global banks, and local unprincipled housing and loan dealers. His testimony noted how house flippers bought foreclosed houses, sold them in poor condition to buyers who got subprime mortgages in amounts greater than the sale price, then immediately defaulted and the process started all over again. Homeowners Barbara Anderson and Audrey Sweet testified about their personal experience with predatory lenders and manipulative mortgage servicers. Rokakis, Brancatelli, and others in the delegation warned Senators that the whole nation was in peril because of irresponsible buying, borrowing, lending, and debt collecting practices that preyed on people in modestly priced houses in ordinary neighborhoods.

None of those testifying realized how drastic the crisis they were reporting would become for the entire nation within months. What their testimony shows is that people in Cleveland recognized the mortgage crisis at the neighborhood level and were sounding alarms well before 2007. Neighborhood community development corporations were the first to see the storm mounting before the turn of the century (Lind 2008, 2). When the storm hit other parts of the nation, it put the Cleveland story in the national spotlight (Kotlowitz).

Clevelanders were not the only ones to recognize this brewing storm. In the summer of 2003, a small group of people from a variety of professional housing and land use disciplines met in Washington D.C. for the announcement of a new coalition called the National Vacant Property Campaign (NVPC).[2] They launched the Campaign in 2003 to help communities prevent abandonment, reclaim abandon properties, and revitalize neighborhoods.[3] Older industrial cities have been dealing with the problems of vacant property for many years. By 2003, however, the sudden rise in mortgage and tax foreclosures along with creeping blight in economically weak neighborhoods had reached crisis levels in many cities, a reality recognized by those gathered for the launching of this new national campaign. This new campaign looked at vacant property in urban neighborhoods not only as a destabilizing disaster, but also as an opportunity to reuse land in new, more efficient, and sustainable ways. Those who launched this campaign saw reuse of abandoned property as an alternative to sprawl, an opportunity to reduce dependence on automobiles and expensive highways, a chance to deliver improved housing opportunities for low-income households, and a way to improve the quality of life for those in economically weak and marginalized communities. In 2010, the NVPC merged with the Genesee Institute, a technical assistance program for land banking in Flint, Michigan, to form the Center for Community Progress (CCP).

National and Local Responses to Abandoned Vacant Property

The NVPC began a program of sending teams of experts to assess vacant housing problems in communities with severe problems. The first of these consultations occurred in Cleveland where in 2004 Neighborhood Progress, Inc., a leading nonprofit community development organization, hired NVPC to do a study of Cleveland's abandoned vacant property problem. Participation in the study and interviews were widespread and included

---

2   Initiators were Smart Growth America, the International City/County Management Association (ICMA), and the Local Initiatives Support Corporation (LISC).

3   See the announcement by LISC at http://www.lisc.org/content/organizations/detail/710. The author participated in this inaugural conference and subsequently participated in several consulting teams until 2011. See also Schilling 2009 112).

public interest community development leaders, academics, neighborhood organizations, and officials in all three branches of municipal government, as well as county agencies. The process climaxed in the spring of 2005 with a series of public presentations of findings and recommendations. These were documented in a report, *Cleveland at The Crossroads: Turning Abandonment into Opportunity* (Mallach et al. 2005).

The Report's most critical findings were that compliance with housing maintenance and neighborhood environmental codes was severely weakened and that enforcement processes were poorly coordinated and so cumbersome that much of the enforcement work was incomplete and ineffective. As a result, it was easy and cheap for investors, speculators, and controlling lienholders to engage in unscrupulous conduct, to manipulate bureaucratic procedures, to disregard housing maintenance regulations, and to ignore or defy law enforcement with impunity. The tidal wave of property abandonment overwhelmed and disabled code enforcement. At the same time, rising mortgage foreclosures pushed cheap, distressed houses into the hands of speculators who helped gullible or fraudulent buyers get loans without hope or intention of repaying those loans. It was observable from the streets that houses in inner city neighborhoods were caught in a repeating cycle of unscrupulous house flipping, fraudulent mortgaging at inflated amounts, early loan defaults, foreclosure, and sheriff sales that often restarted the cycle. At no point in the cycle was code compliance enforcement employed with enough effect to break the cycle. Repair was often limited to cheap cosmetic measures to impress naive buyers. Judges and magistrates presiding over foreclosures were normally unaware of municipal officials' efforts to inspect and condemn houses on their dockets. Private civil nuisance abatement actions by receivers, while innovative, were too few and proceeding at a scale too small to save hardest hit neighborhoods (Lind 2008 9, 2011 128–37).

The Report's list of recommendations included two key items: (1) the establishment of a coordinating group of personnel from various public offices and nonprofit public interest entities who had direct supervision of programs for housing, community development, and code enforcement and (2) the development of a comprehensive parcel-based real property data bank accessible to all those who needed the full array of real property data for their public and public interest work (Mallach et al.). Implementation of these key recommendations is described below.

Similar studies by NVPC consultants, including this author, were conducted in more than a dozen cities around the country facing abandoned property challenges. These studies resulted in a range of findings and recommendations, but they invariably highlighted two concerns: (1) the identification of serious weaknesses in code compliance and enforcement, and (2) the lack of access to real property information needed by those persons evaluating property conditions or tracking ownership, taxation, liens, or transfers of interests. Lacking these resources and laboring under an antiquated code enforcement system, virtually everyone, but especially code compliance officials, were flying blind as the abandoned housing storm spread through vulnerable neighborhoods throughout the country.

A Coalition of the Willing

Reformation of neighborhood code compliance is most evident today where coalitions of people in a variety of institutions, organizations, and departments work together to make change happen. This kind of coalition for reform resembles the dynamic leadership cadre described by urban government political theorist, Clarence Stone, in his book, *Regime Politics* (3–4). Stone sees the leadership cadre positioned outside and distinct from established governmental institutions but consisting of people with knowledge of, and access to, those established institutions in need of political reform.

The reforms considered in this chapter, however, are not political. Rather, they are pragmatic and focused on policies and programs to implement policies. The coalition of the willing is not solely the creature of government; it also includes those who make and administer policies and programs in government and in private public interest positions (Lind 2012 464).

The leadership of a single elected official or institution, no matter how highly placed or powerful, can be useful but is rarely sufficient. There are numerous examples of politicians—mayors, council members, aldermen, and judges—who personally lead the charge to challenge community blight and deterioration only to run into resistance. Some of these politicians lose momentum, some face political opposition, or a campaign for re-election, or still others must shift focus to other competing problems facing a local government. Further, another problem is that other independent (and sometimes competing) elected officials are sometimes charged with aspects of code enforcement. Code enforcement placed in the hands of political officials often presents an impediment to sustained collaboration. Instead, it takes an informal group in public interest positions that are separate from electoral politics, but who have access to institutional resources, to coordinate, develop, and institutionalize meaningful change. Putting together that coordinating leadership group is an initial step toward better code compliance.

In medium-sized and larger cities with departments operating under different elected officials with limited mandates, it takes a sense of acute crisis to trigger the complex and time-consuming task of amending failing policies and practices. The abandoned housing crisis is now widely recognized as a crisis to which localities must respond with whatever means they can muster. Doing nothing is no longer a viable option, and it has adverse political consequences for elected officials. That perceived crisis is a motivating factor for persons who otherwise would be reluctant to allow or actively pursue a new systematic strategy for code compliance.

Formation of a coalition of persons with the means and the will to collaborate may be initiated in various ways. The description of what happened in Cleveland is instructive. A nongovernmental public interest organization with established community credibility commissioned a study culminating in public presentations of findings and recommendations. Outside consultants are generally able to express boldly what they find with more credibility than local advocates, which is what happened in Cleveland. The consultant's recommendation for a collaborative process including public sector and public interest professionals as well as access to an integrated data bank represented a starting point. It also represented a recommendation that could be pursued immediately.

Experience suggests that a strictly public or a strictly private collaboration is less sustainable than a mix of public and private partners. Composition of a coalition can vary but its resilience depends on a strong core of persons who are personally and professionally committed to its purpose and goal. Experience also suggests that the experts from law clinics, urban policy, city planning, and government scholars can provide important resources.

A Well-Studied Example with Ten Years of History

In 2014, an exhaustive study entitled *Cleveland and Cuyahoga County Ohio: A Resilient Region's Responses to Reclaiming Vacant Properties* was published by the Vacant Property Research Network, a Research Initiative of the Metropolitan Institute of Virginia Tech

(Schilling 2014). The study team director and principal author is Joseph Schilling, a practitioner and scholar of housing codes and code compliance for more than two decades. This detailed study of Cleveland's response to its vacant property crisis illuminates what is being done, what worked and what did not, and what still needs to be done in this legacy city's decades of population loss and housing woes.

Schilling's study lists three critical aspects needed for successful vacant property policy systems: (a) Collaborative Networks; (b) Information Systems and Data Driven Interventions; and (c) a Strategic Framework of Vacant Property Policies (2). Each of these elements forms the capacity-building base for thinking and acting strategically as a system. Cleveland provides an impressive example of the development of these critical aspects.

Cleveland's legacy of collaborative networks—nonprofit community developers, nonprofit intermediaries, community development financial institutions (CDFIs), foundations, and public institutions—with decades of experience producing affordable housing and neighborhood improvements has been well–documented (Krumholz and Hexter 1–2; Schilling 2014 68–72). Abandoned houses proliferated faster than new and better housing could be produced. Blight outpaced neighborhood development projects. The community development system and the City of Cleveland needed a greater collaborative effort to take action against the growing surge of vacant abandoned properties. That realization resulted in the Vacant Abandoned Property Action Council (VAPAC). VAPAC was launched in 2005 by Cleveland Neighborhood Progress, Cleveland's leading nonprofit community development entity, in response to the *Cleveland at The Crossroads* report. Drawn together by the access to professional peers and to the best available real property data related to their work, the Council members are able to see over the top of their respective institutional silos and, in many cases, collaborate with each other to achieve objectives impossible for any single nonprofit group, institution, or governmental unit to achieve alone.

Making and Using a Comprehensive Data Base

The second foundational recommendation to build Cleveland's capacity for strategic action on abandoned property was to assemble and provide an integrated information system for real property research, planning, investigation, and accountability. Access to an integrated parcel-based information system is one of the most important tools for any work involving real property. Parcel identification numbers for each parcel of real estate are like social security numbers. All transactions involving a parcel identify it by that code. Records of all kinds, including deeds, liens, taxes, foreclosures and court cases, citations of violations, and permits may be found, tracked over time, and analyzed when placed in a single data system.

Property records are in the custody of public offices and are increasingly available in digital format online in most cities. However, all property data is not generally accessible in any one online source or site. Nor are the official record keepers accustomed to providing more than one record per search. Another big challenge arises from the fact that each public office with the authority to do so obtains its own computerized system from a vendor of its own choosing with designs and protocols only for its own mission. That means these public systems cannot exchange their data with each other. The integration of records from many sources may require translation from one type of program language to another. Integrating data from one data entry point into a collection from other sources requires 'data scrubbing' to account for errors and inconsistencies, such as different spellings of the same name. For these reasons, getting regular and full output of all records in digital form from multiple sources integrated into one database is a daunting task. But only by doing so can information gaps be filled to gain a more nearly complete perspective of what happens with real property.

There is one major divide in the real property recordkeeping world. Records relating to the use and physical condition of real property are nearly always produced and maintained by various municipal administrative offices. Examples of these are building permits; zoning variances; violation notices; public nuisance findings; condemnation orders; environmental conditions; as well as health, police, and fire department actions. Municipal courts and clerks of court may also have separate systems for their public records. Within the municipality, different offices may keep records in a variety of systems using software that does not interact automatically with the systems used by other municipal departments.

Similarly, the records relating to the ownership, legal interests, and taxation of real property, such as the name of the title holders, the identity of those with a lien on the title, property tax assessments and tax payment status, pending bankruptcy proceedings, or trial court judgments affecting title are all in offices that may be in a variety of other government institutions. In some jurisdictions, there are multiple property tax assessing and collection authorities; for example, schools, municipalities, county, regional transportation, water, and sewer authorities.

This separation of property use and condition recordkeeping on one hand, from property title condition and transaction recordkeeping on the other hand, is virtually universal. Municipal government presides over the law and records related to construction, use, and maintenance of real property. County government presides over the law and records related to the ownership, transfer, and encumbrance of real property titles. A title search can reveal who owns a property, the lien holders, and various assessments, but it does not reveal who is in possession, the identity of any occupants, whether its use is lawful, or whether there is a house. Spelling errors, typing errors, and other mistakes are not systematically reviewed and corrected. Therefore, municipal officials who must by law rely on the title records for identities, citations, search warrants, summons, and other information copy incorrect and invalid information from official records. Such problems stop or extend the processes of enforcement. Incorrect information is the basis for a large proportion of housing code enforcement cases dismissed. Such delays often require the process to be started all over again. Code enforcement prosecutors may incorrectly prosecute prior owners of properties because a subsequent owner's deed is not timely recorded. Recording of deeds is not regulated well enough to prevent abusive manipulation of the record system to avoid prosecution for code violations. The separation of the different types of real property records—use and condition from title and encumbrances—wastes time and resources. It is optimistic to expect the array of fragmented government property data sources to be united with a single local government department. Real benefits can be achieved, however, when they can be integrated and housed for access at a single web site.

Besides fully illuminating real property information, the integration of that data in a single, accessible digital system provides a means of improving coordination among the various actors whose work is, or should be, using current and accurate records. A common access point for all available information provides a common interest in cooperation among the users and providers of the data. The integration of data for spreadsheet and mapping display also allows for study and research that could not otherwise be done. Mapping property transfers, ownership, liens, foreclosures, sheriff sales, post–foreclosure transactions, noncompliance with citations, age of housing stock, price fluctuations, and much more can provide new help for working strategically and making the right policy choices to achieve a desired objective.

The mortgage crisis has created an environment that is encouraging more cooperation and collaboration among local public offices on matters related to housing and neighborhood conditions.[4] That includes sharing and using data. There is an urgency about finding ways to act sooner and smarter. Integrating data systems is a key to achieving that goal. Also, simple and inexpensive technology in portable devises is now available not only to the public work force but also to organized resident groups at the neighborhood level. This opens up new information sources and exchanges for engaging concerned citizens at the neighborhood level in the task of neighborhood preservation.

Part 4. Strategic Code Compliance Enforcement

This discussion has proceeded from observing the limitations of historic housing and neighborhood code compliance measures to the recognition of crisis conditions and the need for building capacity to respond more dynamically at the local level. In that regard, I have explored specific examples of community responses to the national vacant property crisis. These locally driven efforts started with the formation of a coalition of willing public and public interest professionals. Those coalitions facilitated the integration of the public real property databases with other available social and demographic data in a unified online system. Now, the discussion turns to the challenge of reorienting housing and neighborhood code compliance and its enforcement. Specifically, this Part focuses on the ways that code compliance and enforcement can be deployed strategically to deal with the critical task of sustaining residential neighborhood character.

Defining Strategic Code Compliance Enforcement

Along with Professor Joseph Schilling, this author has worked for several years on a definition of strategic code compliance enforcement as a term of art distinct from what is commonly meant by code compliance enforcement. It is a concept designed for building effective code compliance capacity in distressed residential communities. Specifically, this system of strategic code enforcement refers to the organization of critical assets and resources into a system of activities with clearly identified goals, principles, and procedures. All elements—personnel, agencies, institutions, policies, procedures, planning, setting priorities, and administration—must have a common purpose in which each plays a contributing part. It focuses resources and skills on problems and dysfunctions most critical to getting desired results. Strategic code enforcement operates pragmatically and proactively. To operate in this fashion requires access to narrowly focused information and data about what is currently happening in the field as well as higher level information and data that measures trends and forecasts problems.

---

4 The National Neighborhood Indicators Partnership is a collaborative effort by the Urban Institute and local partners to further the development and use of neighborhood information systems in local policymaking and community building. *See* http://www.neighborhoodindicators.org/about-nnip. The comments and publications posted by its members demonstrate how integrated real property databases can illuminate critical information about localities.

Thinking and acting strategically to retain historically resilient neighborhoods or to preserve newly restored community assets requires special effort for each of these specific circumstances. It requires the capacity to study and anticipate what is happening over time in the broad picture encompassing the economic, demographic, political, and cultural dynamics of a community, as well as the narrow lens of the history and changing condition of individual parcels and groups of parcels in tax default. (e.g., Ford et al. 2013). It must be able to trace simultaneously the movement of residential property through the various transactions and overlapping jurisdictions of public entities responsible for recording both its physical condition and the condition of its title. It requires perspective beyond the scope of individual government offices, community organizations, or statutory solutions.

Alan Mallach has written about being strategic in the prevention of abandonment in his widely acclaimed book, *Bringing Buildings Back: From Abandoned Properties to Community Assets* (2015). His prescription for strategic prevention of abandonment extends beyond prevention to include preservation of restored housing and rebuilt neighborhoods recovering from abandonment (Mallach 2015 13–15, 314–17). Recovery will not last without sustained maintenance of both houses and neighborhoods. New land uses—gardens, recreational amenities, urban farms, and other innovations—will need maintenance to avoid the fate of their predecessors. Mallach adds that the specific maintenance compliance strategy for recovering and distressed places within the same jurisdiction may vary. Thus, the compliance actions need to be tailored strategically to fit the immediate situation and its specific trends. They need to be flexible and nimble to increase the possibility of timely compliance and to avoid public payment for the failures of private property maintenance.

Distinctive Features of Strategic Policing

Being strategic in code enforcement requires a long-term outlook. It involves looking at the maintenance and condition of homes over time and in the context of the surrounding area, and not merely what happens at a single premises on a particular day. Local governments must see dwellings in the context of their history and future in the surrounding dynamic and consider not only what is wrong with a property, but also what caused a maintenance deficiency and what will be required to remedy the cited defect.

Thinking strategically about how to police housing and neighborhood environments effectively is difficult when policing is divided into separate categories, departments, and institutions. Regulations for residential health, safety, and welfare are usually in separate codified ordinances with enforcement authority assigned to different departments—health, fire, housing, zoning—with the result that no one department looks at the entire set of compliance issues or the relationship of compliance to enforcement actions. Issuing citations for one category of defect at a time by one department is inefficient and frustrating for owners. For instance, a fire department citation for missing smoke detectors in a building that *also* suffers from a leaking roof and water standing in the basement does not accomplish the goal of healthy and safe housing for residents. Similarly, a local government that specially creates and funds a program for lead paint abatement may simultaneously cut personnel in another department that does the inspections to discover lead paint problems. This kind of fragmentation is not uncommon in large municipalities.

Strategic enforcement would look beyond the single violations to the nature and consequence of chronic and widespread failure to meet code requirements. On a street with several chronically vacant and abandoned houses, prosecuting owner-occupants to correct small problems is less significant than protecting the neighborhood from the harm caused by abandoned dwellings and those at risk of abandonment. The true goal of strategic policing is protecting occupants, neighbors, and the public from the harms of noncompliance.

A strategic approach to compliance would also reject the common but unfortunate practice of issuing repeat citations on the same property for the same issue. Instead, strategic compliance means treating enforcement as a tool to fix the problem. It is a departure from the traditional code compliance system that repeatedly imposes a single small fine for recurring violations—fines that property owners dismiss as an annoyance and a cost of doing business. This is a common problem where municipal courts cannot, or do not, raise fines to a meaningful level and where prosecutors fail to litigate with compliance as the goal. Nor does it make sense to cite the same corporate owner for code violations in one nuisance case at a time when the corporation has numerous properties in the same or similar condition. The increase of large-scale investor-owners of residential dwellings has become a new and overwhelming challenge for compliance officers and courts whose enforcement tools are designed to deal with individual consumers rather than high-volume corporate investors owning to sell. What might work to achieve compliance from consumers does not work for profit-seeking owners and servicers. A strategic approach for high volume owners with deep pockets involves working for an alteration of their business practices as well as the repair of houses they neglect to maintain.

Strategic policing also entails careful consideration of the costs incurred and of how those costs are allocated. It means designing procedures to protect taxpayers from being stuck with the cleanup costs resulting from failed or abusive private business transactions—specifically, bad borrowing, lending, debt collecting practices, and maintenance. Municipalities are in no financial position to let wealthy individuals or corporations evade the millions of dollars it costs to cite, secure, maintain, and demolish their abandoned housing. After all, real property rights do not include the right to ruin a neighbor's property interests or the public's legally protected health and safety by unlawful neglect of property ownership responsibilities (Lind 2011 112–22).

When work is organized into discreet departments with differentiated skills, there is often a silo effect, a department operating in a confined space with no view of the surrounding terrain. Strategic thinking and acting abhors tunnel vision, demanding a broad outlook and consideration of a wide range of related matters. The sense of an integrated system of code enforcement and compliance instead of a set of separate programs for code compliance enforcement requires getting out of silos and into relationships. That happens when local government departments and non-profit organizations collaborate to do different albeit interconnected work.

For code compliance enforcers, that list of critical relationships begins with legislators and policy makers, law departments and prosecutors, the courts that adjudicate real property issues, clerks who handle court records, taxing authorities and assessors, other policing and firefighting officials, as well as public and private community and economic development agencies. Getting managers of these operations into sync with each other is hardly possible as an isolated exercise. Instead, sharing data and studies about real properties among these various working groups has proved to be a catalyst for further understanding and collaboration. Acting strategically is animated and guided by a common interest in knowing more than what can be seen from only one perspective (Lind 2015).

As the aforementioned Cleveland experience illustrates, bringing a dedicated group of local government and public interest managers and experts together with an integrated real property data facility enables sophisticated analytical problem solving that can inform strategic neighborhood preservation actions. Those actions involve not only the municipal code enforcement institutions but also an array of nonprofit community developers, county and municipal land banking organizations, and county agencies and departments. In the past 10 years, much of what has brought national attention to Cleveland's resilient response to the mortgage crisis has been influenced by strategies that emerged from this coalition of public and private actors (Schilling 2014 68–75).

Part 5. Demonstrating Strategically Applied Code Compliance Enforcement in Practice

In the previous Parts, strategic code compliance is defined and the concept described. The following offers a look at how it works in practice. This overview has been crafted to show the dynamics involved when practitioners from different disciplines who are armed with information from numerous sources work with real problems. This hypothetical scenario, which is based on the author's firsthand experience, expounds the dynamics of strategic code compliance as described in this chapter.[5] Following an introduction of the hypothetical Neighborhood Stabilization Team and the data available to it, this Part examines the actions on several representative team meeting agenda items. Although the following scenario is fictional, it is based on typical fact patterns and problems.

Neighborhood Stabilization Team Members

A community-based nonprofit development corporation (CDC) located in an urban neighborhood convenes this hypothetical team which meets each month. Regular attendees include community code compliance volunteers as well as CDC code compliance specialists, other staff members, including residential property managers, property acquisition and construction managers, and executive leadership. Legal counsel and data management consultants are also part of the team.[6] Occasionally, a city council member, city housing inspection official, a county official, land bank staff, or other invited guests may participate regarding a specific problem. It is not unusual for local developers to seek information or help from the team. The core interests shared by all members of the team are the health, safety, and stability of the community for its current and future residents.

The Data Available

The indispensable tools at team meetings are maps and spreadsheets displaying the most complete assembly of data available on property parcels within the area of the team's

---

5   Readers unfamiliar with real property law and practice procedures may find some details obscure. Dealing with esoteric title and transaction issues is, however, an inescapable reality in the work of neighborhood stabilization. Law student readers who have not yet taken a bar examination may find the potential bar questions in these hypotheticals terrifying.

6   Some of the discussions in the meetings may be confidential in nature as when attorneys are counseling about prospective or pending litigation, or when information about plans, transactions, or research in process would violate client confidentiality obligations. As meetings convened by private non-profit organizations acting in the public interest, discussion and decisions taken in meetings are generally not subject to sunshine laws; yet discussion and documents can be subject to discovery in litigation or, in the case of material in the hands of public officials, Freedom of Information Act (FOIA) requests.

purview. The data is derived from an integrated data facility located at and managed by a local university.[7] The data covers public information on real property parcels. It is organized by permanent parcel numbers or addresses and includes tax records, title records, sales records, liens, court records, permits, licenses, property inspections and citations. Added to this are census data, criminal data, and health data in the university's data library. The data provided may be as current as the day of the meeting and may extend back a decade or more.

Team members can look at categories of data, including lists of completed and pending foreclosure sheriff sales within their jurisdiction by range of appraised and purchase prices and sorted into categories—creditors, plaintiffs, or buyers. The spreadsheets also provide information about special relationships, including the concentrations of various types of risk factors like code violations, building permits, property tax defaults, occupancy status, ownership characteristics, pending foreclosures, and more. The data can be analyzed to reveal trends, make connections, and comparisons. These are functions critical for planning policies and for measuring performance of persons and programs.

The Agenda for this Stabilization Team Meeting

Good data and the best-credentialed staff cannot revitalize a neighborhood without effective implementation. To reclaim and maintain a once-blighted neighborhood is a methodical battle that proceeds parcel-by-parcel. This effort requires highly coordinated execution and regular monitoring and follow-up. The following team meeting narrative is organized according to a typical team meeting agenda. In five vignettes this fictional agenda details the granular nature of a strategic code compliance strategy.

Agenda Item 1: The meeting starts with an urgent demand from the CDC's executive director for an update on the acquisition of three parcels. The community development corporation needs these specific parcels to demonstrate site control for a 20-unit, low-income-tax-credit-subsidized, housing development. Only 60 days remain to meet the application deadline for tax credit applications and such financing is essential to this project. The three parcels are unoccupied and two of them are vacant lots. The CDC's lawyer reports that the debt collector, a limited liability company in Nevada, is holding the title and mortgage on the first vacant lot. It is not willing to give a quit-claim deed for less than $2,000, even after receiving a copy of the tax duplicate showing that a city nuisance abatement assessment for $8,500 was placed on the title along with liens for unpaid taxes and the water utility totaling another $2,500. The total fair market value for tax assessment of the lot is $1,500. Based on the urgency of the acquisition in this situation, the CDC's lawyer is directed to send the creditor an offer of $1,500 immediately with a photocopy of a certified check for that amount. Then she is instructed to ask city officials for removal of the city's liens in consideration of the intended purpose of the land for new, affordable housing.

The second parcel presents title problems. The parcel's deceased owner died without a will. Although the heirs the CDC has managed to locate are willing to convey the second

---

7   The maps and spreadsheets are designed and maintained by the data management staff based at a high-capacity information technology facility. The National Neighborhood Indicators Partnership (NNIP) is a web-based exchange for the technological and programming for this type of work. *See* http://www.neighborhoodindicators.org. *See also* the property database at the Case Western Reserve University Center on Urban Poverty and Community Development http://neocando.case.edu/cando/housingReport/interface.jsp.

vacant lot, not all of the heirs have been found. The family is scattered and the known heirs are out of touch with younger siblings who drifted away. Property taxes are owed but there are no recorded liens. The lot is too small to permit construction of a house under the present code. It is essentially unmarketable. No legal action is available to cure defects or clear the title within the available time. A risk-management strategy that entails acquiring the defective title from the known heirs will be necessary for this parcel. The objective is to secure payment in escrow for unknown heirs at a future time either upon their request or in an action to quiet title to be initiated by the CDC.

The third parcel presents the biggest challenge. The house is owned by an elderly woman living in a nursing home. She has Alzheimer's disease and lacks legal capacity to convey it. The house has been vacant for the past three years after the death of the owner's daughter who previously occupied it. Now, no one can convey the deed. The house is uninhabitable, stripped of copper and porcelain, and structurally damaged. It was in mortgage foreclosure with a judgment lien for $30,000; but it did not sell at the sheriff's sale held one year ago. The foreclosure case remains open still but nothing new is on the docket. A litany of outstanding debts further burden the parcel: property taxes amounting to $4,500; an unpaid water bill for $1,063; and a judgment lien for unpaid medical bills amounting to $37,347. The city condemned the house as a public nuisance six months ago, but the city will not have funds to demolish it within the coming year.

The team's objective is to get legal control of the elderly woman's property, demolish the house, and hold a clear title. Paying the liens is out of the question because of the cost. None of the creditors can obtain financial benefit from the property and it is unlikely that any of them will spend money to pursue their legal interest in the property. This information leads the team to conclude that the best available solution is for the CDC to quickly file a civil nuisance abatement action in court, seek an expedited hearing to have the building declared a public nuisance, and request the court appoint the CDC as receiver to abate the nuisance by demolition. Under applicable law, a receiver's court-authorized costs for nuisance abatement are a super-priority lien with precedence over all other liens and taxes. Since the CDC qualifies under the nuisance abatement statute as both a plaintiff and as a receiver, and has done this in prior cases, it knows the procedure well. Although the process may take more than 60 days, the court-appointed receiver will have a demonstrable claim for control of the title by virtue of its priority lien in an amount far in excess of the market value of the land. Even though the demolition will cost far more than the land is worth, the ability to establish site control over this parcel is absolutely necessary for the CDC to file its tax credit application.

Item 2: Complaints from residents in the neighborhood about bad conditions are gathered at each meeting for research, analysis, and team action. The spreadsheets and maps document the critical information about each property on the complaint list and the surrounding properties. The most serious complaints are then emailed to the city code enforcement officers with recommendations for enforcement action. Not every complaint warrants the same level of urgency. Some require policing by city officials while others may need assistance from the CDC or its corps of volunteers.

Although it is routine procedure for this CDC to help owner-occupants avoid citations and sentences for housing code violations by intervening on their behalf, more often the CDC uses its resources to establish the existence and urgency of code compliance failures that are deemed to adversely affect the stability, health, and safety of the surrounding neighborhood.[8] In this regard, the CDC prepares and delivers evidence to assist the court in understanding the most critical cases. That evidence typically involves photographs, videos, research, and testimony about the nature and impact of violations. Residents and staff professionals who participate in compliance cases are an important extension of the research and analysis done by the stabilization team. Giving testimony at sentencing hearings enables the neighborhood residents to ask the court for a remedy that makes defendants' full compliance with maintenance codes a priority over fines.

Item 3: A research report requested at a prior team meeting is the next item on the agenda. Noting that the number of abandoned vacant houses has been steadily increasing, a research question was posed whether recent numbers of abandoned homes were exceeding the number of new or rehabilitated houses that were marketed by the CDC. The report confirmed the team's suspicions. It showed that the number of vacant houses demolished in the neighborhood exceeded the number of new or rehabilitated houses put on the market by the CDC. In addition, new and restored houses were staying on the market for an average of 11 months and sellers were selling at a loss. The research further showed that the number of houses condemned for demolition by the city each year was more than twice the number actually demolished.

This report precipitated an intense discussion over whether the neighborhood stabilization strategy should be changed. The data confirmed that most of the abandoned housing in the hardest hit neighborhoods was or had been owned by mortgage creditors following a foreclosure. The increase of abandonment by a small group of large-scale institutional creditors appeared to be undermining the CDC's efforts to fight blight house-by-house and block-by-block. The evidence showed that more and more public funds were spent to clean up the abandoned structures left by financial institution creditors and speculators whose business practices evaded compliance with local public law. Housing prices were so low now that large subsidies were required for rehabilitation of restored dwellings.

These observations led the team to consider what additional strategy they need to pursue in order to protect neighborhoods from the increase in abandonment and neglect. Securing the costs of abating the harmful waste left by commercial absentee homeowners' failed investment decisions became an important priority. One option suggested was large-scale litigation using nuisance abatement law against multiple abandoned houses in a single suit against a corporate owner for reimbursement of public expenditures for nuisance abatement. Another idea was to ask for an ordinance requiring registration of full owner identification of vacant houses in order to facilitate more effective code compliance enforcement while properties were vacant. A more radical idea was to suspend development of new housing while demand was low and concentrate more on shoring up the neighborhood's code compliance and quality of life for current residents. It was decided to do further research on policies and programs being employed in other cities before taking action on a change of strategy.

Item 4: The team noted that the GIS map of the property data portrays two locations in the community with a cluster of distress indicators. In both locations there are several new mortgage foreclosure cases, several known vacant houses, and several neglected vacant

---

8   The cases include matters prosecuted by the municipal housing or environmental officials as well as cases relating to abandoned vacant houses.

lots. It is apparent from the title transfer and mortgage records that owner-occupants are mostly elderly people. These conditions prompt a special neighborhood action consisting of door-to-door canvassing and convening block meetings to fully inform all residents of the various city and nonprofit agency services and programs available to them. The senior citizens' office of the city would participate with social services where needed. The CDC's community organizer proposes new efforts to get neighbors to collaborate on some cleanup and repair projects, gardening, and landscaping. Listening to complaints and worries of residents is important. Getting willing landlords to be involved is a difficult but important component of the program. The ultimate objective is to encourage and empower residents to be engaged and proactive in taking care of their neighborhood in a time of distress. Data and experience demonstrates that neighborhood stability and resilience correlates with good communication and cooperation among neighbors. That is why the stabilization team always looks carefully for indications of where reinforcement of neighborliness is needed to shore up resilience against blight. Early blight prevention with voluntary and assisted code compliance pays off in the long run in most neighborhoods.

Item 5: Advocacy before federal, state, and local legislatures and administrative agencies is also a critical part of a comprehensive code compliance strategy. The Stabilization Team's meeting ends with announcements about requests for help with two such advocacy efforts. First, a national fair housing organization has filed administrative complaints against a number of financial institutions claiming racial discrimination in the way houses in their ownership portfolios are maintained. The charge is that houses in predominantly minority neighborhoods received less and inferior maintenance while those in predominantly white neighborhoods fared better. A complaint is now being prepared regarding that problem in this city. The team is being asked to submit for the plaintiff's review data and other evidence that would support this claim. Since the database is integrated with the wider one for the region, the managers of that system will respond about statistical data. However, CDC staff will review its own case files on code violation complaints made against large financial institutions and field servicers to see if there are racially discriminatory responses or neglect by the institutions and report back to the team.

Secondly, the City Council's Housing and Community Development Committee is considering proposed legislation to require registration and periodic inspection of both interiors and exteriors of single and two-family dwellings that are offered or used as rental property. Current law requires only exterior inspections of dwellings with fewer than four units and then only in response to good-faith complaints. However, some neighborhoods are experiencing a significant increase in rental of single- and two-family housing owned by out-of-state business corporations. There is a concern that the management of these rental properties is sub-standard and that the investor-owners use low standards of care in their maintenance. In anticipation of this kind of legislative proposal, the team is being asked by the Committee chair to prepare testimony based on the data and the experience in the neighborhoods in which this CDC operates.

As the curtain falls ending this scenario, it can be pointed out that keeping neighborhoods resilient and restoring blighted places requires a wide range of actions. Merely citing code violations and enforcing them is not enough. Cities are being forced by budget reductions and reduced staff to use more collaboration to make neighborhoods healthy, safe, and stable. As the hypothetical scenario demonstrates, the complexities of achieving residential stability must be mastered and managed in a collaborative way with both public and private stakeholders taking on new roles.

Conclusion—Toward Resilient Neighborhoods

This chapter attempts to show that strategically designed and deployed code compliance enforcement is essential for recovery of housing blight and stable human habitats of the future. The policies and programs inherited by most cities are weak, fragmented, and broken, especially where the mortgage crisis and its residual effects are reconfiguring city landscapes with disregard for residents. A new approach is both necessary and available to help homeowners and neighborhoods battling blight as well as those in recovery mode.

In the future, code compliance enforcers will have new tools capable of gathering, processing, and communicating information. Neighborhood-level data will be readily available for both officials and residents who value healthy, safe and resilient residency. Hand-held devices linked to web-based data systems connecting people and groups are becoming available. Emerging technology is helping track and instantly transmit information about positive and negative neighborhood conditions. New technology is especially important for connecting community residents to each other and to the code compliance and enforcement processes of their municipal government.

Meanwhile, as municipalities are forced to reduce work forces and provide services with less money, communities will have no choice but to adopt new strategies and new tools for maintaining an acceptable quality of life. Professional, bureaucratic, and institutional silos are relics of an unsustainable past. They inhibit collaboration and divide public servants from the public who depend on municipal services. Experience shows, however, that these silos do not disappear just because they are not working successfully. They need to be overcome with systems that work better than the city's 'old way' of administering code enforcement. Protecting residential health, safety, and welfare is now moving in the direction outlined in this chapter–toward data-driven collaborative strategic planning and actions that use emerging new technology to connect the necessary professional, bureaucratic, and institutional silos with each other and with the residents they serve. The possibility of resilient neighborhoods should not be lost for lack of a better strategy to maintain neighborhoods for the common good.

Works Cited

Alexander, Frank. "Don't Bet the House." *Partners in Community and Economic Development, Federal Reserve Bank of Atlanta* 17.2 (2007): 12–13. Print.

Beckman, Ben. "The Wholesale Decommissioning of Vacant Urban Neighborhoods: Smart Decline, Public Purpose Takings, and the Legality of Shrinking Cities." *Cleveland State Law Review* 58 (2007): 387. Print.

Collins, Brian. "Fair Housing Group Goes After Banks Failing to Maintain REO." nationalmortgagenews.com. n. pag. 27 Feb. 2014. Web. 2 June 2015 <http://www.nationalmortgagenews.com>.

Duke, Elizabeth A. "Addressing Long-Term Vacant Properties to Support Neighborhood Stabilization." Federal Reserve Bank. Web. 5 Oct. 2012. <http://www.federalreserve.gov/newsevents/speech/duke20121005a.htm>.

Eggert, Kurt. "Limiting Abuse and Opportunism by Mortgage Servicers." *Housing Policy Debate* 15.3 (2004): 753–754. Print.

Engel, Kathleen and Patricia McCoy. *The Subprime Virus: Reckless Credit, Regulatory Failure and Next Steps.* USA: Oxford U P, 2011: 43–65. Print.

Federal Reserve White Paper. "The U.S. Housing Market: Current Conditions and Policy Considerations." 4 Jan. 2012. Web. 2 June 2015 <http://federalreserve.gov/publications/other-reports/files/housing-white-paper-20120104.pdf>.

Farkas, Karen. "Housing Conditions May Affect How Well Children Do in School." Cleveland.com, 24 Oct. 2013. Web. 2 June 2015 <http://www.cleveland.com/metro/index.ssf/2013/10/housing_conditions_may_affect.html>.

Ford, Frank. "Cleaning Up After the Foreclosure Tsunami: Tackling Bank Walk-Aways and Vulture Investors." *Shelterforce* (Fall/Winter 2009): 159–160. Print.

Ford, Frank et. al. *The Role of Investors in the One-to-Three-Family REO Market: The Case of Cleveland*. Cambridge: Harvard Joint Center for Housing Studies, 16 Dec. 2013. Web. 2 June 2015 <http://www.jchs.harvard.edu/research/publications/role-investors-one-three-family-reo-market-case-cleveland>.

Garvin, Eugenia, et al. "More Than Just An Eyesore: Local Insights and Solutions on Vacant Land and Urban Health." *Journal of Urban Health* 90.3 (June 2013): 412–26. Print.

Government Accountability Office. *Vacant Properties: Growing Number Increases Communities' Costs and Challenges*. 34th ed. Washington: GAO, 2011: No. 12–34. Print.

Griswold, Nigel G. "Estimating The Effect Of Demolishing Distressed Structures In Cleveland, Ohio 2009—2013: Impacts On Real Estate Equity And Mortgage Foreclosure." Western Reserve Land Conservancy, Thriving Communities Institute, 11 Feb 2014. Web. 28 May 2015 <http://www.wrlandconservancy.org/pdf/Final ReportwithExecSummary.pdf.>.

Harvard Law Review. "Enforcement of Municipal Housing Codes." *Harvard Law Review* 78.4 (1965): 801–60. Print.

Keating, W. Dennis and Kermit J. Lind. "Responding to the Mortgage Crisis: Three Cleveland Examples." *Urban Lawyer* 44.1 (Winter 2012): n. pag. Print.

Kneebone, Elizabeth and Natalie Holmes. "New Census Data Show Few Metro Areas Made Progress Against Poverty in 2013." Brookings Institution. Web. 9 Sept. 2014 <http://www.brookings.edu/research/reports/2014/09/19-census-metros-progress-poverty-kneebone-holmes>.

Kotlowitz, Alex. "All Boarded Up." *New York Times Magazine*. Web. 8 March 2009. <http://www.nytimes.com/2009/03/08/magazine/08Foreclosure-t.html?pagewanted=all>.

Krumholz, Norman and Kathryn W. Hezter. "Re-Thinking the Future of Cleveland's Neighborhood Developers: Interim Report." Center for Community Planning and Development, Cleveland State University, 2012. Web. 2 June 2015 <http://urban.csuohio.edu/publications/center/center_for_community_planning_and_development/Re-thinking_the_Future.pdf>.

Levitin, Adam J. and Tara Twomey. "Mortgage Servicing." *Yale Journal on Regulation* 28.1 (2010): n. pag. Print.

Lind, Kermit J. "The Perfect Storm: An Eyewitness Report From Ground Zero in Cleveland's Neighborhoods." *Journal of Affordable Housing & Community Development Law* 17.237 (2008): n. pag. Print.

——— "Can Public Nuisance Law Protect Your Neighborhood from Big Banks?," *Suffolk University Law Review* 44.89 (2011): n. pag. Print.

——— "Collateral Matters: Housing Code Compliance in the Mortgage Crisis." *Northern Illinois University Law Review* 32 (2012): 445–71. Print.

——— "Perspectives on Abandoned Houses in a Time of Dystopia." *Probate & Property* 52 (March/Apr. 2015): n. pag. Print.

Listokin, David and David Hattis. "Building Codes and Housing." *Cityscape* (2005): 21–67. Print.

Mallach, Alan, et. al. *Cleveland at the Crossroads: Turning Abandonment Into* Opportunity. June 2005. Web. <http://www.communityprogress.net/cleveland-at-the-crossroads--turning-abandonment-into-opportunity--pages-410.php>.

Mallach, Alan. *Bringing Buildings Back: From Abandoned Properties to Community Assets*. 2nd ed. National Housing Institute. 2015. Print.

Marco, Richard J. and James P. Mancino. "Housing Code Enforcement–A New Approach." *Cleveland-Marshall Law Rev*iew 18.368 (1969): n. pag. Print.

Mortgage Foreclosures: Additional Mortgage Servicer Action Could Help Reduce the Frequency and Impact of Abandoned Foreclosures. 93rd ed. Vol. 11. Washington: GAO, n.d. n. pag. Print.

Peck, Don. "Can The Middle Class Be Saved?" *The Atlantic*. Sept. 2011. n. pag. Print.

Pianka, Raymond L. "Cleveland Housing Court—A Problem-solving Court Adapts To New Challenges." *Future Trends In State Courts*, National Center For State Courts, 2012. Web. 2 June 2015 <http://www.ncsc.org/sitecore/content/microsites/future-trends-2012/home/Courts-and-the-Community/~/media/Microsites/Files/Future%20Trends%202012/PDFs/ClevelandHousingCt_Pianka.ashx>.

Pohlman, Jennifer Von. "Share of Seriously Underwater Homes Increases in First Quarter For First Time Since Second Quarter 2012." Renwood Realty Trac. Web. 22 April 2015 <http//www.realtytrac.com>.

Porter, Katherine M. "Misbehavior and Mistake in Bankruptcy Mortgage Claims." *Texas Law Review* 87 (2008): 121. Print.

Schilling, Joseph. "Code Enforcement and Community Stabilization: The Forgotten First Responders to Vacant and Foreclosed Homes." Albany Government Law Review 2.101–162 (2009): n. pag. Print.

———— *Cleveland and Cuyahoga County, Ohio: A Resilient Region's Response to Reclaiming Vacant Properties*. Vacant Property Research Network, Virginia Tech, 2014. Web. <http://vacantpropertyresearch.com/wp-content/uploads/2014/05/VPRN-Cleveland-Case-Study-2014.pdf>.

Schwartz, Mary and Ellen Wilson. "Who Can Afford To Live in a Home? A look at data from the 2006 American Community Survey." U.S. Census Bureau, 2007. Web. 2 June 2015 <http://www.census.gov/housing/census/publications/who-can-afford.pdf>.

Stone, Clarence. *Regime Politics*: *Governing Atlanta, 1946–1988*. Lawrence, Kansas: U P of Kansas, 1989: 3–9. Print.

Theologides, Stergios. "Servicing REO Properties: The Servicer's Role And Incentives." *REO & Vacant Properties: Strategies for Neighborhood Stabilization*. Boston: Federal Reserve Bank of Boston and Federal Reserve Bank of Chicago, 2010. Web. 2 June 2015 <http://www.federalreserve.gov/newsevents/conferences/reo_20100901.pdf>.

United States. Cong. Joint Economic Committee. *A Local Look at the National Foreclosure Crisis: Cleveland Families, Neighborhoods, Economy Under Siege From the Subprime Mortgage Fallout*. 110th Cong., Washington: GPO, n.d. Web.

United States. Cong. Senate. U.S. Senate Permanent Subcommittee on Investigations, Committee on Homeland Security and Government Affairs. *Wall Street and the Financial Crisis: Anatomy of a Financial Collapse*,. S. Rept., 2011: 48–153. Print.

Whitaker, Stephan and Thomas James Fitzpatrick, IV. "The Impact of Vacant, Tax-Delinquent, and Foreclosed Property on Sales Prices of Neighboring Homes." Cleveland, Ohio: Federal Reserve Bank of Cleveland, Dec. 2011: No. 11–23. Print.

Weinstein, Alan C. "Essay: Current and Future Challenges to Local Government Posed by the Housing and Credit Crisis." *Albany Government Law Review* 2 (2009): 259. Print.

White, Alan M. "Deleveraging the American Homeowner: The Failure of 2008 Voluntary Mortgage Contract Modifications." *Connecticut Law Review* 41 (2009): 1107–19 124–129. Print.

White, Brent T. "Underwater and Not Walking Away: Shame, Fear and the Social Management of the Housing Crisis." *Wake Forest Law Review* 45 (2010): 971. Print.

Williams, Timothy. "Blighted Cities Prefer Razing To Rebuilding." *New York Times* 12 Nov. 2013: A15. Print.

Chapter 12
Renegotiating the Constituent's Role in Urban Governance: Participatory Budgeting in New York City

Celina Su

In 2011, at age 24, Corin Mills was not confident that he was capable of completing long-term projects, let alone attend college. He had dropped out of high school and served a brief jail sentence. Then, through a New York community-based organization called Getting Out Staying Out, Mills became involved in participatory budgeting (PB)—a process in which community members, rather than elected officials, decide how to allocate public funds—in New York.

Mills researched the need for and feasibility of project ideas pitched by his neighbors in the fall of 2011. Over the winter, he helped to develop a proposal for a mobile laptop lab to be shared by nine public schools. In the spring of 2012, the proposal Mills worked on won $450,000. His experiences with PB had already been transformative. Before, Mills had only interacted with administrative elites and authority figures, such as local school principals and elected officials, in surveilling and punitive ways; now, he built upon his newfound skills and sense of accomplishment to pursue an even more ambitious goal—to apply to and attend college. Mills was even able to partly cover his college costs with an online crowdfunding campaign that movingly related his struggles. In May 2014, he recounted these experiences at a White House convening (that I attended as well).

Mills' story speaks to PB's potential to engage traditionally marginalized constituents to help them inform policy-makers of their priorities and concerns, bring constituents closer to the governments they elect, hold states accountable, and, put bluntly, (re)enfranchise them. It also speaks to the PB's potential to reallocate public money towards what constituents state they need most, in turn spark critical dialogues on public priorities and, ideally, begin to address America's widening economic inequalities. Indeed, PB has received tremendous attention since it first began in Porto Alegre, Brazil, in 1989. Community organizing coalitions like Right to the City have advocated for PB as one means of reclaiming the commons, and President Obama recently announced PB as a key element of his latest "Open Government" initiative.

As PB continues to gain traction, there remain questions as to whether it can sustain engagement among the traditionally disenfranchised and help engender a more equitable reallocation of public funds, as in well-known past cases (Wampler). Some researchers have argued that PB now runs the risk of becoming a buzzword-turned-fuzzword, an empowering and democratizing process that is diffused and watered down into a politically malleable, innocuous set of procedures (Baiocchi and Ganuza).

In this chapter, I discuss the largest PB process in North America—that of New York City, which just completed its fourth cycle in summer 2015. I examine PB's contested role as an empowering, pro-poor tool for social justice, and traces the difficulty of implementing meaningful collaborative governance. In the New York case, PB has successfully broadened notions of stakeholdership and citizenship for many constituents (especially youth and undocumented citizens), with impressive outcomes thus far. Higher rates of political participation among traditionally marginalized constituents represents an important, intermediary step towards more equitable redistribution of public resources, as well as an outcome in its own right. Still, PB has not necessarily triggered, just yet, a re-prioritization of budget allocations or changes in power dynamics between city agencies and constituents. In order to further address economic inequalities in New York City, the PB process must expand substantially, treat city agency representatives as political stakeholders, and enhance capacity-building for all constituents.

The Promise of Participatory Budgeting

In Porto Alegre, Brazil, a city of roughly 1.5 million people, a disproportionate percentage of government funds historically went to middle- and upper-class neighborhoods. This was true even as slums continued to lack access to potable water and other amenities.

The participatory budgeting process, introduced with a newly democratically elected government in the city in 1989, forced elected officials and roughly 50,000 residents to meet with one another and justify their budget priorities in public, deliberative assemblies. After hearing residents' concerns, delegates translated these concerns into specific program and policy proposals. City officials, in turn, worked with these delegates to make them technically and financially feasible. The resulting budgets from this process are binding. After the process, the proportion of the city budget that went to poor districts and to basic public services rose dramatically (Baiocchi, 2003).

Many of the middle- and upper-class residents who attended neighborhood assemblies voted for projects in slum neighborhoods rather than their own. As a result, sewer and electricity rates rose from 75 percent to 98 percent, and the number of schools quadrupled. Health and education budgets increased from 13 percent to almost 40 percent (Bhatnagar, Rathore, Torres, and Kanungo, 2003). Because of such outcomes, PB's popularity transcended political parties, and it quickly spread to other Brazilian cities in the 1990s. An analysis of Brazil's 220 largest cities suggested that PB there is statistically significantly correlated with lower rates of extreme poverty (Boulding and Wampler).

Since 2000, PB has spread to over 1,500 cities worldwide. Since 2010 alone, PB has spread from one American city to a projected 45 this year. By building more inclusive communities and a more equitable distribution of lifesaving resources—such as water, sanitation, education, and public transit—PB ideally helps communities to better tackle the not only short-term interests, but longer-term policy problems, like economic inequality and poverty (Wampler and Hartz-Karp, 2012; Su, 2012).

A Decline in American Political Participation

The US is a relative latecomer to the PB party. That the US has chosen to implement this democratic experiment might surprise those who primarily perceive the country as an exporter of democracy, rather than an importer of governance ideas from the Global South. At the same time, the US is also fertile soil for PB because of growing disenchantment with our democratic institutions, especially in the Great Recession and its aftermath.

Indeed, American political participation of all sorts—voting, writing to elected officials—has steadily declined since World War II. Although there was a small uptick in voting in 2012, overall voting rates remain lower than they were in 1960 (McDonald). Further, participation is not evenly distributed among demographic groups. For example, Latinos and Asian Americans, women, and low-income constituents tend to vote at lower rates than other racial groups, men, and higher-income constituents, respectively. As income and education levels increase, so does participation in a wide range of political activities, such as working with fellow citizens to solve community problems, making financial contributions, or getting in touch with public officials. Over 90 percent of those with annual incomes of $75,000 and above vote, while only 50 percent of those with household incomes of $15,000 vote (Macedo et al.). Immigrants, youth, and low-income people tend to vote at lower rates for a wide range of reasons, including perceptions that their engagement would not be welcome at civic institutions, a lack of resources and contacts to contribute to political campaigns, and a lack of opportunities that adequately reflect their concerns and make use of their skills (Godsay et al.). Comparatively, among more than 30 high-income countries, the United States reported the highest difference in voting rates between those who have completed high school and those who have not (OCED). This highlights the need for research on innovative attempts to revitalize American democracy, especially amongst those most disenfranchised.

At the same time, new laws in the United States often result in the further disenfranchisement of Americans. In 2011 alone, just before the 2012 Presidential election, 41 American states introduced over 180 legislative bills to restrict voting for laws demanding photo IDs, proof of citizenship, banning same-day registration and university IDs. These laws, all in the name of combating voter fraud, tend to disproportionately discourage eligible voters with fewer resources, such as youth, non-white citizens, lower-income citizens, and immigrants, those who report lower rates of participation to begin with (Weiser and Norden).

These trends suggest that low participation rates in the United States reflect more than apathy or narcissism, popular lines of discourse about Americans (and American youth in particular) in the mainstream media (Bergman et al.; Kaklamanidou and Tally). Most recently, Occupy Wall Street and protests regarding police brutality and other social issues further deepen the puzzle. Americans do engage in politics, but they still appear to remain hesitant to do so via institutional channels.

A Proliferation of Discourse on and Initiatives for Participation

Meanwhile, a decades-long distress regarding the decline and professionalization of American political participation parallels a rise in academic interest in participation worldwide (Skocpol; Putnam). Often, ordinary members in social movements can do little but send in their membership fees or make donations each year, while paid lobbyists from Political Action Committees form the core of these organizations. While these social movement organizations play an important role in politics, they hardly contribute to a deep, healthy democracy at the local level. Some scholars paint a more optimistic picture about American democracy, especially in the field of community organizing, but others emphasize the limited success of attempts at larger-scale social movements, especially in the context of a $1.3 trillion activist "sector" and industry (Warren; Graeber; Incite! Women of Color Against Violence).

In recent years, the domestic and international literatures on political participation, previously largely parallel, have begun to intersect more consistently and substantively. Much of the international participation literature was inspired by the need for constituents throughout the Global South to have a say in the mass-scale dam projects, economic policies, and other governmental (or government-binding) decisions being made by elites, whether domestically or by international institutions like the International Monetary Fund and World Bank. Prominent titles in this literature included *Whose reality counts?* and the World Bank Sourcebook on templates for stakeholder analyses and the limits of relying solely on technical experts in the 1980s and 1990s (Chambers; Carnemark et al.).

By the early 2000s, practitioners and scholars had already begun to call "participatory frameworks" the "new tyranny," a way for funders and institutions to pay lip service to participation while perpetuating the received wisdom (Hickey and Mohan; Cooke and Kothari). Such critiques in international development were especially acute because calls for a strengthening of civil society came in the post-Cold War context, accompanied by policies meant to transition societies and economies away from socialism. Just as "civil society" helped to topple communist regimes in Eastern Europe, perhaps it would battle corruption and keep government in check elsewhere as well (Chandhoke). (Indeed, all other papers in the same World Bank series as the *Participation Sourcebook* focused on combating state corruption.) Some of the most recent works on community participation explicitly tie War on Poverty-era Community Action Programs not just to global post-colonial community development campaigns, but also to current trends in grassroots initiatives both domestically and abroad (Immerwahr). These works emphasize that when participatory programs are run poorly, they reify economic inequalities by benefiting local elites, or helping the "have-somes" while continuing to exclude the "have-nots."

Because so much of the discourse on participation accompanied policies aiming at decentralizing and devolving budgets and policy-making, "participation" became a call of the neoliberal right as well as the left. An emphasis on "community participation" by politicians might bring policy-making closer to the people, rendering the state accountable in a bottom-up way, but it might also highlight the burden on individuals to assume responsibilities that had traditionally been those of the state, especially the welfare state (Paley).

In response, both scholars and practitioners emphasized the need for analyses of alternatives to individual-focused, electoral, market-based models of decision-making and governance, across a range of contexts in both the Global North and the Global South (de Sousa Santos; Fung and Wright; de Souza Briggs; Cornwall and Coelho). Further, as there is no universal template for participation, more recent works have focused on context-specific analyses on tensions such as consultative versus binding decision-making, the risks of rent-seeking and clientelism, and the transformative and contested innovations along the way, especially in diverse and unequal landscapes (Lerner; Baiocchi; Fung; Silver, Scott and Kazepov).

Public Administration and Community Engagement

Finally, as participatory democratic experiments become institutionalized, the role of public administrators—bureaucrats, city managers, city agency representatives, and other non-elected public servants—has also come to the fore. Put bluntly, can government really change, even with constituent pressure? As underscored by a recent review of more than

250 articles and books on designing participatory processes from a public administration perspective, previously top-down practitioners are now also preoccupied with how they can better design participatory processes (Bryson et al.).

At first glance, this seems antithetical to popular views of public administration, long filled with jokes about unfriendly, inflexible service at the local DMV. Such a reputation was partly a reflection of the traditional public administration model, a historical product of a century that began with the progressive era and became epitomized by post-war modernist thinking (with a deep faith in scientific progress, and high trust in government) (Stoker). Public managers were assumed to implement policies in response to well-defined political objectives, and hierarchies in government service were assumed to hold everyone accountable. Politics and administration were seen as separate spheres, even if real life was inherently more complicated.

In the 1980s, the United Kingdom's Prime Minister at the time, Margaret Thatcher, and President Reagan of the United States forwarded neoliberal policies with the assumption that "government is the problem," including a reliance of market forces and competition in the public sector (Salamon). By the 1990s, their successors, Blair and Clinton, respectively, had further institutionalized New Public Management through large-scale implementation of performance reviews, metrics, and incentives, for a "customer-driven government" (Thomas 788). In this model, public managers were not just implementing programs developed by elected officials, but developing a combination of insurance, subsidy, tax, and other products to deliver government services to recipients deemed "customers," rather than citizens. Public managers should be "entrepreneurial," developing service products that beat out the competition from other governmental sectors.

While the New Public Management model remains dominant, widening inequalities, the financial crisis, and deep distrust of government have prompted calls for yet another model of public administration, one that asks public managers "to *govern*, not just manage, in increasingly diverse and complex societies facing increasingly complex problems" (Bryson, Crosby and Bloomberg 447). In this emerging model, citizens are not just customers or voters but partners and co-creators in policy-making, and public managers must act as collaborators, conveners of networks of deliberation, and capacity builders. Yet, public servants and those with private sector experience continue to be invested with perhaps too much trust, whereas community participants are consistently perceived as "emotional, illogical, and lacking in credibility" (Bryer 263). Without careful implementation and bureaucrat retraining, experiments in democratizing public administration may further discredit community input and further marginalize constituents (Durant and Ali).

Conversations Converging: The Case of Participatory Budgeting

Still, even among those who lament the overuse of "more participation" as a prescription for revitalizing democracies, there are no calls to give up the struggle. Rather, participation by non-elite stakeholders remains important for social transformation. Still, in order to work towards equity as well as transparency, it must be "re-articulated to serve broader struggles" and accompanied by governmental restructuring (Leal 96).

This cross-fertilization of perspectives and literatures is significant in several ways. First, the lines between government, civil society, and business continue to blur, with ever more private-public partnerships, social enterprises, and co-production of traditionally public services (Harvey; Bryson, Crosby and Bloomberg). Now, neither community nor public servants can be adequately defined without the other, and PB could be a way to help constituents and local governments to better work together in this new world of shared governance.

Second, as attested by both community participants' experiences on the ground and articulated in the literature, democratic experiments like PB cannot be plopped into a larger context of ossified market-based or traditional top-down logics, for instance, and still make a significant difference in policy-making (Dahl and Soss; Baiocchi and Ganuza). They are only meaningful when accompanied by significant administrative reforms.

Participatory budgeting is ripe for an analysis of the changing social contract between an individual and the state, and how together, we might tackle problems like economic inequality. By far the most prominent example of participatory democracy globally and in North America, PB is usually conducted at the municipal level, but various institutions—such as Toronto Community Housing—have also devoted a portion of their funds to PB (Wampler and Hartz-Karp).

Because of its popularity, there is now a large literature examining its results in terms of deepening participation by the poor, increasing efficiency, and redistributing resources (Wampler; Baiocchi; Lerner). PB attempts to give stakeholders an opportunity to draw upon their knowledge of local needs, articulate proposals, interact with neighbors, deliberate over priorities, and select—not just consult on—which proposals receive funding. Closer examinations of especially more recent cases of PB—where the process was not introduced by new administrations or accompanied by wholesale regime changes— can help us to not only better understand how folks can better participate in democracy, but also to examine the conditions that might lead to more meaningful outcomes in equity.

New York's Participatory Budgeting Process

In New York, four City Councilmembers devoted a portion of their discretionary funds (roughly one million dollars each) to PB in 2011. The New York process is co-conducted by district committees, city councilmembers and their staff, the two lead organizations, and a bevy of volunteers. By the 2014–2015 cycle, roughly half of City Councilmembers participated in PB. This meant that the process involved around $25 million, and roughly four million city residents. The PB-eligible funds can only be used towards capital expenditures, those exceeding $35,000 and officially lasting for more than five years. Thus, in the New York process, additional tutors at an after-school program, hours at a public facility, and murals all currently fall outside the bounds of PB.

In the fall of each year, each city councilmember hosts neighborhood assemblies throughout his or her district, and hundreds of New Yorkers attend to pitch proposals for community projects. Over each winter, residents volunteer to become budget delegates, conducting feasibility and needs assessments to curate the proposals that will end up on the ballot, and working with city agencies to develop ideas into full-fledged proposals. Each spring, residents turn out to vote for the proposals that win funding via PB (Kasdan and Cattell).

New York's PB was designed to be a bottom-up process in two key ways. First, two lead organizations helped to coordinate efforts by the City Councilmembers. They also helped to organize neighborhood assemblies, budget delegate meetings, and outreach with enough public education and information, so that these activities would not easily be manipulated or used as political tools for voter consolidation. The Participatory Budgeting Project (PBP) acts as the lead organization providing technical assistance. A grassroots group with over two decades of experience, Community Voices Heard (CVH), acts as a core partner focused on outreach.

As a non-academic, non-governmental, membership-based organization with deep-seated ties with low-income residents throughout the city, CVH lent the PB process grassroots legitimacy. CVH helps to ensure that everyday constituents (including those with fewer resources and lower levels of educational attainment) participated in and informed the year-long process. CVH organizers know, for instance, the number of people they would need to reach by telephone or via door-knocking in order to guarantee a turnout of 200 people at a neighborhood assembly, which strategies were more likely to yield low-income participants, and how to translate technical information on policies and governmental structures into clear, accessible presentations.

Second, New York's PB structure was designed in a bottom-up way; a wide range of participants wrote the rules that would govern the process for the rest of the year. Each participant proved essential to the process—community board members, for instance, could help to ensure that PB rules did not conflict with theirs, pinpoint which governance institution or bureaucracy had jurisdiction over implementation of likely project proposals, and help to evaluate whether there would be enough local political support for certain initiatives. The principles of participatory decision-making were applied not just to the budgeting process, but to the preceding PB rule-making process as well. A steering committee continued to provide input on rules and strategies along the way. Since 2014, the City Council Office of Policy & Innovation has also worked to coordinate efforts city-wide, and to host the steering committee.

I have been a member of the city-wide steering committee and the research board headed by the Urban Justice Center's Community Development Project (CDP) since the inaugural process, helping to evaluate patterns in participation and outcomes. Each year, the research board collects information on the demographics, civic experiences, and opinions of participants. In 2014, the board collected 8,000 surveys, dozens of interviews, and observation fieldnotes on both experiences with PB and potential barriers to participation (Kasdan, Markman and Covey). I drew upon notes and transcriptions for more than 70 semi-structured interviews conducted with budget delegates, coordinated by me and other research board members, especially Ron Hayduk, Alexa Kasdan, Erin Markman, and Rachel Swaner. Of these, 45 interviews were conducted with current budget delegates, and 30 were conducted with past budget delegates. This chapter also draws upon four years of participant observation at steering committee meetings, neighborhood assemblies, and information sessions, and over 20 1- to 3-hour interviews with PB participants and allies, including outreach staff and representatives of all city agencies involved in New York's PB process, that I conducted. For this chapter, I coded observation and interview data according to thematic codes, engaging in several interpretive iterations of fieldwork and data analysis to explore themes grounded in the data, such as changing perceptions about government and governance in general, and the role of city agencies in budgeting and community development. I read the analytical memos of other research board members, based on the same budget delegate interviews, but I also reviewed original notes and transcripts myself. The names of all fieldwork participants and affiliated agencies or organizations have been withheld for confidentiality reasons. All of the quotations in the remainder of the chapter come from these interview data or my observation fieldnotes.

Broadening Stakeholdership on an Uneven Terrain

New York's PB process has dramatically broadened notions of stakeholdership, engaging traditionally disenfranchised constituents in the city. For instance, the first rulebook dictated that anyone over age 16 who lives, works, attends school, or is the parent of a student in a district could participate in neighborhood assemblies and project-vetting, and residents over age 18, including undocumented immigrants, could vote on the allocations. Immediately, the notion of who should count as a citizen, whose voice and opinion might be valuable in deliberating problems and solutions, was more expansive than in our usual political and legal frameworks. (Notably, Argentina passed a law allowing 16-year-olds to vote in national elections in 2012, and quite a few other countries, such as Chile, allow foreign nationals to vote in local elections after a few years of residency.) Although questions about who should be allowed to hold the rights and responsibilities of voting—college students who live in campus town for only part of the year, or formerly incarcerated citizens who have served their sentences—arise on a regular basis in the US as well, they are often framed in limiting ways, *i.e.*, regarding who should be *denied* the right to vote. PB provides a focused, potentially expansive way in which to engage such questions with specific individuals in mind, and in conversation.

Enthusiastic youth participation in neighborhood assemblies was instrumental in convincing adults to lower the PB voting age to 16, and the participation age to 14, in 2012. Some districts further lowered the voting age to 14 in 2015. At the neighborhood assemblies, it became clear to adults that although the youth sometimes lacked technical expertise, they had the capacity to think about neighborhood needs and draw upon local knowledge, *i.e.*, what infrastructural improvements were most needed in their classrooms and school buildings, what spaces were most welcoming to students after school, and which areas were in dire need of such spaces. Youth from poor communities are often labeled "troublemakers" if they speak critically about local institutions and policies (Alonso et al.). It is significant, then, that PB neighborhood assemblies gave youth opportunities to not only articulate local problems and weaknesses, but to also generate concrete ideas for project proposals and provide constructive feedback.

According to the survey data collected by the Community Development Project of the Urban Justice Center, constituents from traditionally marginalized subpopulations participated in PB at much higher rates than in traditional elections in every cycle thus far. For example, in District 8, the very poor—those with incomes of $10,000 or less—constituted 4 percent of voters in 2009 City Council elections but 22 percent of PB voters (Kasdan and Cattell). In 2014, 39 percent of PB voters had household incomes of $35,000 or less, compared to 21 percent of voters in the 2013 elections. Along lines of race and gender, PB also engaged traditionally underrepresented stakeholders, with Hispanics voting in PB at twice the rate of regular elections. One-third were foreign-born. In one district, over two-thirds of distributed ballots were in languages other than English. Notably, half of 2014 PB voters had never worked with others on a community issue before (Kasdan, Markman and Covey).

Survey data also suggest that strong outreach efforts appear to pay off; lower-income and foreign-born constituents were more likely to learn about PB through word-of-mouth or targeted campaigns, rather than online or through governmental-institutional channels. Districts that hosted assemblies specifically catering to youth or non-English-speaking constituencies saw, in turn, much higher voting rates by those constituents {Kasdan, 2014 #983}. In many ways, then, PB in New York has succeeded in engaging traditionally marginalized constituents, even as more intensive forms of political participation are usually and paradoxically practiced by those with the most resources (Stolle and Hooghe).

Indeed, budget delegates spoke repeatedly about how the PB process allowed and compelled them to engage in discussions with neighbors they may not have met otherwise, and to listen to members of their community who are not property or business owners, parents, and formal citizens, and thus not typically represented by local associations. As one delegate remarked,

> At the meetings you meet people from all over the community. You may walk by them on a daily basis and not know what they're into. It was a good gathering for people to interact, those for women's rights, LGBT, you have all these organization and representatives and say "Oh, cool you're trying to do that, maybe I'll participate in that" and they're like "Oh, yeah, sure, come on down" or whatever. So it really helped me understand how I can help my community.

Delegates also noted the substance and tenor of their conversations through PB, and the fact that they now worked with neighbors from distinctly different background "repeatedly, on a serious but still social level" that felt different from organizing street fairs as a block association. Several participants noted that they especially learned a lot about local community priorities, both because of formal needs assessments and through conversations generally. One delegate from East Harlem noted,

> I was really able to see the needs [of] the community in a way I've never seen before ... I didn't know how bad of an asthma cluster there was in public housing. I don't have kids, so I don't know about needs at school. I don't have any relatives that live in senior housing, so I didn't know about the issues they faced.

In another neighborhood assembly hosted by then-City Councilmember, now Council Speaker Melissa Mark-Viverito, constituents (not just residents, but also business owners and local parents) from lower-income East Harlem engaged in sustained conversations about local needs with constituents from the higher-income Upper West Side. More than once, participants explicitly articulated a change of opinion about district priorities. One Upper West Side resident, for instance, stated that he had attended the assembly to forward his idea for a project near his daughter's school, but decided that laundry rooms for senior citizens in public housing should receive higher priority (Su).

Overall, budget delegates emphasized how PB has a very different, inclusive, and deliberative—encouraging the exchange of ideas and compromise—qualitative feel that contrasted the tenor of electoral politics, even for those already politically active. For one delegate, the combination of working with others unlike herself and working towards binding budgetary decisions gave the PB process a sense of impact lacking in her usual civic engagement: "Every four years, I make an informed choice and vote ... being in this district, with the diversity of the community and feeling like this process is really inclusive and responsive to community input, which you don't often feel .... That is pretty powerful."

One lesson lies in the potential for cross-cutting alliances of groups of residents or organizations, who might usually lobby for funds independently. Budget delegates spoke to the ways in which the PB deliberations allowed them to emphasize more than one aspect of their lives and identities—e.g., as African Americans, as Harlemites, as parents, as public housing residents, as sports fans, etc.—and emphasize issues of intersectionality, rather than a single identity—by race, gender, or other social axis. For instance, one delegate spoke about how he was at first disappointed to be the only person of color, and the only young person, on his PB parks committee, but that to his surprise, he then quickly bonded with the other adults by emphasizing different aspects of his lived experience and background with each. Along the way, these delegates develop proposals and begin to address inequalities they may not have otherwise, such as a playground fit for both local public housing residents and disabled children from around the neighborhood, mobile laptop labs or cooking vans co-sponsored by several schools or nonprofits, and a mobile audio-visual equipment van co-sponsored by three arts organizations that might usually compete for funds.

Toward Models of Civic Education and Critical Pedagogy, Especially Vis-à-vis City Agencies

Many delegates also testified to the skills they learned by participating in the process, as well as the need for additional training. The formerly incarcerated youth I interviewed, for instance, testified to the terror they felt in speaking to school principals during site visits to research project proposals, since they were only used to state institutions as policing, surveilling, and punitive. In fact, one youth mentor spoke of PB as a *de facto* leadership development program:

> In terms of interacting with the community, it's more difficult in an institutionalized setting, such as a school. To kind of go in and act under an authority that you're not necessarily granted by being [a budget delegate] … I can see how intimidating that is. Not only just to set it up but then to follow though. There's a security officer. You have to check in. You have to hand their ID. And then go and actually explain [why you're there, what you're working on] … That was definitely a challenge for them … .

Some of the training likely to edify and empower budget delegates might seem surprising to the uninitiated, yet commonsensical in retrospect. Trainings in public speaking and poster-making, during preps for the pre-vote expos, for example, were taken for granted by middle-class parents active in Parent Teacher Associations, but were just as important as literacy and quantitative needs assessment training for the formerly incarcerated youth I interviewed. The youth mentor continued,

> And then the next phase, making the project poster boards, I actually think that was the portion that most of the delegates had the most fun with. It's basically … almost a big arts and crafts project, but also at this point, they've become more invested after going to all these meetings, participating in some of the site visits. It was really nice to see them liven up for that portion … . Once they made the poster board, I think, that … kind of solidif[ied] their investment. It was tangible as something they could point to and take ownership of.

Such training helps constituents to break out of a "hegemonic discursive code," full of technical jargon that many find difficult to master, and to instead use visual or performative languages to present their arguments in compelling ways (Hajer). It is instrumental in helping constituents to break through (and simultaneously learn) technocratic jargon, so that they can be taken seriously by policy-making, and meaningfully affect governance.

Indeed, many of the budget delegates spoke of the need for critical pedagogy on governmental structures, budgeting, and planning. City agency representatives, who help to define which project proposals are "feasible" and "appropriate," also figured prominently in interviews, and end up shaping the participants' experiences in profound ways. Outside of time constraints, interactions with city agencies were by far the most commonly cited challenge to meaningful participation.

This is a real issue. Between the first and second cycles of New York PB, for example, the number of project ideas proposed plummeted by 70 percent in some districts–and in some categories, the number of proposed ideas went from 200 to 0 in one year. It remains unclear whether this happened because stakeholders have become less interested on those issues, or if they have decided to censor themselves, and to focus their energies elsewhere.

Delegates stated that city agencies were sometimes difficult to work with because many representatives were slow to respond to inquiries, and because so many organically developed project proposals could not move forward, or had to be adjusted dramatically to become eligible. Others expressed that they felt frustrated by the lack of coordination between agencies, or felt confused because the same project idea—a greenroof, for instance—might involve different agencies depending on who owned the parcel of land.

In such situations, delegates noted that more specific criteria, guidelines, and explanations for city agency decisions would have helped them to not only develop better proposals, but to accept rejections as well. They asked for stronger, instant and direct contact with city agencies and different actors in local governance, including an appeals process. As these remarks suggest, budget delegates did not view city agency representatives as neutral parties, but as simultaneous facilitators and gatekeepers with distinct stakeholder interests.

In response, the City Council has worked hard to coordinate information sessions and manage the expectations of both budget delegates and city agency representatives. Indeed, with each year, there have been fewer reported frustrations with the process.

De-Stigmatizing Politics and Popular Governance in Public Administration

PB automatically introduces new tasks and roles for citizens that are not usually discussed, and forces stakeholders to articulate new guidelines for effective and meaningful collaborations. This innovation may be even more difficult for government insiders than for newly enfranchised constituents. In interviews with representatives from all city agencies involved with PB in New York, an emerging theme concerned how agency representatives felt caught between a hierarchical, top-down, bureaucratic culture and new, bottom-up attempts at democratization, collaboration, and co-production. They also reported a range of constraints and conditions to meet alongside PB imperatives. For instance, the agency representatives spoke to a range of organizational structures, funding streams (especially regarding the level of government funding them), and missions.

At first, these interviewees also stated that they are "not really interested in taking their money to do things we're supposed to be doing. So they were all, at the end of the day, mostly pleasant interactions, if they had issues." Such statements, while friendly, also worked to keep delegate expectations contained or perhaps redirected to other agencies: "To be honest, I'd say spend your money on other things, just tell me what you need done, and we'll go fix them if it's an issue." Such comments dovetail well with scholars' findings that regulators "feared commitment to ideas or decisions reached by non-expert ... stakeholders," and that contested politics inevitably distort definitions of the "public good"

(Bryer 264). Over the past few years, however, more agencies have come to embrace PB. Because of the civic education and training embedded in the process, as discussed above, PB de-stigmatizes politics and popular participation in city governance.

Thus far, the agencies have largely fallen into three main categories of agency-budget delegate interactions. First, some agencies worked to contain budget delegates' expectations of PB-funded projects. Although several of these agencies hosted sessions with community members during the PB process, these were largely informational; when budget delegates asked questions about potential project ideas, representatives emphasized technical rules or funding restrictions in their answers. Agency representatives also worked to immediately address pressing needs outside of the PB process, and to announce planned and strategic changes to the agency's overall funding priorities.

A second group of agencies prepared helpful flyers for budget delegates, detailing eligibility criteria in accessible ways, including typical projects that are or are not eligible for PB funds. For several agencies, these presentations and flyers graphically dimmed more ambitious projects, as to manage PB participant expectations. Several agencies also presented at neighborhood delegate meetings lists of hyper-local, large-scale projects already under way, ones which needed a relatively small amount of additional funding to be completed; representatives portrayed these as giving budget delegates "extra bang for [their] buck." Later in the cycle (after project proposals were developed and before voters determined which would be funded), some of the agencies with discrete, neighborhood brick-and-mortar outposts (like parks, libraries, schools, or community centers that were easily contained within a district) also conducted outreach on behalf of the PB process to local patrons; as per campaigning guidelines, all were also careful to state that voters can choose up to five projects on each ballot.

Third, one agency stood out as enthusiastically devoting significant staff time and resources to the PB process—not out of a sense of obligation or to garner PB funds, but as part of its mission. Greater civic engagement, a sense of individual agency, and community participation were seen as larger goals of the Department of Health's mission, broadly conceived. Thus, the representative there stated that, "We could not *not* get involved." This was the only agency that did not track whether agency-related projects won PB votes, but simply worked to promote PB overall.

Balancing Local Knowledge with Technical Expertise.

City agency representatives acknowledged that PB allowed community members to draw upon local knowledge, and to identify local priorities. One, for instance, stated that, "the community is good about identifying the need and saying, 'Yes, this is a really dangerous intersection.'"

At first, a far more common sentiment in city agency interviews concerned the relative lack of technical and budgeting expertise of budget delegates, especially regarding technical rules around capital funding. One specific frustration lay with the cost of proposed projects, and what representatives felt were outsized expectations by the PB delegates. In those cases, these bureaucrats felt as if they were easily villainized, and that the PB process punished the messengers of high-cost figures. These remarks speak to the challenges public managers face in striking delicate balancing acts, acting as liaisons between different technical, political, and local or experiential perspectives in their work (Feldman and Khadermian).

Over the past few years, however, the training for budget delegates has improved, and representatives, too, have stepped up to prepare helpful presentations on eligibility criteria. Thus, the most common mistakes—thinking that air conditioners and iPads might be capital-eligible, for instance, when the city states that they are not—are now largely avoided. Further, even the agency representatives that complained about the PB process feeling like "a waste of time" appeared to enjoy this educative role, and to converse with the infrastructure "geeks" who know a lot about their policy arena.

At the same time, interviews and observations also suggested that city agency representatives, too, might benefit from training and coordination. To what extent should budget delegates—many of them volunteering amidst one or more other full-time jobs, and many of them from traditionally marginalized constituencies—be expected to limit their proposals to ones they could easily develop to perfectly fit individual agencies' respective norms and criteria? Might agency representatives be trained to take the spirit of project proposals, to better incorporate local knowledge and priorities, rather than mandate changes?

Interestingly, one representative stated that she began to do so after serving as a budget delegate herself, in her home neighborhood and working on a committee that worked with an agency other than those she worked with. Other representatives had some past experiences working with community groups or constituents via non-PB processes, and they cited resource ally groups as an essential resource. Such resource ally groups helped lay constituents understand planning regulations or architectural designs, for instance, or translated public focus group data and technical documents into accessible, readable documents into several languages other than English, and with smart graphic design. Without such experiences, agency representatives had difficulty knowing the types of data visualization, exercises, and documents that would make the greatest difference in facilitating deliberative back-and-forth conversations with budget delegates on project proposals.

Remaining Questions of Scale and Equity

New York's experience thus far has demonstrated great potential in tackling persistent and widening economic inequalities; its intermediary outcomes regarding political participation have been particularly impressive. PB has also begun to change some dynamics between constituents and public administrators, leading to new dynamics of contestation and deliberation on the public interest. Because many of the selected projects take years to build, the more long-term outcomes regarding distribution—and especially *re*distribution of funds, as in the original model in Brazil—remain unclear. Some preliminary analyses are quite promising. One analysis shows that census block groups served by PB allocations had a 20 percent higher percentage of housing minority residents and families with income below the poverty level (Goldberg and Finkelstein). Another analysis highlights a correlation between a higher Gini coefficient by income (and thus more income inequality) and likelihood to participate in PB; two-thirds of PB-participating City Council districts have a higher Gini coefficient than the city median, 27 percent (Nakkas).

Despite such promising news, there remain questions of breadth and scale. Both constituents and city agency representatives have lamented PB's limited scope. Roughly $25 million is a tiny fraction of the city's $75 billion operating and $6 billion capital budgets, and significant capital projects—new building developments, sewer systems, public transit—inevitably require cross-district coordination. Thus, agencies that worked on city-wide infrastructure or large geographical areas had trouble mobilizing specific groups of patrons to vote for related projects, or to develop small-scale projects that could be funded by the relatively small pots of PB funds. Representatives from these agencies were also more likely to express views that PB process exacerbated inequalities, since the eligible pots of money are so small and divided by district, when proper reprioritizing would involve much larger economies of scale and coordination across districts in a racially and economically segregated cityscape. They were also less likely to receive city funds, and to express conflicts between local, state, and federal demands.

By contrast, agencies that worked with buildings, intersections, or identifiable spots contained in single districts were more likely to become actively involved in PB. Still, the relatively small amounts of money assigned to PB make project proposals vulnerable to tighter technical restrictions as well as tighter budgets, and make it more likely that city agencies dominate the project-vetting process. One city agency, for example, has been clear that "existing capital projects are ... more attractive than projects [constituents] conceived on their own ... so we come to *them* with projects we know– that are capital, that we can do." This approach appeared to resonate profoundly with several agencies because they have been experiencing severe budget cuts in this era of austerity, and PB allows them to address specific budget shortfalls.

In addition to more funds and fewer restrictions, the current system in NY—organized by City Council district—also limits PB's potential to address issues of economic inequalities in the city. For example, the anecdote above involving high-income Upper West Side and low-income East Harlem residents speaks to PB's ability to facilitate constructive dialogues between people living in disparate socioeconomic conditions. But this exact exchange—and the small-scale attendant redistribution of funds from one area to another—was an exception to the rule, and is not possible under new configurations. District lines have been redrawn, and East Harlem is now districted with higher-poverty stretches of the South Bronx. Because City Council districts tend to be relatively small areas with concentrated wealth or poverty, PB's potential for redistributive impact remains limited unless city-wide portions of the budget become subject to PB.

A Spark Towards Something Bigger, and the Importance of Spillover Effects

As both everyday residents and city agency representatives become adept at navigating the PB process, it becomes tempting to simply forward whatever proposals they now know to be most palatable to city agencies, even when these proposals sideline the concerns and local knowledge that compelled them to participate in the first place. One budget delegate described a hypothetical but, to him, archetypal set of responses:

> "I want new street lights. My street lights aren't broken, but I want prettier street lights" ...
> that often doesn't affect the quality of life for constituents as much as other things. But
> that's often how those things get done because ... funds are available for that purpose.

Another delegate complained that he was told to "think small": "I put [my idea] out there. It was shut down! Even by my peers, those fools around me! They say why it can't, why it won't ... it's ... too big! ... We don't think; we behave small."

Without critical dialogues, PB runs the risk of serving as a technocratic tool for "good governance," romanticizing the role of the individual and reifying a neoliberal logic, enlisting "citizens in measuring, auditing and monitoring ... in a depoliticized technical process that defuses conflicts and treats them as consumers," rather than as political stakeholders (Hickey and Mohan, 2004). After all, should it really be the job of busy, working New Yorkers to research and address which schools need basic repairs, or to "choose" which curbs require extensions to be safe, and by extension, which do not? Discussions regarding taxation and the shape and size of the figurative public budget pie, in such a scenario, are sidelined by competitive exercises determining the slices of the pie. In a neoliberal order of budgeting, lay constituents compete with other budget delegates to develop state-sanctioned projects, including those that were previously paid with core rather than discretionary funds. Therefore, policymakers and constituents must continue to work on maintaining the deliberative aspects of PB integral to the process, and allowing room for critical questions, so that PB votes do not end up as the community development equivalent of school "choice."

Fortunately, there is evidence that the experience of participating in budgeting has educated, empowered, and even outraged constituents to demand more, and to hold government more accountable—not only regarding local discretionary expenditures, but regarding the municipal budget overall. Because the New York PB-eligible funds remain rather limited, many of the most interesting and profound outcomes of PB thus far take the form of spillover effects. For instance, renovations for elementary school bathrooms won PB funds consistently, and PB participants were livid over the fact that they were devoting discretionary funds for what they deemed to be a basic need. In response, Councilmember Brad Lander, of Brooklyn, launched a new initiative to set aside other, non-PB funds for such bathroom renovations. Councilmember Lander similarly began a new public art fund co-administered with two local arts organizations when local residents complained about the lack of PB-funded arts programming. In almost all districts, City Councilmembers have worked with residents to find funding for more projects than the formal PB process allowed, and to work with disgruntled residents on finding alternative ways to implement unfunded proposals.

Further, at the city-wide level, at least one agency has developed a new web portal to enable anyone to more easily search through its recent expenditures and planned projects, and the City Council is developing a project tracker that will enable lay constituents to more easily analyze the progress and patterns found in PB-funded projects. These promising developments are likely to have ripple effects beyond PB, in other forms of public administration and governance.

Discussion

The New York case thus far has been impressively successful in broadening notions of stakeholdership and who is counted in the "community," working against a deficit model of formerly incarcerated youth as troublemakers or undocumented immigrants as a drain on the welfare state, and in facilitating face-to-face deliberations between different stakeholders, allowing them to voice grievances and ideas. Of course, there remain challenges in shifting dynamics of power between political stakeholders and governmental agencies, with contestations over who represents the public here, and how equity ought to be best defined and achieved. Still, it takes time and a lot of tweaking to change cultures and power dynamics in governance, and PB has already begun to help constituents and policy-makers focus on issues of economic inequality in focused, concrete, and collaborative ways.

The constituents' experiences in PB have thus far been quite complex; many are voicing community concerns as best they can within the confines of PB, even as they wish for more power. Rather than blindly internalizing the city agencies' priorities, delegates self-consciously adopt pre-existing limits and criteria—sometimes resisting them, sometimes reifying them, and sometimes simultaneously doing both. Here, participation acts as a double-edged sword.

At its extreme, a dynamic focused on voting for local projects embodies almost a consumer choice model, rather than a deliberative one, with representatives giving pitches for PB funds, and telling delegates exactly what ready-made projects need funding in their neighborhoods. Delegates might then go through the motions of "choosing" the very projects the state would have forwarded in the first place, and well-organized, upper-income parent groups predictably launch winning proposals for their children's schools. But in New York, no dominant order has yet taken hold. Indeed, the tensions that do exist underscore the need for accompanying administrative reforms, without which PB becomes depoliticized as a communicative strategy for transparency (rather than equity or empowerment) (Baiocchi and Ganuza).

Along these lines, another potential lesson lies in how city agency representatives and administrators need more resources and capacity-building as well, and how their material conditions, funding streams, and economies of scale might be better served by larger pots of funds. Both agency and delegate interviewees repeatedly asked for cross-district coordination and expense funding for staffing, alongside capital funding for infrastructure.

Further, even within a single city, no one PB process fits all policy arenas. While libraries, parks, schools, and some other agencies have consistently worked with delegate groups with winning projects, sanitation, environmental protection, public transit, and other agencies have participated a bit less. It may be that running two, concurrent PB tracks—including one that is less intensive, but larger-scale—might help currently marginalized delegate groups and agencies to become more engaged. For instance, in helping city agencies to approve and implement projects that abide by the spirit of community needs, rather than judging proposals by the letter of the laws and dismissing them because of technicalities, the New York PB process might also host neighborhood assemblies in which constituents list community priorities, and then let agencies hash out the details. This would also help PB to more directly address economic inequalities in a segregated city, and to work towards redistribution between, say, the Upper East Side and East Harlem.

Indeed, the New York process has garnered the most attention from the popular press because of its size, and because the city has a history of first piloting policy innovations then modeled by others around the country—but it should by no means be seen as *the* model of PB in the US. In fact, other cities around the country have begun to surpass New York in using PB to reconsider revenues, as well as expenditures, in their public budgets. Vallejo, California, was known as the first city in the state to declare bankruptcy, until it became known as the first local to implement city-wide PB. It uses a sales tax to fund not just pothole repairs, but also internship programs, veteran assistance, and community service-oriented college scholarships. The campaign to implement PB in Buffalo, New York, was spearheaded by organizing groups working on environmental justice issues and lawsuit settlements in the area; they turned to PB in thinking about better ways to allocate these funds. A pilot PB process there has just been approved by its City Council. Boston has focused on youth ages 12 to 25 in its citywide PB process, and several schools around the country have begun to turn over a portion of their budgets to matriculated students.

In all of these settings, PB acts as a laboratory of democracy for conversations essential to our common economic prosperity and social justice. And, despite the legacy of New England hall meetings in smaller towns, PB has the greatest potential and impact as an innovation in urban areas. As debates around gentrification continue to rage around the nation, lower-income residents continue to face risks of displacement and root shock (Fullilove). Neighbors of differing economic strata might engage in friendly banter at the corner bodega, but seldom do they come together to discuss their commonalities and differences, and strategize on how to thrive together. As the New York case suggests, even limited PB process can prompt constituents and policymakers to interact and work across difference in new ways, and to demand more, even in an era of austerity. Across contexts, such meaningful deliberations and dialogues are difficult to facilitate but crucial in making PB truly innovative, substantive, and transformative.

Works Cited

Alonso, Gaston, et al. *Our Schools Suck: Students Talk Back to a Segregated Nation on the Failures of Urban Education*. New York: New York University Press, 2009. Print.

Baiocchi, Gianpaolo. *Militants and Citizens: The Politics of Participatory Democracy in Porto Alegre*. Palo Alto: Stanford University Press, 2005. Print.

Baiocchi, Gianpaolo, and Ernesto Ganuza. "Participatory Budgeting as If Emancipation Mattered." *Politics & Society* 42.1 (2014): 29–50. Print.

Bergman, Shawn M, et al. "Millennials, Narcissism, and Social Networking: What Narcissists Do on Social Networking Sites and Why." *Personality and Individual Differences* 50.5 (2011): 706–11. Print.

Boulding, Carew, and Brian Wampler. "Voice, Votes, and Resources: Evaluating the Effect of Participatory Democracy on Well-Being." *World Development* 38.1 (2010): 125–35. Print.

Bryer, Thomas A. "Public Participation in Regulatory Decision-Making: Cases from Regulations. Gov." *Public Performance & Management Review* 37.2 (2013): 263–79. Print.

Bryson, John M, Barbara C Crosby, and Laura Bloomberg. "Public Value Governance: Moving Beyond Traditional Public Administration and the New Public Management." *Public Administration Review* 74.4 (2014): 445–56. Print.

Bryson, John M, et al. "Designing Public Participation Processes." *Public Administration Review* 73.1 (2013): 23–34. Print.

Carnemark, Curt, et al. *The World Bank Participation Sourcebook*. Washington, DC: World Bank, 1996. Print.

Chambers, Robert. *Whose Reality Counts?: Putting the First Last*. Intermediate Technology Publications Ltd (ITP), 1997. Print.

Chandhoke, Neera. "Civil Society." *Development in Practice* 17.4–5 (2007): 607–14. Print.

Cooke, Bill, and Uma Kothari. *Participation: The New Tyranny?* Zed Books, 2001. Print.

Cornwall, Andrea, and Vera Schatten Coelho. *Spaces for Change?: The Politics of Citizen Participation in New Democratic Arenas*. Vol. 4: Zed Books, 2007. Print.

Dahl, Adam, and Joe Soss. "Neoliberalism for the Common Good? Public Value Governance and the Downsizing of Democracy." *Public Administration Review* 74.4 (2014): 496–504. Print.

de Sousa Santos, Boaventura. *Democratizing Democracy: Beyond the Liberal Democratic Canon*. New York: Verso Books, 2005. Print.

de Souza Briggs, Xavier. *Democracy as Problem Solving: Civic Capacity in Communities across the Globe*. MIT Press, 2008. Print.

Durant, Robert F, and Susannah Bruns Ali. "Repositioning American Public Administration? Citizen Estrangement, Administrative Reform, and the Disarticulated State." *Public Administration Review* 73.2 (2013): 278–89. Print.

Feldman, Martha, and Anne Khadermian. "The Role of the Public Manager in Inclusion: Creating Communities of Participation." *Governance* 20.2 (2007): 305–24. Print.

Fullilove, Mindy. *Root Shock: How Tearing up City Neighborhoods Hurts America, and What We Can Do About It.* One World/Ballantine, 2005. Print.

Fung, Archon. *Empowered Participation: Reinventing Urban Democracy.* Princeton University Press, 2004. Print.

Fung, Archon, and Erik Olin Wright. *Deepening Democracy: Institutional Innovations in Empowered Participatory Governance.* New York: Verso Books, 2003. Print.

Godsay, Surbhi, et al. *"That's Not Democracy": How out-of-School Youth Engage in Civic Life & What Stands in Their Way.* Medford, MA: CIRCLE, the Center for Information & Research on Civic Learning and Engagement, Tufts University, 2012. Print.

Goldberg, Leo, and Rachel Finkelstein. "Comparing Area Demographics for Participatory Budgeting Projects and City Capital Budget Allocations." Cambridge, MA: MIT Department of Urban Studies and Planning, 2014. Print.

Graeber, David. *The Democracy Project: A History, a Crisis, a Movement.* New York: Spiegel & Grau, 2013. Print.

Hajer, Maarten. "Setting the Stage: A Dramaturgy of Policy Deliberation." *Administration & Society* 36.6 (2005): 624–47. Print.

Harvey, David. "From Managerialism to Entrepreneurialism: The Transformation in Urban Governance in Late Capitalism." *Geografiska Annaler Series B, Human Geography* (1989): 3–17. Print.

Hickey, Samuel, and Giles Mohan. *Participation – from Tyranny to Transformation?: Exploring New Approaches to Participation in Development.* London: Zed Books, 2004. Print.

Immerwahr, Daniel. *Thinking Small: The United States and the Lure of Community Development.* Cambridge: Harvard University Press, 2014. Print.

Incite! Women of Color Against Violence. *The Revolution Will Not Be Funded: Beyond the Non-Profit Industrial Complex.* Boston: South End Press, 2007. Print.

Kaklamanidou, Betty, and Margaret Tally. *The Millennials on Film and Television: Essays on the Politics of Popular Culture.* McFarland, 2014. Print.

Kasdan, Alexa, and Lindsay Cattell. *A People's Budget: A Research and Evaluation Report on the Pilot Year of Participatory Budgeting in New York City.* New York: The Community Development Project at the Urban Justice Center, 2012. Print.

Kasdan, Alexa, Erin Markman, and Pat Covey. *A People's Budget: A Research and Evaluation Report on the Participatory Budgeting in New York City in 2013–2014 (Tentative Title, Forthcoming).* New York: The Community Development Project at the Urban Justice Center 2014. Print.

Leal, Pablo Alejandro. "Participation: The Ascendancy of a Buzzword in the Neo-Liberal Era." *Deconstructing Development Discourse: Buzzwords and Fuzzwords.* Eds. Cornwall, Andrea and Deborah Eade. Oxford: Oxfam, 2010. 89–100. Print.

Lerner, Josh. *Making Democracy Fun: How Game Design Can Empower Citizens and Transform Politics.* Cambridge: MIT Press, 2014. Print.

Macedo, Stephen, et al. *Democracy at Risk: How Political Choices Have Undermined Citizenship and What We Can Do About It.* Washington, DC: Brookings Institution Press, 2005. Print.

McDonald, Michael. *2014 November General Election Turnout Rates*. Gainesville, FL: United States Elections Project, 2014. Print.

Nakkas, George. "Participatory Budgeting and Economic Inequality." Brooklyn, NY: Brooklyn College, 2015. Print.

OCED. *Society at a Glance 2011: Social Indicators*. Paris: OECD Publishing, 2011. Print.

Paley, Julia. *Marketing Democracy: Power and Social Movements in Post-Dictatorship Chile*. Berkeley: University of California Press, 2001. Print.

Putnam, Robert. *Bowling Alone: The Collapse and Revival of American Community*. Simon and Schuster, 2001. Print.

Salamon, Lester M. *The Tools of Government: A Guide to the New Governance*. Oxford University Press, 2002. Print.

Silver, Hilary, Alan Scott, and Yuri Kazepov. "Participation in Urban Contention and Deliberation." *International Journal of Urban and Regional Research* 34.3 (2010): 453–77. Print.

Skocpol, Theda. "Voice and Inequality: The Transformation of American Civic Democracy." *Perspectives on Politics* 2.01 (2004): 3–20. Print.

Stoker, Gerry. "Public Value Management a New Narrative for Networked Governance?" *The American review of public administration* 36.1 (2006): 41–57. Print.

Stolle, Dietlind, and Marc Hooghe. "Shifting Inequalities: Patterns of Exclusion and Inclusion in Emerging Forms of Political Participation." *European Societies* 13.1 (2011): 119–42. Print.

Su, Celina. "Whose Budget? Our Budget? Broadening Political Stakeholdership Via Participatory Budgeting." *Journal of Public Deliberation* 8.2 (2012): 1–14. Print.

Thomas, John Clayton. "Citizen, Customer, Partner: Rethinking the Place of the Public in Public Management." *Public Administration Review* 73.6 (2013): 786–96. Print.

Wampler, Brian. *Participatory Budgeting in Brazil: Contestation, Cooperation, and Accountability*. University Park: Pennsylvania State Press, 2010. Print.

Wampler, Brian, and Janette Hartz-Karp. "Participatory Budgeting: Diffusion and Outcomes across the World." *Journal of public deliberation* (2012). Print.

Warren, Mark. "Communities and Schools: A New View of Urban Education Reform." *Harvard Educational Review* 75.2 (2005): 133–73. Print.

Weiser, Wendy R., and Larence Norden. *Voting Law Changes in 2012*. New York: Brennan Center for Justice, 2011. Print.

Chapter 13
Financing Urban Transportation Infrastructure: Old and New Approaches for Funding Urban Regeneration and Resiliency: The Atlanta Example

M. Becht Neel and Julian Conrad Juergensmeyer

I. Introduction

When the mayors of Canadian cities that rank among the world's most functional, attractive cities gathered to lobby their national government for an even more explicitly pro-urban agenda, one mayor was quoted as saying, "[g]rowth shines a spotlight on a number of problems around how cities are financed and in particular how cities invest in infrastructure" (Church et al.). No words could ring more true for the great metropolitan areas of the world—and America's cities in particular.

The United States entered the 21st century with essential infrastructure barely adequate to meet current demands, not to mention future burdens associated with increased population, heavy continuing use, diminished government investment in upkeep, or climate change. Road, bridge, electrical grid, and water supply failures are, unfortunately, not unfamiliar news headlines. The challenge of replacing or rebuilding infrastructure looms larger every year. What is at stake, however, is not just the task of carrying out construction projects so that the nation's cities can maintain the status quo. The stakes are potentially much higher, and the payoff to urban communities is potentially much greater: infrastructure construction and redevelopment is not merely a band aid, it can also be part of the solution to the most pressing challenges that cities will face over the next century, including population growth, climate change and the ever more urgent imperative to connect economically disadvantaged communities to resources and opportunities.

The current outlook for US cities is not good. They already face enormous price tags for completing updates to existing infrastructure. Most have no viable plan for how to pay for infrastructure improvements required to accommodate growing population, increased development, and climate change. One of the chief reasons for this bleak outlook is that US cities lack forward-thinking approaches to infrastructure financing. Current funding streams largely dependent on local government property taxes, state support, and federal grants are too small to fill the need. It is imperative that local and state governments have appropriate financing tools at their disposal. This chapter outlines the range of financing mechanisms that local and state governments must consider adopting to support transportation infrastructure redevelopment and construction using the Atlanta, Georgia, metropolitan area as a case study.

Atlanta's dynamic growth and evolution have exposed the inadequacy of the city's transportation infrastructure and the enormous amount of funding needed to keep the city on a prosperous path. Unfortunately, the numbers do not add up in Atlanta's favor. From 2000–2015, the 2 percent population growth for the 29-county metropolitan Atlanta area makes Georgia the third fasting-growing state in the US. At the same time, Atlanta ranks 49th in transportation spending per capita (Atlanta Regional Commission 2010). Even in 49th place, the Atlanta region spends around $2 billion annually on transportation infrastructure (Atlanta Regional Commission n.d.). In 2013, cities and counties in metro Atlanta spent approximately $1.3 billion on transportation projects (Norton). An average of 35 percent of those funds came from the Federal Highway Trust Fund. However, Atlanta is estimated to need an additional $1.5 billion in annual revenue to fund the present and long-term infrastructure the city needs to stay a competitive international city (Lindsey 18 Jan 2015). The Statewide Transportation Plan for 2005–2035 has an estimated cost of $160 billion, but present revenue resources are expected to bring in only $86 billion over this period, leaving a funding gap of $74 billion (Lindsey 18 Aug. 2014).

Atlanta's glaring infrastructure funding shortfall raises the question of how the city can manage the challenges and reap the opportunities promised by the coming decades. With the federal government's primary source for bridge, road, and transit funding, the Federal Highway Trust fund, slated to run out of money in the near future (Fuentes) and infrastructure only receiving attention from leadership in Washington when there is failure and tragedy, it is our cities and regional governments, as well as the private and non-profit sectors, that are left to figure out innovative ways to rebuild their cities and provide their inhabitants with the reliable and modern transportation infrastructure needed to keep their cities alive. But the infrastructure challenge looms so large that it will not be enough for any single city or county to channel resources to meeting the enormous funding gap. Even development of the most ingenious sources of infrastructure funding will not save Atlanta and other cities unless they embrace change and adopt metropolitan-wide urban planning and development regulation.

II. What Needs to Change for Cities to Have Adequate Infrastructure Funding?

Despite the economic storm the Great Recession unleashed on American cites, it is hard to deny that the vast majority enjoy a relatively stable government and economy. This is particularly true when juxtaposed to the turbulent politics and economic strife experienced by many countries throughout Europe, including Italy and Spain. With unemployment hovering well above 20 percent and tens-of-thousands of families losing their homes, the years since the 2008 crash have precipitated profound social and political upheaval. But the hardships these countries have experienced following the 2008 worldwide economic shock mask the striking resilience of some of these countries' major cities. As Atlanta and its sister American cities search for possible solutions to addressing their current and future infrastructure crisis, they should not ignore the strategies employed by major European cities, including cities in Spain and Italy. Although these cities faced dire economic straits, a number of them have deployed creative legal strategies to sustain their cities.

Italy and Spain found themselves with weak economies and thus fading tax revenue funds to maintain and grow their infrastructure and their cities. In the midst of their fiscal dilemma, they identified a new approach to sustaining their cities. They found that "regional unification" of local governments was an essential strategy for resuscitating their cities. Barcelona, internationally admired for overcoming waves of economic, social and political adversity, provides an excellent model for large US cities like Atlanta. In 1976, Barcelona implemented a region-wide "General Metropolitan Plan" with the purpose of providing a legally binding metro-wide plan, encompassing Barcelona, seven counties, and 164 surrounding municipalities (Esteban). That 1976 plan, which established a legal foundation for a supportive relationship among metro-Barcelona's local governments, also provided a platform for ensuing urban revitalization and funding plans. Among the most noteworthy of these regional coordinating plans is 22@ Barcelona, which establishes and funds "Innovation Districts" supported by regional legitimacy, strength, and interconnectivity. Italy has gone even further toward solving the problem by amending its constitution to consolidate and absorb most of the 8,000+ cities and towns (each of which had a separate civic, economic, and political center and agenda) into 14 newly created "metropolitan cities" or "areas" that are centered on Italy's current largest city centers in which 1 in 3 Italians will now live (Griffith). The goal of the historic change was to encourage better coordination on urban problems that do not stop at city borders and to augment cites' authority by cutting out the provincial layer of local government and giving that power to the new "metro areas" in order to eliminate bureaucratic red tape and support the efficiency and growth of the larger regions that make up Italy's metropolitan cities.[1] It is also worth noting that a somewhat similar approach has been initiated in France pursuant to MAPTAM (Club Metropolitan).[2]

The exact opposite situation can be found in American cities like Atlanta (Juergensmeyer and Roberts). Fulton County, which encompasses a major portion of the City of Atlanta, is home to 14 different cities, three of which are among Georgia's top 10 largest in population (Crane). Instead of regional unification, Atlanta has become even more fractured in the past decade with the formation of new cities (accompanied by additional local agencies) such as Sandy Springs (June 2005), John's Creek (July 2006), Milton (Dec. 2006), Chattahoochee Hills (June 2007), Dunwoody (Dec. 2008), Peachtree Corners (Nov. 2011) and Brookhaven (July 2012) (Williams). More new cities are on the horizon for Fulton County—each with the objective of detaching themselves from a larger encompassing governmental entity. These proposed communities include Tucker (Georgia House Bill 515) and LaVista Hills (Georgia House Bill 520), whose city formation has been submitted to referendum (Niesse). This lack of metropolitan-wide jurisdictional control is one of the main challenges local government officials and planners face worldwide when seeking to plan, coordinate, and implement innovative strategies for funding transportation and other infrastructure. The different approaches of dozens of jurisdictions within a metropolitan area to infrastructure funding lead to conflict, competition, and inefficiency. Without substantial region-wide coordination and cooperation, neither traditional nor innovative infrastructure funding approaches will be able to provide adequate infrastructure —transportation or other—for metropolitan areas (Ziegler).

What follows are ideas about various current and future sources of transportation infrastructure funding for the Atlanta metropolitan area. However, these infrastructure-funding sources are only going to work effectively if there is a metro Atlanta regional "unification" that allows for a single governing body to have principal power and authority

---

1   Legge 7 aprile 2014, n. 56, NORMATTIVA. This resource may is available in English at http://www.loc.gov/lawweb/servlet/lloc_news?disp3_l205403979_text. It is also available in Italian at http://www.normattiva.it/uri-res/N2Ls?urn:nir:stato:legge:2014–04–07;56.

2   This reference is available in French at: LOI n° 2014–58 du 27 janvier 2014 de modernisation de l'action publique territoriale et d'affirmation des métropoles (1) NOR: RDFX1306287L Version consolidée au 26 avril 2015.

over the 29-counties and 236 local government entities' transportation infrastructure-related decision.

The purpose of this chapter is not to address the logistics or specifics behind "how" metro Atlanta should go about uniting the region (the international examples above suggest such unification is more than viable), but rather to demonstrate that if you solve the jurisdictional problem, there are ways to generate annually the additional $1.5 billion Atlanta needs for infrastructure. None of these alternative-funding sources, no matter how innovative they are, are going to work well unless implemented on a metropolitan-wide level.

III. Transportation Infrastructure Funding Sources

Before exploring the details of the many different ways to fund transportation infrastructure, it is helpful to categorize generally the wide range of tools cities and states like Atlanta and Georgia have at their disposal. Some of these funding mechanisms may already be in place and working well, others are current practices that warrant improvement, and still others have yet to even be considered for implementation by US metropolitan jurisdictions. For every alternative source of infrastructure funding there are advantages and disadvantages. Implementation assessments discussed later in the chapter will focus on some of the more innovative, encouraging and profitable funding possibilities for metro Atlanta.

The five (5) primary funding mechanism categories are as follows (State Smart Transportation Initiative (SSTI)):

1. General income and consumption taxes. An unsophisticated means of revenue for infrastructure, the following broad-based taxes on residents and businesses could help fund metropolitan infrastructure: (i) payroll taxes; (ii) sales taxes; (ii) property taxes; (iv) income taxes; (v) "sin" taxes (lottery, tobacco, marijuana, alcohol revenue); (vi) capital gains taxes; and (vii) real estate transfer taxes.

2. Activity-based user fees. A popular but by no means widespread mechanism for funding infrastructure, local governments can collect revenue from charges on the users of transportation facilities through: (i) fuel tax; (ii) fare-box revenue; (iii) tolling; (iv) carbon fees; (v) weight-mile fees; (vi) mileage-based user fees (MBUF); (vii) tire and battery fees; (viii) passenger facility charges; (ix) right-of-way leasing; (x) terminal use fees (land fees, berthing fees); (xi) parking space taxes; (xii) energy use taxes; (xiii) hotel taxes; (xiv) rental car taxes; (xv) bicycle user fees; and (xvi) mobile source emissions credits.

3. Administrative fees and fines. As a public agent, cities can derive revenue through their authorization, administration, and enforcement of activities including: (i) motor vehicle registration fees; (ii) driver's license fees; (iii) vehicle transfer fees; (iv) dedicated traffic violation revenue; (v) utility or franchise fees; and (vi) regional vehicle ad valorem taxes.

4. Value capture. This approach to funding transportation infrastructure offers an equitable means of recouping value and therefore revenue from the private sector in proportion to the benefit received from transportation services and facility improvements (Atlanta Regional Commission). Value capture (VC) mechanisms include: (i) tax increment financing (TIF); (ii) transportation benefit districts or special resident assessment districts (SRAD); (iii) joint development or air rights development; (iv) exactions; (v) developer impact fees (DIF); (vi) land value or split

rate taxes; (vii) transportation utility fees (TUF); (viii) sponsorship, advertising, and naming rights; and (ix) systems development charges (SSTI).
5. Public-private partnerships or joint participation. Contractual agreements between public and private entities for the direct provision of transportation facilities or services allow market demand to improve infrastructure without public funds. These funding schemes include: (i) business improvement districts, (ii) developer financing; (iii) joint development; (iv) private ownership (highways or transit); (v) private donation, (vi) negotiated exactions; or (vii) urban service boundary expansion windfall taxes.

Currently, Atlanta's transportation and infrastructure moneys are derived from three of these five general funding mechanisms. The metro region's relatively unsophisticated revenue generators include: the motor fuel and motor carrier taxes (activity-based user fees), car title registration and license tag fees (administrative fees), personal property tax and the MARTA sales and use tax (general income and consumption taxes) (Atlanta Regional Commission 16). A truly viable future for the Atlanta metropolitan region depends on a jurisdictionally unified metro Atlanta that can cooperate not only to restructure these archaic revenue generators but to adopt more creative and sources of infrastructure revenue.

IV. Traditional Transportation Infrastructure Funding "Reform"

The transportation infrastructure funding techniques available to fund metro Atlanta's infrastructure needs are blunt tools for raising funds. They tend to charge transportation system users costs that are not proportionate to the frequency and intensity of their transportation system uses and they frequently impose outsized burdens on lower income families forced to drive long distances to work due to inadequate public transit or lack of affordable housing options proximate to their workplace. Despite these significant drawbacks, it is difficult to envision meeting the metro Atlanta region's enormous transportation needs without continuing to rely on the existing menu of transportation infrastructure funding options. This part of the chapter reviews the current funding mechanisms and suggests ways these existing tools could be further refined to help meet the needs of the metro Atlanta region.

Of the six current revenue sources that help support metro Atlanta transportation infrastructure, the two motor fuel taxes (excise and prepaid sales taxes) are the primary funding mechanism for transportation at the State level. In 2009, 96 percent of the Georgia Department of Transportation's (GDOT) budget stemmed from the motor fuel taxes ($775 million) (Atlanta Regional Commission 2010, 16). In Georgia, the reliance on State rather than federal funding leads to a severe inability for a city like Atlanta to address its transportation needs when the overall economy contracts and motor fuel tax revenues inevitably decline. An examination of transportation funding mechanisms shows how a united Atlanta region could swiftly cure these disorders.

A. Motor Fuel Excise Tax Rate Increase

The motor fuel excise tax is a tax based on the volume (in gallons) of fuel purchased. Revenues from this tax are strongly interrelated with vehicle-miles traveled and the fuel economy of the motor vehicles traveling in the state because the tax is based solely on the volume of gasoline sold. Today, this tax, which is a relic of the 1930s, is no longer efficient.

There is a negative correlation between tax revenue generated and the Miles Per Gallon (MPG) of automobiles because of exponential technological advancements in vehicle fuel efficiency, as well as changing land use patterns that increase the selection of modal choices for households (Atlanta Regional Commission 2010, 39).

However, recapitalization of the Georgia Transportation Infrastructure Bank (GTIB) by the increase of the gas tax from its 1971 level at 7.5 cents per gallon up a single cent or as much as 4 cents would generate up to $240 million per year (Lindsey 18 Jan. 2015). A 3-cent increase would equate to 40 percent additional revenue for Metro Atlanta by 2045 (Atlanta Regional Commission 39). Adding just 2 cents to the 7.5 cents yields a 26.7 percent increase in transportation revenue generated by 2045. Even a 1 cent excise tax rate increase would generate approximately 13.3 percent more revenue over a 30-year period. This increased revenue, devoted exclusively to the GTIB, could be used to create a loan or grant fund to match the funds for transportation projects in the Atlanta region (Lindsey 18 Jan. 2015).

Additionally, the provisions of the Georgia Constitution authorizing both motor fuel taxes prohibit any of the revenue from being spent on public transportation, limiting the allowable expenditures to roads and bridges.[3] Lifting this use restriction to allow the revenue to be split between the two forms of transportation infrastructure would empower the region to act in a flexible manner to address transportation needs (Atlanta Regional Commission 16).

B. Prepaid Motor Fuel Sales Tax Increase

Georgia also collects a 4 percent sales tax on the average retail price of fuel, of which only 3 percent of that 4 percent is dedicated to transportation. The Prepaid Motor Fuel Sales Tax is the more effective of the two motor fuel taxes because it is less dependent on the gallons of fuel consumed and more dependent on the price of fuel itself. Since revenues from this tax increase and decrease with the price of gasoline (Atlanta Regional Commission 2010, 17), the revenue stream from this fuel sales tax has been declining since 2008 due to the decrease in gas prices.

One possible solution is to pass legislation that phases out the practice of allowing state and local governments to siphon off gasoline sales tax to their general funds for spending on non-transportation items and enables the remaining 1 percent of the sales tax to go toward (and earmarking that 1 percent for) public transit infrastructure.

More effective legislation would recapture the approximately $180 million and $500 million in *annual* state and local gasoline taxes respectively that were diverted into general funds and dedicate the entire 4 percent of the fuel sales tax collected toward transportation infrastructure (Lindsey 18 Jan. 2015). The Atlanta Regional Commission (ARC) forecasts that increasing the prepaid motor fuel sales tax to 5 percent amounts to $447 million in additional annual revenue, and an increase in the tax rate up to 7 percent yields around $894 million in additional *annual* transportation revenue (Atlanta Regional Commission 2010, 41).

---

3   "All money derived from motor fuel taxes received by the state … is hereby appropriated … for all activities incident to providing and maintaining an adequate system of public roads and bridges in this state … " Ga. Const. art. III, § 9, ¶ VI.

C. Regional Vehicle Ad Valorem Tax

In Georgia, the Vehicle Ad Valorem Tax (VAVT) is a type of personal property tax based upon the vehicle's value that generates revenues for the general funds of state and local governments (distinguishable from a "user fee" that collects money for transportation only) (Atlanta Regional Commission 2010, 23). The State, county, and local millage rates are levied on registered motor vehicles in the form of the ad valorem tax and must be paid when an application is made for tag renewal (Atlanta Regional Commission 2010, 24–25). A 1-millage and 2-millage VAVT in the inner 18-country region (not the whole 29-county region) would generate annual revenue of about $16 million and $31 million, respectively.

Some states like Arizona saw the potential for transportation funding from VAVTs and now dedicate a portion (1/3) of the revenue to be used towards state highways and mass transit (Joint Legislative). The Atlanta region should follow Arizona's example and adopt this new alternative to help fund transportation infrastructure.

The advantage to using an increased VAVT is that tax assessment and collection during the vehicle registration process translates into low collection and administrative cost (Atlanta Regional Commission 51). Also, the progressive nature of the tax based on vehicle value correlates to household income, and tax revenue would increase as the cost of vehicles increase. Additionally, the tax is highway related and would be responsive to economic fluctuations. Finally, using existing VAVT as Arizona does, or in the alternative, imposing a small millage rate on motor vehicles, has the potential to be more acceptable to voters and constituents because it can be justified as a user fee. On the other hand, VAVTs are placed solely on local residents to bear the cost, with no impact on non-residents and/or tourists. Further, the potential of generating just $16 million or $31 million in additional revenue does little to alleviate the annual $1.5 billion of increased transportation spending needed by metro Atlanta (Lindsey 2 Feb. 2015).

D. Regional Vehicle Registration/ License Plate Fees.

Many states already implement types of license plate fees to help create additional transportation infrastructure revenue. The number of vehicles registered in the Atlanta region is forecasted to increase by 79 percent over the next 25 years: an additional 3.2 million vehicles. The massive increase means implementation of a regional $10 per license plate fee would generate approximately$40 million in 2016 and nearly $45 million by 2040.

The disadvantage of the fee is its regressive nature. It also places another burden on vehicle owners who are already burdened by existing ad valorem taxes placed on motor vehicles by the State and local governments (Atlanta Regional Commission 2010, 42). Also, similar to the VAVT, $40 million is only a fraction of the billions of additional annual transportation infrastructure revenue that Atlanta needs to keep pace with mounting demand.

E. Past Proposals: Regional Transportation Special Purpose Local Option Sales Tax (T-SPLOST).

Once regarded as the Atlanta region's most promising solution to its transportation problems, metro Atlanta voters defeated a proposed Regional Transportation Special Purpose Local Option Sales Tax (T-SPLOST) program in 2012 partially due to public perception that the money from the 1 percent sales tax increase would fail to provide real regional "gridlock relief." (Georgia House Bill 227). The goal of T-SPLOST was to allow multiple counties to create SPLOSTs to meet particularly critical road and transit needs in their region. Although this particular regional solution to transportation infrastructure spending would have represented a significant steps toward meeting the metropolitan area's unfunded need, it is important to note that a metro Atlanta region united with force of law into one whole "metropolitan city" would accomplish what T-SPLOST attempted to do and much more.

V. Innovative Alternative Transportation Infrastructure Funding Mechanisms

The last section of this chapter focuses on innovative ways to fund transportation infrastructure in addition to the six longstanding revenue generators that currently attempt to support the metro Atlanta region's transportation infrastructure. While current infrastructure funding mechanisms will have an important role to play in almost any future comprehensive infrastructure funding reform plan, these existing tools are not on their own a firm foundation for a sustainable future for one of the worlds' leading cities. If international mega-cities, like Atlanta, are going to meet their own expectations, then they must adopt innovative solutions for bridging the current transportation infrastructure gap— solutions that have a realistic expectation of garnering public support.

As discussed above, a unified Atlanta region that supports the entire Southeast needs a transportation infrastructure overhaul to achieve those aspirations. This section addresses four funding approaches that would help fund Atlanta's transportation infrastructure and perhaps offer viable alternatives to other metropolitan areas struggling with transportation infrastructure deficiencies. These options include: (1) increasing the property tax for different property types; (2) value capture methods; (3) user fees; and (4) parking fees.

A. Property Tax of Different Property Types: Regional Millage Rate Increase

Over the past half-century, the metro Atlanta region's explosive growth, unimpeded by any significant geographic barrier, has manifested itself as dense ripples of sprawling development extending outward more than 60 miles in every direction from the city center. It is this unchecked growth that has led to the transportation and infrastructure problems crippling the region today. However, the Atlanta metro region's growth, which is almost unprecedented in the annals of American urban development, may also present an opportunity for addressing the dire transportation needs that the region now faces. The Atlanta region's booming real estate development sector, fueled by local governments readiness to rezone former farms and cow pastures for tens of thousands of homes and commercial buildings, as well as infill and brownfield development of former industrial sites in the city center, have increased the value of metro Atlanta counties' tax digests pools by 200–700 percent (Atlanta Regional Commission 2010, 22). Mindful of the political challenge of securing increases in ad valorem property taxes, an increase in property tax, whether on all property use types or specific property types, represents one of the more significant revenue generating options for funding the Atlanta region's transportation infrastructure.

1. Property tax & millage rate increase
Despite the struggle for property values to return to pre-2008 levels, a 0.5-mil increase in property tax throughout the inner 18 counties that comprise metro Atlanta could yield around $100 million for transportation funding in 2016 and a 1-mill increase could result in an additional $225 million contribution to Atlanta's pool of transportation funding, with

that number rising if all 29 counties in the region were subject to the tax increase (Atlanta Regional Commission 2010, 46).

The property tax increase on all types of real property brings a significant boost in infrastructure funds from a broad tax base at a minimal impact per resident. Furthermore, the burden new development places on infrastructure would be partially funded by the increased property tax. The tax increase could also be justified as a modest "capture" of a portion of the rise in property values of lands benefitted by their adjacency to new transit infrastructure.

The political prospects are currently grim for local approval of these property tax increases. Real estate taxes are among the most unpopular types of taxes. In addition, citizens perceive the tax as placing an unfair burden on them to pay for facilities that should be shared with non-residents. Further, as the nation emerges from the recent real estate crash that battered the Atlanta region, it is important to consider that the revenue generated from this tax is subject to volatile fluctuations in the real estate market. After decades of influx and growth in metro Atlanta, this tax could also have a negative impact on residential and commercial relocation decisions because a low cost of living is a main source of attraction to the Atlanta area.

2. Industrial and/or commercial property tax increase

To avoid public opposition a tax increase could cause, the 18 (or 29) county Atlanta region could consider increasing the property tax on commercial or solely industrial properties by 1-mil. This increase would annually generate up to $79 million (if adopted in all 29 counties) and $15 million (if adopted in only 15 counties) in additional revenues by 2040. A 2-mil tax increase on commercial or industrial properties would lead to $160 million or $30 million in revenue respectively for transportation funding (Atlanta Regional Commission 2010, 47).

This type of commercial or industrial property tax increase is also attractive because it provides additional monies to address transportation needs and issues that arise from new large-scale commercial development projects. Of course it is likely that a tax focusing on commercial or industrial properties would have negative impacts on Atlanta's businesses and its economic climate. Despite the outspoken need to find additional sources of transportation funding emanating from prominent members of the metro Atlanta business community, a tax that appears to target them and result in elevated commercial rents would be met with substantial opposition. This type of tax would also have more profound effects on regional commercial location decisions and regional economic development.

3. Land value or split rate taxes

Land value taxes are a different way of calculating property taxes. They constitute a general area value capture strategy. Whereas typical property taxes lump together the value of both land and buildings, land value taxes focus only on the value of the land's location (Atlanta Regional Commission 2010, 49).

A related method, known as a split rate tax, provides separate taxing rates for the land and the buildings on it. The value of the land is determined primarily by its access to transportation and proximity to major destinations and amenities, regardless of whether a small one-story retail establishment or a multi-story mixed-use development occupies the property. This tax structure encourages landowners to develop the land. Land value taxes are calculated based on the benefit provided by the transportation network, whether the property actually uses the transport amenity or not. Although in wide use in several Asian, Eastern

European, and South American nations, Pennsylvania is the only US state with extensive experience using a split rate tax formula (SSTI 21). Implementing one of these innovative taxing strategies in the 29-county Atlanta region could be the cornerstone for driving smart growth and development centered on accessible transportation in metro Atlanta.

B. Implement Value Capture

Transportation infrastructure improvements like transit stations, roadway networks, or interchanges add value to nearby lands and the location of these transportation facilities are key decision factors for employees, employers, and the general traveling public. The unsustainable arrangement of government providing transportation improvements as a public good is shifting as transportation budgets trickle dry. There is now an increasing movement to ask users of transportation improvements, who receive the greatest share of the added value these facilities produced, to assume the greater part of the burden in funding transportation improvements (SSTI 19). Value capture offers an equitable means of recouping value from the private sector in proportion to the benefit received from transportation improvements.

Just as we may have to look outside our domestic borders to Italy for inspiration to pursue regional governments, we can also learn from Spain and the Latin American region where land value capture strategies are implemented as a major source of infrastructure funding. The Spanish consider value capture as the public's recovery of "the land value increments (unearned income ...) generated by the actions other than the landowner's direct investments ... The objective is to draw on publicly generated land value increments to enable local administrations to improve the performance of land use management and to fund urban infrastructure and service provisions (Smolka)." In other words, the Spanish appreciate that value capture achieves the equitable objective of making sure "that benefits provided by governments to private landowners should be shared fairly among all residents."

A future metro Atlanta central transportation department could develop entrepreneurial policies that would encourage investment by the specific neighborhoods or businesses receiving the majority of the benefit from improved transportation facilities. Value capture mechanisms range from district-based value-capture methods to common project-based value-capture strategies and general area, value-capture strategies.

1. District-based value capture mechanisms

One type of district-based value-capture tactic is tax increment financing (TIF). TIF anticipates additional tax revenues from rising property values associated with new transportation infrastructure and borrows against the expected increase to provide up-front financing for the transportation project. Capturing this incremental increase is usually accomplished through the issuance of bonds at the beginning of a project. TIF districts are premised on the "but for" notion—enhanced development value, and the resulting higher tax proceeds, would not be possible "but for" the provision of the enhanced transportation. TIFs generally expire over a period of years or a few decades.

Another district-based instrument the Atlanta region could use to capture transportation infrastructure funds is transportation benefit districts (TBDs) and/or special resident assessment districts (SRADs). These apply a special fee on properties located near a new transportation project or service based on the benefit they receive from their proximity (SSTI 20). Many states utilize a special or resident assessment program to finance part or all of proposed improvements to local roads or transportation facilities (Atlanta Regional

Commission 2010, 51). These assessments are charges to the owner of a property that benefits from an improved transportation facility. The charge can be based on frontage, value, or a combination of factors. They can also be used to support bond issues, although special legislation is usually required (SSTI 20).

Special assessment districts have been used to fund modern streetcars in Portland and new infill metro transit stations in Washington, DC (American Association 2014). While special assessments can cover the whole cost of new investments, they most often cover all or part of the state or local portion of a project (SSTI 20). The advantages of SRADs allow the burden of the tax to be placed upon the landowners who benefit from the infrastructure improvement. Thus, costs are shifted to a group of property owners in return for special benefits that accrue to their property as a result of the nearby, publicly constructed infrastructure improvements, and the tax would be equally connected to the benefit (Atlanta Regional Commission 2010, 51). Another major advantage is that, unlike TIFs, these districts can run indefinitely, supporting not just construction but also operation of the system (SSTI 20). Unfortunately, SRADs require enabling legislation in GA, and despite the logical fairness with the tax on its face, it can be perceived by the public as unjust or inequitable (Atlanta Regional Commission 2010, 51).

2. Common project-based value capture strategies
Instead of capturing value for transportation infrastructure funds from a specific district impacted with transposition improvements, metro Atlanta could look to a common project-based value capture strategy. These include more well-known approaches like joint development/air rights development, exactions, and development impact fees (DIF).

Joint development, or air rights development as it is sometimes referred to, is a promising option for Atlanta to raise significant transportation improvement funds. This value-capture method involves publicly or authority-controlled properties above, below, or adjacent to a piece of infrastructure or right-of-way that are sold or leased to developers. The proceeds are reinvested in the transit or transportation system or bonded out to leverage an even greater funding stream. Some states or transit authorities have created special accounts to manage revenues from these properties to ensure they are used for asset maintenance or alternative transportation investments (SSTI 20). Many large, fixed-rail transit systems such as Washington, D.C.'s Metro or BART in San Francisco have well-established joint development programs, and Boston's central artery project ("The Big Dig") resulted in several air rights projects (Bechtel).

MARTA in Atlanta has implemented this value-capture tool to raise revenue in the past. The transit system's Lindbergh Station is a good example of the smart growth that joint development can stimulate. Recently, MARTA revealed plans to jointly develop a dozen acres of underutilized parking lots at the Brookhaven Station into large-scale mixed-use development. Furthermore, MARTA has received attention from national developers for air rights development around and on top of its valuable Midtown-Arts Station. MARTA would not sell these properties or rights, but rather would partner with developers to share in the proceeds from the leases to then allow them to bond out that revenue stream. The result of this strategy is that instead of a one-time income from the sale of property rights of say $10 million, MARTA has a joint partnership with a development bringing in $100,000 of annual revenue that can be bonded for $100 million. So instead of remodeling a single transit station, MARTA can lay new track and build three new stations.

Exactions and developer impact fees (DIF) also capture value, but not with large-scale impact. Narrow in scope, exactions are contributions negotiated with individual development projects. They are typically used for specific on-site improvements to an area being developed, such as the dedication of rights-of-way and the construction of new roadway networks, new traffic signals, sidewalks, and intermodal stations. Exactions are often in-kind contributions, but may be fees or contributions paid to the locality.

Development impact fees (DIF) are charges levied by local (or potentially regional) governments to pay for the new or expanded transportation facilities or services necessary to support a new development. They are often used in conjunction with adequate public facility ordinances (APFOs) or concurrency requirements, but do not require them (SSTI 20). To date, approximately 27 states have enacted impact fee enabling legislation (Nelson et al.) but they are most prevalent in fast-growing areas and have been utilized extensively in California, Texas, and Florida (Transportation Cooperative). DIFs are generally applied at the county or municipal level. They are similar to development exactions, except that the transportation improvements they support are commonly located outside of the specific property. In the Georgia Development Impact Fee Act, a leading impact fee enabling statute, an "impact fee" is defined as "a payment of money imposed upon development as a condition of development approval to pay for a proportionate share of the cost of system improvements needed to serve new growth and development."[4] A regional DIF would take advantage of the growth booms in metro Atlanta and charge developers a cut of their significant profits to mitigate the burdensome effects of the population influx on an area.

3. General area value capture strategies

General area approaches for capturing value in the Atlanta region include transportation utility fees (TUFs) and also land value/split rate taxes discussed earlier in the chapter. As previously noted, the revenue drawn from fuel taxes is inadequate to fully fund the transportation infrastructure needs for the Atlanta region so money from general funds are diverted to maintain transportation services and assets. TUFs are fees for service based on a user's estimated use. Employing TUFs could potentially eliminate the need to use general funds by treating transportation the same as any other public utility–water, sewer, and electric. TUFs shift the burden of supporting the transportation network from the residential base to commercial and industrial businesses because commercial uses tend to impose greater impacts on transportation networks than residential uses. All properties that are "transportation users"-including nontaxable properties such as nonprofit institutions-must pay the utility fee. TUFs (and Land Value Taxes) may be a good fit for the metro Atlanta region because they are applied area-wide and thus create the largest revenue base (SSTI 21).

C. User Fees –Regional Vehicle Miles of Travel Tax

A mileage-based user fee (MBUF) is a distance-based tax levied on miles driven whose revenues can be suited to fund transportation cost. Compared to the current motor fuel tax system, a system based on mileage traveled will strengthen the effects urban form has on travel behavior and thus improve the ability of planners to use land use planning to moderate single occupancy vehicle travel demand. In addition, charging higher fees for peak-hour travel and for travel in highly congested designated areas could effectively

---

4   Georgia Development Impact Fee Act, Ga. Code Ann. § 36–71–2(8). System improvements, also called non-site related improvements, are to be distinguished from project improvements, also called site related improvements.

reduce congestion where it is the worst. MBUF's approach of charging by the mile more efficiently allocates the cost to frequent road users because of the wear and tear a vehicle imposes on a transportation system (SSTI 26). Alternatively, the Atlanta region or State of Georgia could create a new per-miles method of taxing alternative fuel vehicles to ensure that they also pay their fair share toward transportation.

However an MBUF most likely would be implemented on a national level because of logistics and interstate travel. Estimates show a one-cent-per-mile fee nationally would raise $32.4 billion per year (American Association 2015). For the Atlanta region, adding a 1.5–2 cent per mile MBUF tax (indexed to inflation) would have the potential to generate $1–1.4 billion a year on average over the next 25 years (Atlanta Regional Commission 2010, 55).

Potential drawbacks associated with MBUFs must also be considered (Humphrey School). The cost of imposing and collecting the new fees could equal up to 6 percent of revenues received compared with the less than 1 percent expenses associated with the motor fuel tax (I-95 Coalition). In addition, there are interstate jurisdictional issues in the absence of a federal MBUF, including how to charge out-of-state drivers using an in-state roadway. The additional hurdle of privacy and public acceptance of the government using technology to monitor and charge travel behavior must be vetted (SSTI 27).

D. Parking Fees

The extensive use of parking in public spaces and facilities (whether in office parking decks or supermarket parking lots) in metro Atlanta makes parking facilities a logical point-of-collection for fees from users of the region's transportation infrastructure. Atlanta also has one of the lowest costs of parking in an urban center in the nation at an average of $90 per month. Parking fees are most useful at major regional activity centers in metro Atlanta and help to incentivize people to drive less and live closer to areas of importance to them–reducing the overall burden on the transportation system.

One way to receive revenue from parking fees is a Transactional Tax. Similar to a "sales tax," this is a fee collected on every transaction made for parking as a percentage of the overall parking cost and is the most commonly used collection technique in the US. Unfortunately, the number of free parking spaces reduces, not only the incentives motorist would have for utilizing another mode of transportation, but also the amount of money the fees could generate for infrastructure funding would not be substantial—only $181.1 million annually by 2030 for a $1 daily surcharge levied on transactions for the City of Atlanta's 200,000 parking spaces (Atlanta Regional Commission 2010, 58).

The Atlanta region could also charge an ownership tax, which basically taxes an owner of a parking space (parking garage/deck) through yearly billing on a per-space basis. The likely result of such a tax would unfortunately be the owner of the space shifting the cost to the user to generate additional revenues the owner needs to pay the tax (Atlanta Regional Commission 2010, 58). The City of Atlanta estimated that instituting a 10 percent tax rate on 50,000 spaces in the city that average $90 per month could generate $13.4 million annually by 2030. Other advantages to implementing parking fees to fund transportation infrastructure in the unified Atlanta region are that the revenues generated place a burden of maintenance and operation that is widely dispersed among all users of the local transportation network rather than solely being placed on local residents, and this makes it more of a user-based fee than a controversial conventional tax increase (Atlanta Regional Commission 2010, 59).

VI. Conclusion

Exactly what transportation infrastructure will be needed by the cities of the future is difficult to predict although some themes seem apparent. For example, there is little doubt that public transit of all types is inadequate in Atlanta and most other large American metro areas. It seems equally clear that nearly all cities have not even begun to assess the need for infrastructure to enable the use of transportation innovations such as driverless cars, platooning, and "people movers" nor to adopt the land use policies necessary to encourage or require alternative transportation approaches rather than the current automobile-centric model.

What is clear is that funding all types of transportation infrastructure is becoming more and more difficult and this has made it more important than ever to identify innovative new approaches and implement reform of traditional funding sources. The dilemma faced by Atlanta and its peers is that governmental reform creating regional or metropolitan-wide jurisdictions is a prerequisite to effective and efficient provision of infrastructure. When governmental consolidation and multifaceted and innovative funding are combined, Atlanta and its fellow metropolitan areas will have the potential to provide the quantity and quality of transportation infrastructure needed to ensure urban regeneration and resilience.

Works Cited

American Association of State Highway and Transportation Officials. "VMT Fees." 2015. Web. 12 Mar. 2015 <http://www.transportation-finance.org/funding_financing/funding/proposed_funding_sources/vmt_fees.aspx>.

American Association of State Highway and Transportation Officials, Center for Excellence in Project Finance. *"New York Avenue—Florida Avenue—Gulladet University Metro Station: A Case Study."* 2014. Web. 2 Feb. 2014 <http://www.transportation-finance. org/pdf/funding_financing/funding/local_funding/New_York_Avenue_Case_Study. pdf>.

Atlanta Regional Commission. *Bridging the Gap 2010: Investigation Solutions for Transportation Funding Alternative in the Atlanta Region.* Jul. 2010. Print.

Atlanta Regional Commission. "Financing Transportation." N.d. Web. <http://www. atlantaregional.com/transportation/financing-transportation>.

Bechtel. *Boston Central Artery, Massachusetts, USA: The Most Complex Urban Transportation Project in U.S. History."* Bechtel. 2015. Web. 20 February 2015 <www. bechtel.com/boston_central_artery.html>.

Church, Elizabeth, et al. "Canada's Big City Mayors Ready to Push Urban Agenda." *The Globe and Mail* 1 Feb. 2015. Web. 8 Feb. 2015.

Club Metropolitan. "France Adopts First Law of the Territorial Organisation Reform Project." 21 Feb. 2014. Web. <http://www.clubmetropolitan.ro/WP/france-adopts-first-law-of-the-territorial-organisation-reform-project/?lang=en>.

Copeland, Larry. "Georgia Scraps Over Creation of New, Mostly White Cities." *USA Today* 30 Jul. 2012. Web.

Crane, Bill. "GA View: A New-City Epidemic." *Georgia Trend* 1 Jul. 2012. Web.

d'Antonio, Simone. "'Metropolitan Cities' Are Born in Italy." *Cityscope.* n. pag. 11 Dec. 2014. Web. <http://citiscope.org/story/2014/metropolitan-cities-are-born-italy>.

Esteban, Juli. "The Planning Project: Bringing Value to the Periphery, Recovering the Centre." *Transforming Barcelona.* Ed. Tim Marshall. London: Routledge, 2004. 111–15. Print.

Figueroa, Dante. "Italy: Law Reordering the Territorial Organization of the Country." *The Law Library of Congress* 14 May 2014. Web.

Fuentes, Robert. "America's Infrastructure Woes No Joke: John Oliver Takes on Infrastructure." *Brookings.* The Brookings Institution, 2 Mar. 2015. Web. <http://www.brookings.edu/blogs/the-avenue/posts/2015/03/06-americas-infrastructure-woes-puentes>.

Georgia Constitution. Art. III, Sec. 9. Print.

Georgia House Bill 227. 2010. Web. <http://www.legis.ga.gov/Legislation/20092010/106194.pdf>.

Georgia House Bill 515. 2015. Web. <http://www.legis.ga.gov/legislation/en-US/Display/20152016/HB/515>.

Georgia House Bill 520. 2015. Web. <http://www.legis.ga.gov/legislation/en-US/Display/20152016/HB/520>.

Georgia Development Impact Fee Act. Ga. Code Ann. § 36–71–2(8). 2009. Print.

Georgia Power. Metro Atlanta Overview -- Atlanta: The Growth Engine of the Southeast. n.d. n. pag. Print.

Griffith, Janice. "Regional Governance Reconsidered." *Journal of Law and Politics* 21 (2005): 505. Print.

Humphrey School of Public Affairs at the University of Minnesota. *Report of Minnesota's Mileage-Based User Fee Policy Task Force.* 2011. Web. 30 March 2015.

I-95 Coalition. *The I-95 Corridor Coalition Concept of Operations for the Administration of Mileage-Based User Fees in a Multistate Environment.* Apr. 2012. Web. 25 Mar. 2015.

Joint Legislative Budgeting Committee. *State of Arizona Tax Handbook.* 2010: 163–176. Print.

Juergensmeyer, Julian and Thomas Roberts. *Land Use Planning and Development Regulation Law.* 3rd ed. St. Paul: West, 2012. Print.

Lindsey, Edward. "Now is Time to Act to Fix Transportation." *Atlanta Journal Constitution* 18 Jan. 2015. Web.

Lindsey, Edward. "Transportation Bill Pros and Cons: Help Wanted: Critics Need Not Apply." *Atlanta Journal Constitution* 2 Feb. 2015. Web.

Lindsey, Edward. "Transportation Trouble: Find a Fix, or Georgia Loses." *Atlanta Journal Constitution* 18 Aug. 2014. Web.

Nelson, Arthur C., James C. Nicholas, and Julian C. Juergensmeyer. *Impact Fees: Principles and Practice of Proportionate Share Development Fees.* Chicago: APA, 2009. Print.

Niesse, Mark. "Senate Approves Cityhood for LaVista Hills and Tucker." *Atlanta Journal Constitution* 25 Mar. 2015. Web.

Norton, Lamar. "Transportation Bill Pros and Cons: Dream Big on Statewide Revenue." *Atlanta Journal Constitution* 2 Feb. 2015. Web.

Pelham, Thomas. *Transportation Concurrency, Mobility Fees, and Urban Sprawl in Florida.* Chicago: ABA, 2011. Print.

Smolka, Martim. *Implementing Value Capture in Latin America: Policies and Tools for Urban Development.* Cambridge: Lincoln Institute of Land Policy. 2013. Print.

State Smart Transportation Initiative (SSTI). *The Innovative DOT: A Handbook of Policies and Practice. Focus Area 1: Revenue Sources,* 2015. Web. 9 Jan. 2015.

Transportation Cooperative Research Program. *Use of Fees or Alternatives to Fund Transit.*
      Dec. 2008. Print.
Williams, Dick. "Brookhaven Looks at City Options." *Dunwoody Crier* 27 Jul. 2007. Web.
Ziegler, Edward. "Sustainable Urban Development and the Next American Landscape:
      Some Thoughts on Transportation, Regionalism, and Urban Planning Law Reform in
      the 21st Century." *The Urban Lawyer* 43.2 (2011). Print.

Index

Note: figures and tables are denoted with italicized page numbers; footnote information is denoted with an n and note number following the page number.

Printed and bound by CPI Group (UK) Ltd, Croydon, CR0 4YY

22/10/2024

01777623-0017